T0140472

Risk Assessment Methods for Biological and Chemical Hazards in Food

Risk Assessment Methods for Biological and Chemical Hazards in Food

Edited by
Fernando Pérez-Rodríguez

CRC Press
Taylor & Francis Group
Boca Raton London New York

CRC Press is an imprint of the
Taylor & Francis Group, an **informa** business

First edition published 2021
by CRC Press
6000 Broken Sound Parkway NW, Suite 300, Boca Raton, FL 33487-2742

and by CRC Press
2 Park Square, Milton Park, Abingdon, Oxon, OX14 4RN

First issued in paperback 2022

© 2021 Taylor & Francis Group, LLC

CRC Press is an imprint of Taylor & Francis Group, an Informa business

Library of Congress Cataloging-in-Publication Data

Names: Perez-Rodriguez, Fernando, editor.
Title: Risk assessment methods for biological and chemical hazards in food / edited by Fernando Pérez-Rodríguez.
Description: Boca Raton : CRC Press, 2021. | Includes bibliographical references and index.
Identifiers: LCCN 2020024661 (print) | LCCN 2020024662 (ebook) | ISBN 9781498762021 (hardback) | ISBN 9780429083525 (ebook)
Subjects: LCSH: Food contamination--Risk assessment. | Food industry and trade--Risk management. | Food--Microbiology. | Food--Toxicology.
Classification: LCC TX535 .R57 2021 (print) | LCC TX535 (ebook) | DDC 363.19/26--dc23
LC record available at https://lccn.loc.gov/2020024661
LC ebook record available at https://lccn.loc.gov/2020024662

ISBN 13: 978-0-367-52358-9 (pbk)
ISBN 13: 978-1-498-76202-1 (hbk)
ISBN 13: 978-0-429-08352-5 (ebk)

Typeset in Palatino
by Deanta Global Publishing Services, Chennai, India

CONTENTS

SECTION II Microbial Risk Assessment

SECTION III   Chemical Risk Assessment

CONTENTS

PREFACE

Risk assessment is a set of techniques and methods that have been further developed in several scientific fields, such as environmental science, economics, civil engineering, etc. More recently, as a consequence of the Sanitary and Phytosanitary Measures agreement and the General Agreement on Tariffs and Trade on the use of the risk analysis framework, international organizations and governments have adopted risk assessment as a science-based process to support food safety decision making, thereby protecting consumer health and ensuring fair practices in the food trade. Also, the food industry can benefit from the use of these techniques for food process optimization and quality assurance. The present book will introduce the reader to food risk assessment methods, considering both general aspects and those specific to the type of hazard, that is, biological and chemical hazards. In addition, specific sections are intended to show more technical details on key elements of risk assessment, such as separation of uncertainty and variability, risk metrics or application of sensitivity analysis. New trends and more innovative methods, such as the application of quantitative risk assessment technology for quality aspects in foods, risk-benefit assessment or risk ranking, are also addressed in the present book. Finally, specific examples are developed in order to inform readers concerning how risk assessment techniques can be applied in practice.

MATLAB® is a registered trademark of The MathWorks, Inc. For product information, please contact:

The MathWorks, Inc.
3 Apple Hill Drive
Natick, MA 01760-2098 USA
Tel: 508 647 7000
Fax: 508-647-7001
E-mail: info@mathworks.com
Web: www.mathworks.com

EDITOR

Fernando Pérez-Rodríguez, PhD, earned his degrees in Biological Science and in Food Science and Technology from the University of Córdoba in 1999 and 2002, respectively. He earned his PhD in Risk Analysis from the University of Córdoba (2007), dealing with microbiological risk assessment and cross contamination in cooked meat products. He has published more than 120 peer-reviewed papers, book chapters and books concerning predictive microbiology, and microbial and chemical quantitative risk assessment in foods. Dr. Pérez-Rodríguez has been involved in several research projects at the national and international levels, conducting quantitative risk assessment studies. Due to his expertise, he has participated as a scientific advisor in several expert panels at the Spanish Food Safety Agency and as a Risk Assessment Expert of the European Food Safety Authority (EFSA) and Food and Agriculture Organization (FAO) providing scientific opinion and reports. Dr. Pérez-Rodríguez is the leading manager and designer of the online risk assessment supporting tool "MicroHibro" for microbial hazards and spoilage microorganisms in foods. He has also taught several national and international master and other training courses in predictive microbiology and he is a professor of Food Microbiology and Hygiene at the University of Córdoba (Spain).

CONTRIBUTORS

Verônica O. Alvarenga
Department of Food Science and
　Nutrition
University of Campinas
Campinas, Brazil

and

Department of Food Science
Federal University of Minas Gerais
Belo Horizonte, Brazil

Michael B. Batz
Center for Food Safety and
　Applied Nutrition
U.S. Food and Drug
　Administration
College Park, Maryland

Araceli Bolívar
Department of Food Science and
　Technology
University of Córdoba
Córdoba, Spain

Fernanda B. Campagnollo
Department of Food Science and
　Nutrition
University of Campinas
Campinas, Brazil

German Cano-Sancho
Laboratoire d'Etude des Résidus
　et Contaminants dans les
　Aliments (LABERCA)
Oniris, INRAE
Nantes, France

and

INRA Centre de recherche Pays de
　la Loire
Nantes, France

Rafael D. Chaves
Department of Food Science and
　Nutrition
University of Campinas
Campinas, Brazil

Jean Carlos Correia Peres Costa
Department of Food Science and
　Technology
University of Córdoba
Córdoba, Spain

and

Department of Food Engineering
Federal University of Rondônia
Ariquemes, Brazil

Amélie Crépet
Risk Assessment Department
French Agency for Food,
    Environmental and Occupational
    Health & Safety (ANSES)
Maisons-Alfort, France

Brecht Devleesschauwer
Department of Epidemiology and
    Public Health
Sciensano
Brussels, Belgium

and

Department of Veterinary Public
    Health and Food Safety
Ghent University
Ghent, Belgium

Mariem Ellouze
Food Safety Research Department
Nestlé Research
Lausanne, Switzerland

Gláucia Maria Falcão de Aragão
Department of Chemical
    Engineering and Food
    Engineering
Federal University of Santa
    Catarina
Florianópolis, Brazil

M. Filter
German Federal Institute for Risk
    Assessment (BfR)
Department 4–Biological Safety
Berlin, Germany

Sandrine Fraize-Frontier
Risk Assessment Department
French Agency for Food,
    Environmental and
    Occupational Health &
    Safety (ANSES)
France

Marios Georgiadis
European Food Safety Authority
    (EFSA)
Parma, Italy

Laurent Guillier
Risk Assessment Department
French Agency for Food,
    Environmental and
    Occupational Health & Safety
    (ANSES)
Maisons-Alfort, France

Letícia Ungaretti Haberbeck
Nestlé Waters Product Technology
    Center
Vittel, France

Tina B. Hansen
National Food Institute
Technical University of Denmark
Kgs. Lyngby, Denmark

Arie H. Havelaar
Food Systems Institute
Emerging Pathogens Institute and
    Department of Animal Sciences
University of Florida
Gainesville, Florida

Melinda M. Hayman
Center for Food Safety and
Applied Nutrition
U.S. Food and Drug
Administration
College Park, Maryland

Lea Sletting Jakobsen
Risk-Benefit Research Group
National Food Institute
Technical University of
Denmark
Kgs. Lyngby, Denmark

K. Koutsoumanis
Laboratory of Food Microbiology
and Hygiene
Department of Food Science and
Technology
Aristotle University of
Thessaloniki
Thessaloniki, Greece

Barbara B. Kowalcyk
Center for Foodborne Illness
Research and Prevention
Department of Food Science and
Technology
Translational Data Analytics
Institute
The Ohio State University
Columbus, Ohio

Daniel Angelo Longhi
Food Engineering
Federal University of Paraná
Jandaia do Sul, Brazil

Sonia Marín
Food Technology Department
University of Lleida
Agrotecnio Center
Lleida, Spain

Caroline Merten
European Food Safety Authority
(EFSA)
Parma, Italy

Winy Messens
European Food Safety Authority
(EFSA)
Parma, Italy

Cleide O. de A. Møller
University of Copenhagen
Department of Food Science
Frederiksberg, Denmark

Maarten J. Nauta
National Food Institute
Technical University of Denmark
Kgs. Lyngby, Denmark

Juliana Lane Paixão dos Santos
Laboratory of Food Microbiology
and Food Preservation
Department of Food Safety and
Food Quality, Ghent University
Ghent, Belgium

Fernando Pérez-Rodríguez
Department of Food Science and
Technology
University of Córdoba
Córdoba, Spain

Maria Persson
Risk-Benefit Research Group
National Food Institute
Technical University of Denmark
Kgs. Lyngby, Denmark

Sara M. Pires
Risk-Benefit Research Group
National Food Institute
Technical University of Denmark
Kgs. Lyngby, Denmark

C. Plaza-Rodriguez
German Federal Institute for Risk
    Assessment (BfR)
Department 4–Biological Safety
Berlin, Germany

Arícia Possas
Department of Food Science and
    Technology
University of Córdoba
Córdoba, Spain

Régis Pouillot
Independent Consultant

Leonardo do Prado-Silva
Department of Food Science and
    Nutrition
University of Campinas
Campinas, Brazil

Jukka Ranta
Risk Assessment Unit
Finnish Food Authority
Helsinki, Finland

Chris Roth
Risk Assessment Department
French Agency for Food,
    Environmental and
    Occupational Health &
    Safety (ANSES)
France

Fernando Sampedro
Environmental Health Sciences
    Department
University of Minnesota
Minneapolis, Minnesota

Moez Sanaa
French Agency for Food
Environmental and Occupational
    Health & Safety (ANSES)
France

Anderson S. Sant'Ana
Department of Food Science and
    Nutrition
University of Campinas
Campinas, Brazil

Sofia M. Santillana Farakos
Center for Food Safety and
    Applied Nutrition
U.S. Food and Drug
    Administration
College Park, Maryland

Robert L. Scharff
Department of Human Sciences
The Ohio State University
Columbus, Ohio

Niko Speybroeck
Institute for Health and
    Society (IRSS)
Université catholique de Louvain
Brussels, Belgium

Sofie Theresa Thomsen
Risk-Benefit Research Group
National Food Institute
Technical University of Denmark
Kgs. Lyngby, Denmark

Section I

General Aspects

1

Food Risk Assessment Framework
Foundations and Concepts

Arícia Possas, Letícia Ungaretti Haberbeck, and Fernando Pérez-Rodríguez

Contents

1.1 INTRODUCTION

Globalization of trade is having a big impact on food systems worldwide, resulting in greater food availability and diversity. A food commodity produced on one side of the world can be available on the other side in a matter of days. The increase in food productivity is driven by scientific and technological advances, genetic improvements, development of fertilizers and pesticides, and use of antibiotics and growth promoting substances (Cummins 2017; FAO 2004). As consequences of the globalization of trade, the transmission of harmful bacteria with increased resistance and the presence of chemicals with toxicological effects are of big concern for human health. Thus, the reduction in barriers to the cross-border movement of foods has serious implications for food safety.

To face the issues resulting from globalization, the risk analysis approach has been created and is used as a dominant process to ensure food safety. Risk assessment is the scientifically based component of risk analysis and consists of a systematic framework conducted with the goal of achieving a full understanding of the nature, magnitude and probability of potential hazards in foods (Kavlock et al. 2018).

A microbial risk assessment (MRA) is performed to describe the risk and the potential adverse health effects of microbial hazards in the whole farm-to-fork food production chain or the part that is relevant to the problem (Nauta 2008; Codex Alimentarius Commission 1999). Chemical risk assessment (CRA) can be described as the characterization of potential hazards and the associated risks to life and health resulting from exposure of humans to chemicals present in food over a specified period (EFSA 2018).

The results of a risk assessment are an important management tool that can help in the detection of critical points in the food chain, in the assessment of intervention strategies and in the elaboration of standards for food in international trade (FAO/WHO 2008). In this chapter, the foundations and concepts of food risk assessments are described.

1.2 HAZARD VERSUS RISK

Based on the need for uniform terminology, the Codex Alimentarius committee defined and published terms of risk assessment related to food safety according to the recommendations of the Food Agriculture Organization (FAO)/World Health Organization (WHO) (FAO/WHO 2013). Among the definitions, the terms "hazard" and "risk" are fundamental, since in many languages these terms are not differentiated. According to the Codex, a hazard is a biological, chemical or physical agent that can cause an adverse effect on health, while risk is the probability of occurrence of an adverse health effect (i.e. death or illness) as a consequence of the presence of a hazard in foods (FAO/WHO 2013).

The definitions of hazard and risk published by the Codex, which cover chemical, biological and physical agents, differ from the definitions of bodies that deal specifically with CRA. In a CRA, the hazard is not the chemical itself but a property associated with it. According to the International Programme of Chemical Safety (IPCS), hazard is an "inherent property of an agent or situation having the potential to cause adverse effects when an organism, system or (sub) population is exposed to that agent". Finally, risk is defined as the "probability of an adverse effect in an organism, system, or (sub) population caused under specified circumstances by exposure to an agent" (IPCS 2004).

1.3 RISK ASSESSMENT AND ITS ROLE IN RISK ANALYSIS

The structural framework of risk assessments was formalized by the development and adoption of the risk analysis paradigm by FAO/WHO, taking the lead of the World Trade Organization (WTO), in 1995 (Pérez-Rodríguez and Valero 2013). Risk analysis is a process comprising three components: risk assessment, risk management and risk communication (Figure 1.1) (FAO/WHO 2013). The development of food standards to ensure global food safety is based on the systematized risk analysis process (Cummins 2017; Pérez-Rodríguez and Valero 2013).

Risk assessment, the central scientific component of the risk analysis process, is the qualitative and/or quantitative evaluation of the adverse effects linked to chemical, physical and biological agents that may be present in foods (FAO/WHO 2013). In the qualitative evaluation, risk is described by descriptive terms, while in a quantitative risk assessment, the risk is estimated in terms of numerical outcomes, typically the probability of illness or death (Cummins 2017). Risk assessment was developed due to the need to make decisions to protect health in spite of scientific uncertainty (FAO/WHO 2009).

The decision on whether a risk assessment is necessary is taken by risk managers, who carry out tasks including the description of the objective and the questions to be answered by the risk assessment. Risk

5

Figure 1.1   Interaction between the three components of risk analysis.

managers also establish the risk assessment policy, set time schedules and provide the resources needed for the risk assessment to be carried out (FAO/WHO 2009).

The risk management component of a risk analysis is in charge of deciding whether a risk is acceptable in the light of the results of the risk assessment and what control measures must be implemented in the case that the risk is not acceptable (FAO/WHO 2013). The risk management team may be comprised of industry, public body representatives and policy makers alike (Cummins 2017).

Risk communication refers to the exchange of information and opinions regarding risk between risk assessors, risk managers and all stakeholders. This risk analysis component is critical to ensure that regardless of scientific understanding, the aims and outcomes of a risk assessment are communicated to all the interested parties in a clear and effective manner (FAO/WHO 2013).

Although the interaction between the three risk analysis components is relevant, they may be functionally separated in order to avoid conflicts of interest or bias in the risk assessment process. Finally, the risk analysis process may be evaluated and reviewed when appropriate.

1.4 RISK ASSESSMENT FRAMEWORK

The development of a risk assessment comprises four well-established components: (i) hazard identification, (ii) hazard characterization, (iii) exposure assessment and (iv) risk characterization (CAC 1999). Although the same structure is adopted for microbial and chemical risk assessments, it is appropriate to subdivide their descriptions into individual sections, since some terms adopted for chemicals are different from terms adopted for microbial hazards. The definitions of risk assessment in the context of chemicals have been developed as a part of the project "Approaches to the Assessment of Risk from Exposure to Chemicals" (IPCS 2004), while the definitions for microbial risk assessment (MRA) are the ones established by the Codex Alimentarius (CAC 1999).

1.4.1 Microbial Risk Assessment Concepts

Figure 1.2 depicts the four components of an MRA and briefly summarizes its main outputs and the type of information it describes. The scope of the MRA is defined between the risk question and the hazard identification.

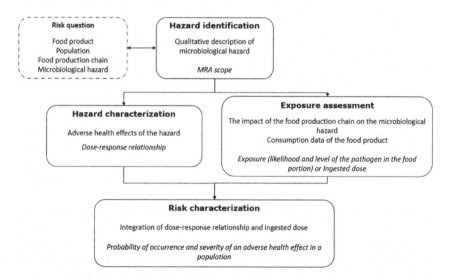

Figure 1.2 The main outputs (in italic) and the type of information described in each of the four components of a microbiological risk assessment (MRA).

7

Similarly, the scope of the four components depends on the precise objective of the MRA. Generally, the objective is described as a risk question developed by risk managers in consultation with risk assessors. This question can describe the hazard, food, population and steps in the food production chain to be considered (Dennis et al. 2008; Nauta 2008). Two examples of risk questions from the literature are:

- What is the efficacy of different intervention strategies to reduce the risk of acquiring *Vibrio parahaemolyticus* gastroenteritis from raw oyster consumption? (FDA 2005)
- What is the estimation of the number of listeriosis cases per year in the European Union population from consumption of a meal containing each of the three ready-to-eat (RTE) food categories: heat-treated meat, gravad and smoked fish, and soft and semi-soft cheese? (Pérez-Rodríguez et al. 2017)

The hazard identification aims to identify microorganisms or microbial toxins of concern in the food product or water. It is predominantly a qualitative description of the microbiological hazard of concern as well as relevant related data, such as clinical and surveillance data (FAO/WHO 2003; Lammerding and Fazil 2000). This first component is described further in Chapter 10.

The hazard characterization provides a qualitative and/or quantitative description of the adverse health effects that may result from ingestion of the microorganisms or microbial toxins. When quantitative data are available, a dose–response model is the main output of this component. The dose–response model describes the relation between the dose ingested (e.g. colony forming units per gram or millilitre of a food product) and the frequency of a given effect (e.g. vomiting, diarrhoea, hospitalization or death). Dose–response models are detailed in Chapter 14.

The exposure assessment provides a qualitative and/or quantitative estimate of the likelihood and level of the pathogen in a determined portion of food. Qualitative exposure assessments are descriptive or categorical treatments of information, whereas quantitative assessments are mathematical analyses of numerical data. If the available data are inadequate to develop a quantitative assessment, a qualitative assessment may be developed by assigning descriptive ratings of probability and severity such as "negligible", "low", "medium" or "high" to the exposure factors (FAO/WHO 2008). The use of models and data from the predictive microbiology (PM) research area is often deployed in quantitative exposure assessments. PM models describe mathematically the behaviour of micro-organisms over time and according to environmental factors (Tenenhaus-Aziza and Ellouze 2015). More details on exposure assessment approaches for MRA are given in Chapters 9, 11 and 12.

In the last component, the risk is characterized by combining the exposure assessment and the dose–response relation (Nauta 2008). In an MRA, the risk is the probability of occurrence and severity of known or potential adverse health effects in a given population over a given period (Codex Alimentarius Commission 1999).

It is important to highlight that the estimation of microbial concentration and prevalence in food products by the end of the production process or at the time of consumption is typically more relevant to the industry than the estimation of occurrence of illness. Thus, QMEAs (quantitative microbiological exposure assessments) are usually performed rather than QMRAs (quantitative microbial risk assessments) (Membre 2016).

1.4.2 Chemical Risk Assessment Concepts

Despite being essential building blocks for foods, chemicals can have a variety of toxicological properties, some which can be harmful to human health. CRA forms the foundation of regulatory decisions for a wide range of chemical substances, including those intentionally added or chemical residues that end up in foods by the end of the production process or the distribution chain (EFSA 2018).

The CRA structure provides a mechanism to review all the relevant information necessary to estimate health outcomes in relation to the exposure to chemicals present in foods. Quantitative methods for CRA are presented in Chapter 15, while examples of CRA are available in Chapter 17. The four steps of risk assessment for food chemicals are briefly described in the following.

Hazard identification: the purpose of hazard identification is the evaluation of the weight of evidence that a chemical can cause an adverse health effect according to data available on toxicity and mode of action. These data can come from observations in humans, domestic or laboratory animals, or *in vitro* studies. From the observed data, the toxicity nature or the health effect and the affected organs or tissues are identified (FAO/WHO 2009). Hence, at this stage, two primary questions must be answered: 1) What is the nature of any health hazard to humans an agent may pose? 2) Under what circumstances may the identified hazard be expressed?

Hazard characterization: after confirmation of a cause–effect relationship between the exposure to a chemical and the incidence of an adverse health effect, this relationship is qualitatively or quantitatively described at the hazard characterization stage, including a dose–response assessment where possible. At this stage, dose–response data derived from observations during *in vivo* or *in vitro* studies are essential. Based on these data, the effects of increasing the exposure to a chemical with the increase in incidence of the adverse health effect are characterized at this stage, as well as the first adverse effect resulting from an increase in dose or exposure, i.e. the critical effect (FAO/WHO 2009). Also, the level of exposure to a chemical that does not produce appreciable health effects and health-guidance values such as the ADI (acceptable daily intake) for additives or residues or TDI (tolerable daily intake) for contaminants are established. From these health-guidance values, the maximum legally allowed concentrations of a chemical in food commodities are set (Brimer 2011).

Exposure assessment: according to IPCS (2004), exposure assessment is the "evaluation of the exposure of an organism, system, or (sub) population to an agent (and its derivatives)". The exposure assessment takes into account the occurrence and concentration of the chemical in the diet and food consumption data to estimate average and high level daily intakes (FAO/WHO 2009).

Risk characterization: the information from the exposure assessment and the hazard characterization is integrated to estimate quantitatively or qualitatively the potential health risk associated with human exposure to a chemical hazard present in food. Risk estimates are communicated to risk managers for decision-making, including the clear explanation of any uncertainties derived from the limitations in the risk assessment process (FAO/WHO 2009).

1.5 DETERMINISTIC VERSUS STOCHASTIC RISK ASSESSMENT

Risk assessment models can be characterized as deterministic or stochastic with regard to how input variables are handled (Vose 2008). In the first approach, point-estimate values are used to describe the variables of the model, and only individual scenarios are analysed. Since the worst-case scenario is typically reflected, deterministic approaches are usually unrealistic or "overcautious", and the outcomes are not representative of real situations (Tennant 2012; Pérez-Rodríguez and Valero 2013). In the second approach, variables are defined with probability distributions that encompass all possible scenarios, taking into account uncertainty and/or variability in those variables (Cummins 2017). Hence, stochastic approaches are more reflective of real-life scenarios.

1.6 UNCERTAINTY AND VARIABILITY IN RISK ASSESSMENTS

These components are related to the level of knowledge on risk model inputs. Briefly, uncertainty is the lack of knowledge, for instance, regarding a quantity (Membré and Boué 2018). As uncertainty is usually related to analytical limitations or low precision of measurement methods, it should be minimized whenever possible by further study, for example by increasing the number of samples analysed or by improving measurement methods. The prevalence of a pathogen in a food commodity can be used to illustrate uncertainty: to ascertain prevalence with 100% certainty, 100% of the food products might be tested for the presence of the pathogen, which is not feasible. Hence, we have to rely upon available prevalence data to estimate the prevalence of the whole population, and the greater the number of samples, the higher is our degree of certainty regarding the estimate. On the other hand, variability represents the true heterogeneity in a population (Membré and Boué 2018). For instance, the ability to metabolize or detoxify chemicals can vary from person to person. This variability is not reducible by further study since it is related to natural randomness. Uncertainty and variability are described in more detail in Chapters 7 and 16.

1.7 LIMITATIONS AND CHALLENGES OF RISK ASSESSMENT IN FOODS

A multidisciplinary team that supplies the variety of knowledge to handle the available scientific information is required to carry out a risk assessment. It includes professionals from different fields, such as microbiology, mathematics, epidemiology, food technology and social sciences, among others (Membré and Boué 2018). The complexity derived from this multidisciplinary approach represents a big challenge when performing a risk assessment. The lack of guides or protocols to develop risk assessments and the lack of harmonization in vocabulary or terms employed are also big limitations of the field, since the employment of a common structure would be crucial to compare hazards, risks, management measures, etc. between autonomous regions, and ideally between countries, and over time. Finally, practical guidelines to translate risk-based food safety management for operational use, as well as instructional and training resources to assist in building skills for risk assessments, must be created (Membré and Boué 2018).

1.8 CURRENT DEVELOPMENTS AND FUTURE PERSPECTIVES

The incorporation of omics technology in the exposure assessment component will move towards the next generation of microbiological risk assessment. With this technology, the behaviour of microorganisms in relation to food preservation treatments and environmental conditions will be described with mechanistic cellular information (den Besten et al. 2017; Brul et al. 2012). Njage and Buys (2017) included the potential of gene transfer between strains into the exposure to *Escherichia coli* due to the consumption of lettuce. Fritsch et al. (2018) worked on the refinements of a *Listeria monocytogenes* QMRA by integrating genomic data and considering phenogenotype associations for its hazardous properties, such as growth ability at low temperature and virulence. In addition, whole genome sequencing (WGS) has been frequently used to refine the hazard identification component of MRA (Membré and Guillou 2016).

The QMRA community has invested great effort and time to develop a rich variety of data, databases, models and software (Membré and Guillou 2016; Tenenhaus-Aziza and Ellouze 2015). However, their reusability and the information exchange between the software and the databases may currently be difficult and time consuming (Plaza-Rodríguez et al., 2017). This situation represents an obstacle to the performance of risk assessment using the most up-to-date knowledge. A recent initiative aims to establish a new community resource called Risk Assessment Modelling and Knowledge Integration Platform (RAKIP). This platform will facilitate the sharing and execution of curated QMRA and PM models using a harmonized metadata schema and information exchange format. The aim of RAKIP is to promote knowledge reusability and high-quality information exchange between stakeholders within QMRA and PM modelling (Haberbeck et al. 2018; Plaza-Rodríguez et al. 2017). Chapter 9 describes the principle of knowledge exchange in more detail.

The approaches of chemical and microbiological risk assessment and the nutritional aspects of food consumption are integrated into one of the most recent risk-based methods, the so-called risk-benefit assessments (RBAs). Currently, most RBAs integrate chemical and nutritional assessments; microbial risk is occasionally assessed, and mostly qualitatively (Boué et al. 2015). Some recent examples are the studies of Berjia et al. (2012) that integrated microbiological risks and nutritional benefits in cold smoked salmon and those of Boué et al. (2017) that integrated microbiological and chemical risks with nutritional benefits in infant feeding. RBAs are discussed in detail in Chapter 4.

REFERENCES

Berjia, F. L., R. Andersen, J. Hoekstra, M. Poulsen, and M. Nauta. 2012. "Risk Benefit Assessment of Cold-Smoked Salmon: Microbial Risk Versus Nutritional Benefit." *European Journal of Food Research & Review* 2: 49–68.

Boué, G., E. Cummins, S. Guillou, J. P. Antignac, B. Le Bizec, and J. M. Membré. 2017. "Development and Application of a Probabilistic Risk-Benefit Assessment Model for Infant Feeding Integrating Microbiological, Nutritional, and Chemical Components." *Risk Analysis* 37(12): 2360–88.

Boué, G., S. Guillou, J. P. Antignac, B. Le Bizec, and J. M. Membré. 2015. "Public Health Risk-Benefit Assessment Associated with Food Consumption — A Review." *European Journal of Nutrition & Food Safety* 5(1): 32–58.

Brimer, Leon. 2011. *Chemical Food Safety*. Edited by Leon Brimer. 1st ed. Cambridge: Cambridge University Press.

Brul, S., J. Bassett, P. Cook, S. Kathariou, P. McClure, P. R. Jasti, and R. Betts. 2012. "'Omics' Technologies in Quantitative Microbial Risk Assessment." *Trends in Food Science and Technology* 27(31): 12–24.

Codex Alimentarius Commission. 1999. "Principles and Guidelines for the Conduct of Microbiological Risk Assessment." In: *Joint FAO/WHO Food Standards Programme* (Ed.), CAC/GL-30, Rome.

Cummins, Enda. 2017. "Fundamental Principles of Risk Assessment." In: *Quantitative Tools for Sustainable Food and Energy in the Food Chain*, edited by V. P. Valdramidis, E. Cummins, and J. Van Impe, 151–72. Ostend, Belgium: Eurosis-ETI.

den Besten, Heidy M. W., Alejandro Amézquita, Sara Bover-Cid, Stéphane Dagnas, Mariem Ellouze, Sandrine Guillou, George Nychas, Cian O'Mahony, Fernando Pérez-Rodriguez, and Jeanne Marie Membré. 2017. "Next Generation of Microbiological Risk Assessment: Potential of Omics Data for Exposure Assessment." *International Journal of Food Microbiology* 287: 1–10. doi:10.1016/j.ijfoodmicro.2017.10.006.

Dennis, S. B., K. Kause, M. Losikoff, D. L. Engeljohn, and R. L. Buchanan. 2008. "Using Risk Analysis for Microbial Food Safety Regulatory Decision Making." In: *Microbial Risk Analysis of Foods*, edited by D. W. Schaffner, 137–75. Washington, DC: ASM Press.

EFSA. 2018. "Chemicals in Food." http://www.efsa.europa.eu/en/topics/topic/chemicals-food. Accessed in 02/10/2018.

FAO. 2004. "Globalization of Food Systems in Developing Countries: Impact on Food Security and Nutrition." *FAO Food and Nutrition Paper* 83: 107. doi:10.1186/1475-2891-10-104.

FAO/WHO. 2003. "Hazard Characterization for Pathogens in Food and Water: Guidelines." *Microbiological Risk Assessment Series* 3: 61.

FAO/WHO. 2008. "Exposure Assessment of Microbiological Hazards in Food: Guidelines." *Microbiological Risk Assessment Series* 7(61): 102.

FAO/WHO. 2009. "Principles and Methods for the Risk Assessment of Chemicals in Food." *Environmental Health Criteria* 240. http://www.who.int/foodsafety/publications/chemical-food/en/.

FAO/WHO. 2013. *Codex Alimentarius Commission Procedural Manual*. 21st ed. Rome. http://www.fao.org/3/a-i3243e.pdf.

Food and Drug Administration (FDA). 2005. "Quantitative Risk Assessment on the Public Health Impact of Pathogenic *Vibrio parahaemolyticus* in Raw Oysters." Center for Food Safety and Applied Nutrition, Food and Drug Administration, U.S. Department of Health and Human Services.

Fritsch, L., L. Guillier, and J. C. Augustin. 2018. "Next Generation Quantitative Microbiological Risk Assessment: Refinement of the Cold Smoked Salmon-Related Listeriosis Risk Model by Integrating Genomic Data." *Microbial Risk Analysis* 10: 20–27.

Haberbeck, L. U., C. Plaza-Rodríguez, V. Desvignes, P. Dalgaard, M. Sanaa, L. Guillier, M. Nauta, and M. Filter. 2018. "Harmonized Terms, Concepts and Metadata for Microbiological Risk Assessment Models: The Basis for Knowledge Integration and Exchange." *Microbial Risk Analysis* 10: 3–12.

IPCS. 2004. "IPCS Risk Assessment Terminology." Geneva: World Health Organization, International Programme on Chemical Safety (Harmonization Project Document, No. 1). http://www.who.int/ipcs/methods/harmonization/areas/ ipcsterminologyparts1and2.pdf.

Kavlock, Robert J., Tina Bahadori, Tara S. Barton-Maclaren, Maureen R. Gwinn, Mike Rasenberg, and Russell S. Thomas. 2018. "Accelerating the Pace of Chemical Risk Assessment." Review-Article. *Chemical Research in Toxicology* 31(5). American Chemical Society: 287–90. doi:10.1021/acs.chemrestox.7b00339.

Lammerding, Anna M., and Aamir Fazil. 2000. "Hazard Identification and Exposure Assessment for Microbial Food Safety Risk Assessment." *International Journal of Food Microbiology* 58(3): 147–57. doi:10.1016/S0168-1605(00)00269-5.

Membre, Jeanne Marie. 2016. "Microbiological Risk Assessments in Food Industry". In *Food Hygiene and Toxicology in Ready-to-Eat Foods*, edited by P. Kotzekidou, 337–50, 1st ed. Cambridge, USA: Academic Press. doi:10.1016/B978-0-12-801916-0.00019-4.

Membré, Jeanne Marie, and Géraldine Boué. 2018. "Quantitative Microbiological Risk Assessment in Food Industry: Theory and Practical Application." *Food Research International* 106(August 2017). Elsevier: 1132–39. doi:10.1016/j.foodres.2017.11.025.

Membré, Jeanne Marie, and Sandrine Guillou. 2016. "Lastest Developments in Foodborne Pathogen Risk Assessment." *Current Opinion in Food Science* 8: 120–26. doi:10.1016/j.cofs.2016.04.011.

Nauta, Maarten J. 2008. "The Modular Process Risk Model (MPRM): A Structured Approach to Food Chain Exposure Assessment." In: *Microbial Risk Analysis of Foods*, edited by D. W. Schaffner, 99–136. Washington, DC: ASM Press.

Njage, P. M. K., and E. M. Buys. 2017. "Quantitative Assessment of Human Exposure to Extended Spectrum and AmpC B-Lactamases Bearing *E. coli* in Lettuce Attributable to Irrigation Water and Subsequent Horizontal Gene Transfer." *International Journal of Food Microbiology* 240: 141–51.

Pérez-Rodríguez, Fernando, Elena Carrasco, Sara Bover-Cid, Anna Jofré, and Antonio Valero. 2017. "*Listeria monocytogenes* Risk Assessment Model for Three Ready-to-Eat Food Categories in the EU." doi:10.5281/zenodo.822350.

Pérez-Rodríguez, Fernando, and Antonio Valero. 2013. *Predictive Microbiology in Foods*. New York, USA: Springer.

Plaza-Rodríguez, C., L. U. Haberbeck, V. Desvignes, P. Dalgaard, M. Sanaa, M. Nauta, M. Filter, and L. Guillier. 2017. "Towards Transparent and Consistent Exchange of Knowledge for Improved Microbiological Food Safety." *Current Opinion in Food Science* 19: 129–37.

Tenenhaus-Aziza, Fanny, and Mariem Ellouze. 2015. "Software for Predictive Microbiology and Risk Assessment: A Description and Comparison of Tools Presented at the ICPMF8 Software Fair." *Food Microbiology* 45(February). Academic Press: 290–99. doi:10.1016/J.FM.2014.06.026.

Tennant, D. R. 2012. *Food Chemical Risk Analysis*. Edited by D. R. Tennant. 2nd ed. London: Blackie Academic and Professional.

Vose, D. 2008. *Risk Analysis: A Quantitative Guide*. 3rd ed. West Sussex, England: John Wiley & Sons, Ltd.

2

Risk Ranking
Moving towards a Risk-Based Inspection and Surveillance System

Fernando Sampedro

Contents

2.1 THE NEED FOR A RISK-BASED FOOD INSPECTION AND SURVEILLANCE SYSTEM

A report by the Food and Agriculture Organization of the United Nations (FAO) and the World Trade Organization (WTO) stated that the annual value of trade in agricultural products has grown almost threefold over the past decade, largely in emerging economies and developing countries, reaching US$1.7 trillion (FAO/WTO, 2017). Countries worldwide have also experienced an exponential increase of small-scale farming operations and micro and small food processing and distribution enterprises (formal and informal) in urban areas in response to the fast population growth and consumer demands for locally produced food (Cabannes and Marocchino, 2018). A special case is the growth of artisanal and family-owned cheese-making businesses that have emerged as a result of demands for artisanal dairy products and local economic incentives (see for example the case of México and Brazil in González-Córdova et al., 2016 and Campagnollo et al., 2018, respectively).

It is common to find food safety regulatory agencies struggling to fulfill their routine inspection duties as budgets and work force reduce and small-size establishments and street vendors increase. In addition, inspection activities are often reactive (i.e., respond to specific food safety incidents), based on out-of-date regulations and on a fragmented and weak public health reporting system, making it difficult to identify which foods and hazards pose the highest risk to consumers (Jaffee et al., 2019). As a result, the use of resources is inefficient and often does not address the most critical food safety risks, failing to protect public health (Jaffee et al., 2019).

The current challenges faced by regulatory agencies require a modern and risk-based food inspection and surveillance system that is able to identify the products and establishments that pose the greatest public health risk so that they can be targeted and inspected more frequently (FAO, 2008). Such a risk-based system must be rooted in the risk analysis principles defined by the Codex (CAC, 1997) and can be achieved by using risk ranking approaches to better guide food safety control activities in the country.

2.2 IDENTIFYING FOOD SAFETY RISKS: RISK RANKING

Risk analysis (assessment, management and communication) has emerged as the foundation for developing food safety systems and policies around the world (Hoffman, 2010). The implementation of risk analysis at the country level requires that governments conduct the following steps (adapted from CAC, 2007): 1) identify public health objectives and establish a risk management plan with measurable metrics; 2) identify and prioritize the main food safety risks; 3) allocate resources to collect data relevant to the identified risks or to conduct a risk assessment; 4) analyze and select intervention strategies for implementation; 5) design and implement an intervention plan; and 6) monitor and review the plan to evaluate whether the interventions met the public health objectives.

A common situation encountered by countries when embarking on the implementation of risk analysis is the lack of knowledge of their main food safety risks. Identifying such risks remains the logical place to start thinking about the most effective measures to reduce public health risk (Hoffman, 2010). Food safety risks can lead towards the implementation of risk analysis (i.e., conducting a risk assessment, developing a risk-based inspection and generating a risk communication plan). Conversely, advances in risk analysis have mostly centered around quantitative risk assessment methodology for single food product–pathogen pairs, missing an important aspect of identifying the food safety risks (Newsome et al., 2009). This has resulted in a lack of adequate methodology for risk ranking or at least, national food safety authorities not giving it adequate attention until recently (Hoffman, 2010; Speybroeck et al., 2015).

Food safety risk ranking involves systemically identifying what hazards (mainly biological and chemical) and/or food products pose the greatest risk to public health. Risk ranking is the first step towards prioritization of future risk management decisions, where the ranking based on public health risks is taken into consideration along with other relevant factors (e.g., economic impact, feasibility, and consumer perception) for decision making (FAO, 2017). In decision-making prioritization, all these factors as well as public health are integrated to identify food safety priorities for the country to take action in a way that is structured, transparent and provides national food safety authorities with the basis for making objective and evidence-based decisions on how to better allocate their resources (FAO, 2017). Risk ranking can be used to (EFSA, 2012a; Stella et al., 2013; van der Fels-Klerx et al., 2018):

- Identify food safety issues that have the highest public health impact
- Develop a risk-based disease and hazard surveillance program
- Develop a risk-based food inspection program

- Identify research and data needs
- Develop a risk communication plan targeted to the most significant food safety risks

Risk ranking and prioritization terms have been used interchangeably, but it is important to differentiate them. According to FAO (2017), "risk ranking is the systematic analysis and ordering of foodborne hazards and/or foods in terms of the likelihood and severity of adverse impacts on human health in a target population". Risk ranking is a primary consideration (public health) in the prioritization of risk management decisions, but other types of impact are also considered in prioritization, such as social, economic, and political consequences. Examples can include pressing public or political demands, trade restrictions, reduced revenue from exports and significant impacts on the vulnerable.

Recent review publications have highlighted the need for a harmonized framework for food safety risk ranking (EFSA, 2012a; Mangen et al., 2010; van der Fels-Klerx et al., 2018). A recent survey also stressed the need for more formal training in risk ranking and case-studies among decision makers and stakeholders (Speybroeck et al., 2015). A range of risk ranking (based only on public health) methods have been applied to address a variety of food safety scenarios (Table 2.1). Some of the studies also included prioritization of interventions.

The next section in this chapter will cover the general framework for conducting a risk ranking exercise.

2.3 RISK RANKING FRAMEWORK

There is no internationally recognized framework to rank foodborne risks; however, the European Food Safety Authority (EFSA) (2012a) and an FAO expert panel of risk assessors (FAO, unpublished) have proposed a series of steps to objectively and systematically rank microbial and chemical risks in food, which are described in the following. Figure 2.1 illustrates the seven steps, which will be discussed in detail.

2.3.1 Defining the Scope

Risk ranking can be overwhelming when the number of foods and/or hazards is too large for it to be feasible (EFSA, 2012a). The scope should be defined by the risk managers and decision makers and should include the goals for risk ranking. Defining the scope is a critical step that will

Table 2.1 Risk Ranking Methods Used in Food Safety

Type of Hazards	Topic	Approach	Reference
Biological	Foodborne pathogens	Ranking	Cardoen et al., 2009; Batz et al., 2012; Devleesschauwer et al., 2017; Felicio et al., 2015; Stella et al., 2013
		Ranking and prioritization	FAO/WHO, 2014; Robertson et al., 2015
	Bioterrorism threats	Ranking	Tomuzia et al., 2013
Chemical	All chemicals	Ranking	Langerholc et al., 2018; Stroheker et al., 2017; van Asselt et al., 2018a,b; Vromman et al., 2014
		Ranking and prioritization	Felter et al., 2009; Papadopoulos et al., 2015
	Veterinary drug residues	Ranking	FDA, 2015; van Asselt et al., 2013
	Pesticides	Ranking	Chou et al., 2019; Labite and Cummins, 2012; Melnyk et al., 2016; Nougadère et al., 2011; Tsaboula et al., 2016
	Heavy metals	Ranking	Groth, 2010

guide all risk ranking activities, from identifying the applicable foods and hazards to choosing the risk ranking approach. Risk ranking is rooted in the basic definition of risk by the Codex Alimentarius as the probability of an adverse health effect, and the severity of that effect, consequential to a hazard(s) in food (CAC, 1999). Risk managers are usually interested in the highest-risk outcomes (hazard–food pair combinations that are both severe and likely to occur), and that is generally the scope of risk ranking. Risk ranking focuses on one of three levels: ranking foods (single hazard in multiple foods), ranking hazards (multiple hazards in a single food), or a combined ranking of foods and hazards (multiple hazards in multiple foods). The broader the scope, the more resources and data are needed to complete the risk ranking exercise.

The use of a food classification system whereby food products (products that share similar composition and production characteristics) are grouped into a single food category (e.g., fresh chicken parts) has been

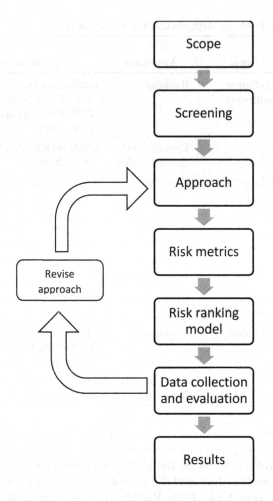

Figure 2.1 Risk ranking framework. (Adapted from EFSA (2012a, *Scientific opinion on the development of a risk ranking framework on biological hazards*) and FAO, *Preliminary Guide to Ranking Food Safety Risks at the National Level*, Rome, forthcoming.)

proposed as a reasonable approach to simplify and reduce the burden of the ranking (EFSA, 2012a). In other instances, the food categorization scheme used to collect data on food consumption or to categorize food production establishments could be used. Examples of food categories can be found in EFSA (2011), FDA (2012) and Richardson et al. (2017). The choice of how to categorize the foods will need to be compatible with the initial goal of the risk ranking as defined by the risk managers and should follow a common set of principles as described by Morgan et al. (2000) and summarized in Table 2.2.

Identifying the hazards of concern is also a critical task when defining the scope. Using the national food safety regulations (e.g., microbiological criteria) to identify potential hazards as the only data source may not be comprehensive when regulations are not up to date with the current

Table 2.2   Desirable Attributes for an Ideal Risk Ranking Categorization System

Attribute	Description
Logically Consistent	• Exhaustive so that no relevant risks are overlooked. • Mutually exclusive so that risks are not double-counted. • Homogeneous so that all risk categories can be evaluated on the same set of attributes.
Administratively Compatible	• Compatible with existing organizational structures and legislative mandates so that lines of authority are clear and management actions at cross purposes are avoided. • Relevant to management so that risk priorities can be mapped into risk management actions. • Large enough in number so that regulatory attention can be finely targeted, with a minimum of interpretation by agency staff. • Compatible with existing databases, to make best use of available information in any analysis leading to ranking.
Equitable	• Fairly drawn so that the interests of various stakeholders, including the general public, are balanced.
Compatible with Cognitive Constraints and Biases	• Chosen with an awareness of inevitable framing biases. • Simple and compatible with people's existing mental models so that risk categories are easy to communicate. • Few enough in number so that the ranking task is tractable. • Free of the "lamp-post" effect, in which better-understood risks are categorized more finely than less well-understood risks.

From FAO Guidance forthcoming and adapted from Morgan, M.G. et al., *Risk Anal.* 20, 49–58, 2000.

scientific knowledge. Risk managers should seek additional sources of data, which may include published literature, national outbreak data, the WHO's global burden of foodborne diseases (WHO, 2015), international food standards such as the Codex Alimentarius, food recall databases such as the European Commission's Rapid Alert System for Food and Feed (RASFF) and the FDA's Reportable Food Registry for Industry, and expert committee sources such as the International Commission for Microbiological Specifications in Food (ICMSF), the National Advisory Committee on Microbiological Criteria for Foods (NACMCF) and EFSA expert panel scientific opinions.

2.3.2 Screening

Risk managers are often tempted to include all foods and hazards that appear to match the purpose of the risk ranking exercise. In reality, this is not practical, and frequently, it is not appropriate. Conducting a filtering process whereby foods and hazards with a weak or non-existent relationship (e.g., no evidence of outbreaks, contamination, or recall events) or in a known, well-described low-risk category (e.g., sterilized canned food) are excluded from further consideration greatly reduces the overall work burden. Decision trees (explained later in this chapter) can be used to make the screening process more transparent and justify the inclusion of specific foods and hazards.

2.3.3 Approach

Food safety risk ranking methodology can use different approaches: qualitative, semi-quantitative and fully quantitative. Risk ranking approaches are not one-size-fits-all, so it is important to consider, at a high level, the type of risk ranking needed to address the objectives of the risk managers. Table 2.3 summarizes the main attributes of each approach.

The qualitative approach generally uses categorical (descriptive) terms to define the likelihood and severity components and the risk level. Examples of descriptive terms are "low", "moderate" and "high". The EFSA's biohazard panel scientific opinion summarized the main drivers for risk managers to use a qualitative ranking approach, which is perceived as a quicker, simpler and easier-to-understand approach when mathematical skills and/or data are lacking (EFSA, 2012a). An important aspect of the qualitative approach is the use of clearly defined terms to categorize the risk, uncertainty and/or variability (e.g., also using descriptive

Table 2.3   Types of Risk Ranking Approach

Type of Risk Ranking	Attributes
Qualitative	Methods include decision trees, decision matrices and deliberative process. Outputs are typically descriptive (e.g., high, medium and low) and do not relate to a numerical value. Useful to perform a high-level screening process. Low level of data and resources required
Semi-quantitative	Methods include risk matrix, risk ratio (exposure/effect) and multicriteria decision analysis. Outputs are ordinal categories that are defined using non-overlapping numerical intervals. Useful to combine qualitative and quantitative variables. Moderate level of data and resources required
Quantitative	Methods include cost of illness, DALY[a]/QALY[b], quantitative risk assessment. Outputs are numeric in form of a single estimate or probability distribution. Useful to conduct a detailed quantitative analysis. High level of data and resources required

[a] Disability-adjusted life years; [b]quality-adjusted life years.

terms such as low, moderate and high). Whereas the qualitative approach has been recognized as a valid method for conducting a risk assessment (CAC, 1999; FAO/WHO, 2006), there is still some controversy over the appropriateness of this approach for risk ranking and risk assessment, as other authors have raised concerns that it is arbitrary and subjective (Cox et al., 2005; Rozell, 2015). While I acknowledge some of the limitations of the qualitative approach, qualitative methods can be useful to conduct an initial high-level screening to exclude negligible risk scenarios (e.g., food–hazard combinations) from the risk ranking exercise or when data is very scarce, thus identifying information gaps that need to be filled by future studies. Examples of qualitative risk ranking methods are decision trees, decision matrices and deliberative processes (Van der Fels-Klerx et al., 2018). Examples of using decision trees for risk ranking can be found in Stella et al. (2013) and van Asselt et al. (2018 b) and the use of deliberative process and expert consultation in FAO/WHO (2008), Hartmann et al. (2018), Siegrist et al. (2018) and Webster et al. (2010).

Quantitative methods require the development of mathematical models, which may be deterministic (inputs and outputs are single values or point-estimates) or stochastic/probabilistic (inputs and outputs are characterized by probability distributions, and calculations are made using computer simulations like Monte Carlo). Quantitative risk ranking can also be approached in one of two broad ways: top-down or bottom-up (EFSA, 2012a) (Figure 2.2).

In the top-down approach, the variables (or inputs) and risk metric (output) are estimated using population-based health metrics (e.g., disease incidence) derived from official epidemiological systems. Developing countries often lack the disease incidence and source attribution data from in-country surveillance systems, making the use of top-down approaches challenging. The use of the WHO's global burden of foodborne diseases (WHO, 2015) can help countries obtain regional disease incidence data, as well as other relevant inputs for the risk ranking, in the absence of country-specific information. In the bottom-up approach, estimates are derived using the classic risk assessment paradigm defined in the Codex Alimentarius (CAC, 1999) and explained in detail in several books (Haas et al., 2014; Shaffner, 2008; Vose, 2008). Basically, quantitative risk assessment evaluates the risk using exposure to the hazard at the time of consumption and dose–response information. Predictive microbiology tools (explained later in this chapter) are very useful in this approach, as they predict the growth and survival of pathogens based on specific food product characteristics. The bottom-up approach can be used to rank chemical hazards due to the lack of epidemiological data for chronic diseases. In general, the quantitative approaches are more complex and require greater technical expertise, resources and data than either qualitative or semi-quantitative methods. Due to the nature of a risk ranking exercise (high number of food products and hazards to assess), such a large investment may not be warranted, unless the focus of the ranking exercise is narrowed and previous resources (e.g., developed quantitative models) are readily available.

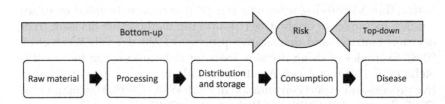

Figure 2.2 Quantitative risk ranking approaches. (Adapted from the European Food Safety Authority, Panel on Biological Hazards, *EFSA Journal*, 10, 2724, 2012.)

The semi-quantitative approach is probably the most common approach used by regulatory and international agencies (CAC, 2012; FAO/WHO, 2014; FDA, 2013, 2015; UK's veterinary residues committee, 2015). It uses a combination of qualitative and quantitative attributes. The main difference is that some sort of quantitative scoring system is used within defined boundaries. The final risk estimate is calculated by an algebraic expression (e.g., summing up or multiplying different risk factors) and is expressed in relative quantitative terms (e.g., using a scale from 0 to 100) or transformed in qualitative terms (e.g., low, moderate and high). It is important that the labeling of the scoring system is unambiguous, non-overlapping and clearly defined (EFSA, 2012a). Examples of semi-quantitative risk ranking tools include multicriteria decision analysis (MCDA), generic scoring systems, risk ratios and risk matrices. MCDA can be used to aggregate and rank food safety issues; however, it often includes more criteria than just public health, being more frequently associated with prioritization efforts. Several publications show the semi-quantitative approach to risk ranking (Chou et al. 2019; CAC, 2013; Collineau et al. 2018; Di Nica et al. 2015; van Asselt et al. 2018 a; van der Fels-Klerx et al. 2017) and ranking and prioritization (Ruzante et al. 2010).

As with any risk assessment tool, lack of data and data gaps result in higher uncertainty around risk ranking estimates, requiring good communication between assessor and manager so that decisions can be made. As mentioned earlier, developing countries often lack epidemiological and national food consumption data, baseline studies on hazard levels and prevalence, as well as consumer handling practices. This precludes them from using fully quantitative approaches when data are lacking, resulting in large uncertainty around the model estimates. The flexibility of the semi-quantitative approach allows qualitative descriptors to be included and the qualitative tools (e.g., expert judgment) to be used on risk factors that are not well characterized, where large knowledge gaps exist, and more quantitative descriptors on factors where more data are available.

2.3.4 Risk Variables and Metrics

The common basis for risk ranking is the estimation of the overall risk (risk metric) based on several variables or inputs. The variables are basically the major factors that should be considered in decision making regarding an issue, and they define what is considered important in ranking one risk relative to another (EFSA, 2012a). Most of the inputs can be classified as belonging to likelihood (e.g., pathogen presence in the final product and disease incidence) and severity (e.g., hospitalization and mortality rates). Table 2.4 lists examples of variables for likelihood and severity as well as the risk metrics that could be used in risk ranking.

It is important to distinguish among the risk metrics (outputs) used in the selected approach. Semi-quantitative approaches commonly use the relative risk (a relative measure not linked to the actual food safety risk or any other biological meaning) as the risk metric to conduct the risk ranking exercise. If a fully quantitative approach is used, then the risk metric corresponds to an absolute risk value (direct measure of risk) (e.g., number of human cases per year). It is imperative to make that distinction clear to risk managers when defining the risk metric to avoid future confusion in the ranking exercise results.

2.3.5 Risk Ranking Model

Once foods and/or hazards are narrowed down and the metric identified, the risk ranking model will be built. In general, the model will combine inputs to produce an estimation of the risk metric. Depending on the approach, the combination of the inputs will increase in complexity. For example, if a qualitative approach is selected, a decision tree can be used to combine inputs from a range of risk variables (yes/no questions) to achieve the final risk estimate (e.g., low, moderate or high). In the case of the semi-quantitative approach, a simple mathematical expression can be built to estimate a relative risk metric (output) that is affected by three risk variables (inputs) in a summative way by using a generic scoring system:

$$\text{Relative risk} = C1 + C2 + C3 \tag{2.1}$$

Each risk variable (C1 to C3) can have its own scoring system with categories that need to be clearly described so that the process is transparent and can be reproduced by different users. It is important to normalize the scores for each factor when the value ranges are very different (e.g., orders of magnitude) or the units are not comparable (e.g., percent vs. milligrams per kilogram), and in the FAO Guidance for risk ranking (forthcoming), it has been recommended that the number of scoring categories should be more than three to allow more differentiation between them.

It is possible to assign weights to each public health variable so that they contribute differently to the overall output. The different weights often reflect the values of risk managers and experts. For instance, decision makers might be interested in controlling diseases that cause a high number of cases of illness versus those that cause a few cases but have

Table 2.4 Risk Variables and Metrics

Likelihood Risk Variables	Severity Risk Variables	Risk Metrics
• Number of outbreak cases or illnesses reported per year • Hazard prevalence (raw commodities or final product) • Growth potential during shelf-life • Effectiveness of control measures • Consumption (per capita, annual, number of servings) • Annual food production (tons, kg)	• DALY[a]/case • Hospitalization rate • Mortality rate • Duration of acute and chronic illness • Probability of sequelae • Cost of illness/case • QALY[b]/case • Case Fatality Ratio[c] (CFR) • Toxicological endpoints (e.g., LD_{50}[d], NOAEL[e], LOAEL[f])	• Incidence of cases/100,000 population • Probability of illness • Total DALY/QALY/Cost exposed population • Relative risk

[a] Disability-adjusted life years; [b] Quality-adjusted life years; [c] Proportion of deaths within a designated population of cases (exposed population) over the course of the disease; [d] Lethal dose for 50% of the study population; [e] No-observed-adverse-effect-level, which represents the maximum dose at which no adverse health effect occurs in test organisms; [f] Lowest-observed-adverse-effect level, which represents the lowest dose that causes an adverse effect.

a high fatality rate. This can be accounted for by assigning different weights to incidence of illness and number of deaths. If different weights are assigned, it is important to describe the rationale behind it so that it is transparent. Equation 2.1 can be modified to include different weights:

$$\text{Relative risk} = C1 * W1 + C2 * W2 + C3 * W3 \qquad (2.2)$$

The relative risk values obtained from Equation 2.2 will then be used to rank the food safety scenarios from high to low risk. This simplified approach is one of several that are part of the MCDA methods. It can be used to aggregate different data across multiple criteria and has been used by several organizations for conducting risk ranking exercises (Codex Alimentarius, US Food and Drug Administration, FAO/WHO, European Center for Disease Control, the UK's veterinary residues committee and Food Standards Australia New Zealand, among others).

More complex models can be developed when using a fully quantitative approach. As stated previously, however, both the top-down and the bottom-up approach are data and resource intensive, so that they may not be the best methodology for broad risk ranking; the advantage is the fact that the risk dimensions are maintained.

2.3.6 Data Collection and Evaluation

After the risk ranking model has been developed, risk managers need to gather the in-country data and with risk assessors, evaluate additional data needed for ranking the food categories or hazards of concern. Data required for risk ranking may come from a wide variety of sources. In general, these sources can be categorized into five main categories (EFSA, 2012a): i) national databases; ii) public health reporting system; iii) inspection and surveillance activities; iv) literature review; v) expert opinion.

As shown in Table 2.4, there are several likelihood and severity metrics that can be employed to estimate the overall risk. National databases on the food import and/or production volume and food consumption and dietary patterns are important sources for exposure estimation. Data from official food surveillance systems (i.e., baseline pathogen prevalence studies or veterinary drug residues), results from official inspections (e.g., imported food testing) and food recall databases are also important likelihood sources to estimate the extent of hazard exposure. The incidence of outbreaks and sporadic cases with source attribution can also be used as sources for likelihood. Whenever one is using epidemiological data, it is important to evaluate potential diagnosis and reporting biases (under- or over-). For example, over-reporting may occur when only the most severe cases are reported or caused by a few severe pathogens, making the attribution of cases to those hazards more frequent.

Systematic reviews and meta-analyses as well as reviews of the literature and published reports are other important sources to obtain relevant published data (e.g., research studies on the prevalence of a hazard in food commodities or consumer handling practices). An EFSA scientific opinion summarizes the methodology used by the agency to conduct literature reviews in feed and food safety assessments (Butcher et al., 2011; EFSA, 2010). There are also several methods to formally elicit expert opinion (Cooke, 1991; Colson and Cooke, 2018), ranging from online surveys to face-to-face consensus-based meetings (i.e., deliberative processes). Expert elicitation is very useful when critical information gaps for the risk ranking have been identified. However, some limitations of using expert elicitation have been expressed by several authors: it does not provide empirical data and due to its subjective nature, can be biased by several factors, including the expert's background and scientific expertise (Pires et al., 2014). Still, if it is well designed, experts are able to produce quantitative information, and uncertainty around their estimates can be calculated and incorporated into the analysis (Hoffman et al., 2007).

2.3.7 Risk Ranking Results

In the presentation of the results, the risk ranking needs to be fully documented. It is imperative that the risk ranking model, steps, data and assumptions undertaken to produce the final output are fully documented in a transparent way (including the uncertainty and variability in the data set used) such that the process of risk ranking is reproducible (EFSA, 2012a; FAO, forthcoming). It is also critical that the strengths and limitations of the approach are explained so that decision makers can take them into consideration. An executive summary with the main takeaways and limitations of the risk ranking exercise is also extremely useful for broad audiences.

There are different ways to reproduce the results of risk ranking. A table listing the overall relative score from high to low of the food–hazard combinations has often been used. Other visualization tools such as color-coded risk matrices (explained later in this chapter) and relative risk ranking charts are an easy way to present the results of the risk ranking exercise. Figure 2.3 shows a relative risk ranking chart used in a FAO/WHO ranking report on foodborne parasites.

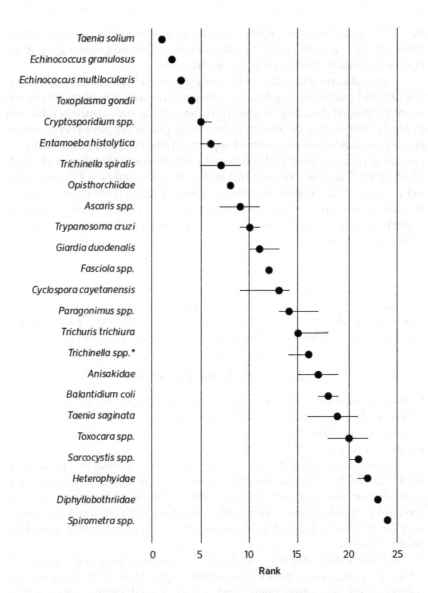

Figure 2.3 Example of a relative risk ranking chart. (From FAO/WHO, *Microbiol. Risk Assess. Series*, 23, 302, 2014.)

2.4 RISK RANKING TOOLS

There are several risk ranking tools available that can be used for the three approaches discussed in this chapter (qualitative, semi-quantitative and quantitative). Table 2.5 shows the available tools and the main approach to their use. The risk ranking tools include decision trees, decision matrices, expert judgment, scoring systems, risk ratios, MCDA, spreadsheet calculators and online software. A detailed explanation of the main ranking tools is given in the following.

2.4.1 Decision Trees

Decision trees are simple visual tools that provide an objective approach to conducting qualitative risk ranking exercises. The tool consists of a flow chart with a series of simple questions (typically with yes/no answers). It is important that each node of the tree results in a clear yes or no decision (i.e., does product support pathogen growth?). Due to their simplicity,

Table 2.5 Risk Ranking Methods and Tools

Method/Tool	Qualitative	Semi-Quantitative	Quantitative
Decision tree	X	X	
Decision matrix	X	X	
Deliberative process	X		
WHO global burden of disease			X
Scoring systems		X	
Risk ratios		X	
Multicriteria Decision Analysis (MCDA)	X	X	X
Risk Ranger		X	
iRISK			X
Predictive microbiology software (ComBase, Predictive Modeling Program)			X
Swift quantitative microbial risk assessment (sQMRA)			X

33

decision trees are easy to customize and can be generic or specific depending on the ranking exercise. The output from decision trees is categorical risk bins (e.g., high, moderate or low) that can be used by risk managers to identify high-risk food categories or hazards (for example). As mentioned earlier in this chapter, decision trees are useful at the prescreening step, when risk managers want to conduct a high-level ranking excluding negligible/low-risk scenarios. They can also be useful when the sources of information are qualitative or consist of poor quantitative data, providing very high versatility (EFSA, 2015b). Examples of using decision trees for risk ranking can be found in EFSA (2012 b), Stella et al. (2013), Stroheker et al. (2017) and van Asselt et al. (2018b). Figure 2.4 shows an example of a generic decision tree that can be used to rank food categories based on the presence of biological hazards.

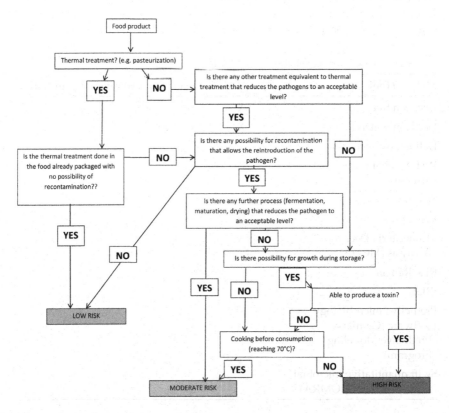

Figure 2.4 Example of a decision tree to rank biological hazards.

2.4.2 Decision/Risk Matrices

The risk/decision matrix is a qualitative or semi-quantitative tool that considers a wide variety of both qualitative and semi-quantitative data to rank risks. Qualitative decision matrices are built by categorizing the risks into bins according to their relative severity and likelihood. For practical use, 5 by 5 risk matrices are often used and based on the assumption that risk equals severity multiplied by likelihood. There is no widely recognized guidance on how to aggregate the different qualitative and semi-quantitative scores for likelihood and severity into different scoring bins. To improve clarity, it has been suggested that binning should be conducted at the last step, and binning categories should be clearly tied to risk outputs. Risk matrices used as the only means to conduct a risk ranking exercise have been criticized as being subjective, lacking resolution (false sense of precision), containing errors, and containing invalid mathematical processes, as mentioned earlier for the qualitative approach (Cox et al., 2005).

2.4.3 Spreadsheet Calculators

Risk Ranger is a publicly available spreadsheet semi-quantitative risk ranking tool developed by Ross and Sumner (2002). The tool consists of 11 risk factors (inputs) in the form of qualitative and quantitative questions affecting the food safety risk of a specific population for selected product–hazard combinations. It uses the bottom-up approach, therefore evaluating risk from harvest to consumption. The spreadsheet converts the inputs to numerical scores, providing an overall risk ranking score (from 0 [no risk] to 100 [highest risk]). Risk estimates also include predicted annual illnesses or probability of illness per day in the target population. Risk Ranger is available online free of charge at https://www.cbpremium.org/RiskRanger

The swift quantitative microbial risk assessment (sQMRA) is another spreadsheet calculator that uses a quantitative deterministic approach to compare the risk of pathogen–food product combinations (Evers and Chardon, 2010; Chardon and Evers, 2017). It is also considered within the bottom-up approach, as it models pathogen growth and reduction from retail to consumption using 11 risk factors. sQMRA is also available online free of charge at http://foodrisk.org/exclusives/sqmra/.

2.4.4 Scoring Systems and Multicriteria Decision Analysis

Generic scoring systems and MCDA are semi-quantitative tools that offer a transparent and objective approach to accounting for all the criteria relevant to the risk manager's decision (Ruzante et al., 2010). MCDA incorporates multiple qualitative and quantitative data types from different dimensions (e.g., public health, social and economic) as well as expert opinion and judgment. As explained in this chapter, risk variables can be combined by simple algebraic expressions including their assigned weights. A typical MCDA exercise contains the following steps (CAC, 2013; Devleesschauwer et al., 2017): identification of hazards/food categories to be ranked; definition of criteria and description of scoring methods; normalization of the values to make them comparable between criteria; elicitation of weights for the different criteria; and scoring and combination of the criteria. It is important to remember that MCDA does not provide an absolute risk estimation; rather, it provides a relative risk estimate (relative importance) of the scenario analyzed (Devleesschauwer et al., 2017). MCDA has been used by several agencies to rank hazards in feed (Codex Alimentarius) (CAC, 2012), rank high-risk food categories and animal drug residues in milk and milk products (FDA, 2013, 2015), rank veterinary residues in milk (UK's veterinary residues committee, 2015), rank foodborne parasites (FAO/WHO, 2014) and has been applied in several research studies (Anderson et al., 2011; Chou et al., 2019; Collineau et al., 2018; Di Nica et al., 2015; Ruzante et al., 2010; van Asselt et al., 2018a; van der Fels-Klerx et al., 2017).

2.4.5 Web-Based Tools

The iRISK tool is a quantitative probabilistic risk assessment tool that integrates data from seven components: the food, the hazard, the consumer population, process models describing the changes in hazard prevalence and concentration up to the point of consumption, consumption patterns, dose–response curves, and health effects (Chen et al., 2013). One of the main capabilities of iRISK is the possibility of evaluating both microbial and chemical risks. The main outputs of the tool are public health metrics (total disability-adjusted life years [DALYs] or cost of illness [COI]) per eating occasion or consumer, total illnesses, and total illnesses per eating occasion or consumer. However, as mentioned earlier, the tools based on quantitative risk assessment are more time-consuming and data intensive, making it impractical for ranking a large number of hazards and foods unless a more targeted risk management question is framed. iRISK is available online free of charge at irisk.foodrisk.org.

ComBase is an online predictive microbiology tool that predicts the growth and survival of pathogens depending on the food characteristics (e.g., pH, a_w) and extrinsic characteristics (e.g., temperature and CO_2) in media solution (peptone water) (Baranyi and Tamplin, 2004). It is important to always validate the predictions against the real food product. The tool also allows dynamic predictions during shelf-life when the temperature is constantly changing. It also includes a module for the growth of *Salmonella* spp. in liquid egg, validation of the cooling phase of a thermal process against *Clostridium perfringens* in meat products, and a database of growth data of pathogens and spoilage microorganisms in different food products. This tool can be useful when the risk ranking exercise contains factors related to the microbial growth potential of the food category or when a more quantitative approach is used and the pathogen growth needs to be estimated. ComBase is available online free of charge at www.cbpremium.org/.

The Pathogen Modeling Program (PMP) is an online predictive microbiology tool that also allows the growth and survival prediction of foodborne pathogens. As with ComBase, a validation in the food product is required when using data obtained in media solutions. It has certain particularities, as it contains specific models for *Clostridium botulinum* (growth and toxin production) and models related to the transfer of *Listeria monocytogenes* during slicing of a deli meat product. PMP is available online free of charge at https://pmp.errc.ars.usda.gov/PMPOnline.aspx.

Microhibro is a quantitative risk assessment online tool that allows the growth of *L. monocytogenes*, *Salmonella* spp. and *Escherichia coli* to be estimated in fresh produce, dairy and meat products, and seafood. Microhibro is available online free of charge at www.microhibro.com.

2.5 CASE STUDY: DEVELOPMENT OF A RISK-BASED INSPECTION SYSTEM

The first section of this chapter has described the framework and tools risk managers can utilize to conduct risk ranking. One of the main goals of a risk ranking is to help develop a risk-based inspection system. A risk-based inspection can be defined as a system that is based on the risk analysis framework and uses risk ranking to establish a priority list of primary production and food processing establishments to be inspected (FAO, 2008). In this particular case, the risk metric is the relative food safety risk of the establishment. Several risk factors will affect how an establishment behaves and the risk level (adapted from FAO, 2008): 1) compliance with the established national food safety regulations; 2) compliance with the established Good Manufacturing Practices (GMPs) or Hazard Analysis and Critical Control Points (HACCP) plans; 3) risk profile of the products processed in the establishment; 4) production volume; and 5) high-risk consumer population (e.g., infants and hospitals).

The implementation of a risk-based inspection system has some similarities with a risk ranking exercise. The framework to develop a risk-based inspection may be as follows:

1. Development of a database of food categories and establishments. An exhaustive database of the current food categories produced in the country and the establishments associated with their production needs to be created. Examples of food category databases were mentioned earlier in this chapter (e.g., EFSA and FDA databases). The number of establishments and food categories to be included is considered the scope of the risk-based inspection program, as countries may decide on developing an inspection system for a particular commodity or food chain.

2. Risk level classification of food categories. Different tools can be used for estimating the risk level of the food categories produced by the establishments within the scope of the inspection program. As mentioned earlier, a decision tree (Figure 2.4), decision matrix or MCDA can be built to estimate the presence of biological and/or chemical hazards in a specific food related to the food processing characteristics and consumer preparation (e.g., fresh or cooked consumption). Variables related to the likelihood and severity need to be developed to account for the public health metrics (see Table 2.4). A relative risk score (1 for low, 2 for moderate and 4 for high risk) can be used to estimate the risk level of each food category.

3. Risk level classification of food establishments. A decision matrix can be designed to assess the food safety level of the processing plants through several factors related to the hygienic design of the plant and the food safety control systems in place:
   i) Compliance with GMPs or HACCP plans: This variable may include the percentage of compliance with current GMPs or the HACCP plan if required at the last inspection. To quantitatively assess the establishment's performance according to current regulations, the country needs to develop a quantitative check-list for inspectors so that the percentage of accomplishment can be estimated from each inspection.

ii) Facility layout and personnel movement flow: The establishment's type of layout and movement of personnel are an important variable to evaluate the potential for cross-contamination and lack of hygienic design and separation of raw material versus finished product zones.

iii) Effective validation, monitoring and verification activities: Effective validation of the CCPs of the HACCP plan and adequate monitoring and verification activities of the sanitation processes and final product testing (along with other activities) need to be in place and be adequately recorded to ensure that the food safety system is working as intended.

iv) History of compliance with inspection: A record of previous inspections, including the number/percentage of non-compliances (moderate and severe), needs to be incorporated into the matrix to evaluate the establishment's performance over time. As mentioned in variable (i), the quantitative nature of check-lists will help take account of the previous performance history.

v) Recall and international market rejection: Recall data involving the establishment or the food category and other data on international markets (if the establishment is exporting) will help to include contamination events or data captured by other countries' surveillance systems.

vi) Traceability plan: The extent to which the establishment is able to track its products (beyond one step back and one forward) will evaluate the establishment's performance in the case of a product recall.

Each variable can be evaluated through a performance score (e.g., 1 through 7), and a summative score for each plant can then be obtained. This score will represent the overall risk score for the establishment.

4. Overall risk score. A simple mathematical expression, multiplying the food category score by the processing plant score, can be built to obtain the final risk score:

Risk score = food category risk × food establishment risk   (2.3)

5. Inspection frequency. The frequency of inspection can be designed according to the final overall risk score obtained for each combination of food category and establishment. The higher

the score (risk), the higher the frequency of inspection. For example, inspection frequency can be broken into three categories (e.g., quarterly, biannually and annually), and for each one, a scoring range will be assigned (minimum score to 33rd percentile value, 34th–66th percentile value and 67th–99th percentile value). If several food categories are produced within a single establishment, the highest-risk food is used to categorize the establishment.

6. Development of strategic goals. The inspection program needs to be reviewed at least annually to ensure its validity. The regulatory agency may want to develop its own performance goals by estimating the percentage of establishments under certain performance categories. For example, the country may decide to use three categories for establishments (low, medium and high performance) and estimate the number of establishments in each category. The agency then develops some strategic goals by improving the overall inspection program performance and increasing the percentages of medium- and high-performance establishments. The incentive for establishments to move to a higher performance is the lower frequency of inspections.

2.6 CONCLUSIONS

Identifying food safety risks by the use of objective and scientific risk ranking tools allows countries to be more efficient and optimize resources by targeting inspection and surveillance activities to high-risk food–hazard combinations. Risk ranking tools and extensive literature are available for countries to perform risk ranking exercises. A case-study on the implementation of a risk-based food inspection system is shown as a practical guide on using risk ranking to develop risk management strategies.

REFERENCES

Anderson, M., Jaykus, L.A., Beaulieu, S. and Dennis, S. 2012. Pathogen-produce pair attribution risk ranking tool to prioritize fresh produce commodity and pathogen combinations for further evaluation (P³ARRT). *Food Control* 22(12):1865–1872.

Baranyi, J. and Tamplin, M.L. 2004. ComBase: A combined database on microbial responses to food environments. *J Food Prot* 67(9):1967–1971.

Batz, M.B., Hoffmann, S. and Morris, J.G. 2012. Ranking the disease burden of 14 pathogens in food sources in the United States using attribution data from outbreak investigations and expert elicitation. *J Food Prot* 75(7): 1278–1291.

Bucher, O., Fazil, A., Rajic, A., Farrar, A., Wills, R. and McEwen, S.A. 2011. Evaluating interventions against Salmonella in broiler chickens: Applying synthesis research in support of quantitative exposure assessment. *EPI Inf* 140(5):925–945.

Cabannes, Y. and Marocchino, C. (eds). 2018. *Integrating Food into Urban Planning*. London, UCL Press/Rome, FAO. https://doi.org/10.14324/111.9781787353763.

Campagnollo, F.B., Gonzales-Barron, U., Cadavez, V.A.P., Sant'Ana, A.S. and Schaffner, D.W. 2018. Quantitative risk assessment of *Listeria monocytogenes* in traditional Minas cheeses: The cases of artisanal semi-hard and fresh soft cheeses. *Food Control* 92:370–379.

Cardoen, S., Van Huffel, X., Berkvens, D., Quoilin, S., Ducoffre, G., Saegerman, C., Speybroeck, N., Imberechts, H., Herman, L., Ducatelle, R. and Dierick, K. 2009. Evidence-based semiquantitative methodology for prioritization of foodborne zoonoses. *Foodborne Pathog Dis* 6(9):1083–1096.

Chardon, J.E. and Evers, E.G. 2017. Improved swift quantitative microbiological risk assessment (sQMRA) methodology. *Food Control* 73:1285–1297.

Chen, Y., Dennis, S.B., Hartnett, E., Paoli, G., Pouillot, R., Ruthman, T. and Wilson, M. 2013. FDA-iRISK-a comparative risk assessment system for evaluating and ranking food-hazard pairs: Case studies on microbial hazards. *J Food Prot* 76(3):376–385.

Chou, W.-C., Tsai, W.-R., Chang, H.-H., Lu, S.-Y., Lin, K.-F. and Lin, P. 2019. Prioritization of pesticides in crops with a semi-quantitative risk ranking method for Taiwan postmarket monitoring program. *J Food Drug Anal* 27(1):347–354.

Codex Alimentarius Commission (CAC). 1999. Principles and guidelines for the conduct of microbiological risk assessment. http://www.fao.org/docrep/004/y1579e/y1579e05.htm (accessed on November 10, 2018).

Codex Alimentarius Commission (CAC). 2007. Working principles for risk analysis for food safety for application by governments. http://www.fao.org/3/a-a1550t.pdf (accessed on December 4, 2018).

Codex Alimentarius Commission (CAC). 2013. Guidance for governments on prioritizing hazards in feed. CAC/GL 81-2013, 7 pp.

Collineau, L., Carmo, L.P., Endimiani, A., Magouras, I., Muntener, C., Schupbach-Regula, G. and Stark, K.D.C. 2018. Risk ranking of antimicrobial-resistant hazards found in meat in Switzerland. *Risk Anal* 38(5):1070–1084.

Colson, A.R. and Cooke, R.M. 2018. Expert elicitation: Using the classical model to validate experts' judgments. *Rev Environ Econ Policy* 12(1):113–132.

Cooke, R.M. 1991. *Experts in Uncertainty-Opinion and Subjective Probability in Science*. Oxford, UK: Oxford University Press.

Cox, L.A., Babayev, D. and Huber, W. 2005. Some limitations of qualitative risk rating systems. *Risk Anal* 25(3):651–662.

Devleesschauwer, B., Bouwknegt, M., Dorny, P., Gabriël, S., Havelaar, A.H., Quoilin, S., Robertson, L.J., Speybroeck, N., Torgerson, P.R., van der Giessen, J. W.B. and Trevisan, C. 2017. Risk ranking of foodborne parasites: State of the art. *Food Waterborne Parasitol* 8(9):1–13.

Di Nica, V., Menaballi, L., Azimonti, G. and Finizio, A. 2015. RANKVET: A new ranking method for comparing and prioritizing the environmental risk of veterinary pharmaceuticals. *Ecol Ind* 52:270–276.

EFSA. 2010. Application of systematic review methodology to food and feed safety assessments to support decision making. *EFSA J* 8(6): 1637, 90 pp.

EFSA. 2011. The food classification and description system foodEx 2 (draft revision 1). *Support Publ* 215, 438 pp. 1–90.

EFSA Panel on Biological Hazards (BIOHAZ). 2012a. Scientific opinion on the development of a risk ranking framework on biological hazards. *EFSA J* 10(6):2724, 88 pp.

EFSA Panel on Biological Hazards (BIOHAZ). 2012b. Scientific opinion on public health risks represented by certain composite products containing food of animal origin. *EFSA J* 10(5):2662, 132 pp.

EFSA. 2015a. *The Food Classification and Description System FoodEx2* (revision 2). EFSA supporting publication 2015:EN-804, 90 pp, Parma.

EFSA. 2015b. Scientific opinion on the development of a risk ranking toolbox for the EFSA BIOHAZ panel. *EFSA J* 13(1):3939, 131 pp.

Evers, E.G. and Chardon, J.E. 2010. A swift quantitative microbiological risk assessment (sQMRA) tool. *Food Control* 21(3):319–330.

FAO. 2008. Risk-based food inspection manual. *FAO Food Nutrition Paper*, No. 89. Rome. 92 pp.

FAO/WHO. 2008. Microbiological hazards in fresh leafy vegetables and herbs. *Microbiol Risk Assess Series*, No. 14. 135 pp, Geneva and Rome.

FAO/WHO. 2014. Multicriteria-based ranking for risk management of food-borne parasites. *Microbiol Risk Assess Series*, No. 23. 302 pp, Rome.

FAO. 2017. Food safety risk management: Evidence-informed policies and decisions, considering multiple factors. *Food Safety and Quality Series*, No. 4. Rome, 106 pp.

FAO. 2020 (forthcoming). *Preliminary Guide to Ranking Food Safety Risks at the National Level*. Rome.

FAO/WTO. 2017. Trade and food standards. https://www.wto.org/english/res_e/booksp_e/tradefoodfao17_e.pdf (accessed on January 20, 2019).

FDA. 2012. *Reportable Food Summary Report RFR COMMODITIES Definitions*, pp. 9, Silver Spring, MD.

FDA. 2013. *FDA's Draft Approach for Designating High-Risk Foods as Required by Section 204 of FSMA*, Silver Spring, MD.

FDA. 2015. Multicriteria-based ranking model for risk management of animal drug residues in milk and milk products. https://www.fda.gov/Food/FoodScienceResearch/RiskSafetyAssessment/ucm443549.htm (accessed on December 5, 2018).

Felicio, M.T.D.S., Hald, T., Liebana, E., Allende, A., Hugas, M., Nguyen-The, C., Johannessen, G.S., Niskanen, T., Uyttendaele, M. and McLauchlin, J. 2015. Risk ranking of pathogens in ready-to-eat unprocessed foods of non-animal origin (FoNAO) in the EU: Initial evaluation using outbreak data (2007–2011). *Int J Food Micro* 195:9–19.

Felter, S., Lane, R.W., Latulippe, M.E., Llewellyn, C.G., Olin, S.S., Scimeca, J.A. and Trautman, T.D. 2009. Refining the threshold of toxicological concern (TTC) for risk prioritization of trace chemicals in food. *Food Chem Toxicol* 47(9):2236–2245.

González-Córdova, A.F., Yescas, C., Ortiz-Estrada, A.M., De la Rosa-Alcaraz, M., Hernández-Mendoza, A. and Vallejo-Córdoba, B. 2016. Invited review: Artisanal Mexican cheeses. *J Dairy Sci* 99(5):3250–3262.

Groth III, E. 2010. Ranking the contributions of commercial fish and shellfish varieties to mercury exposure in the United States: Implications for risk communication. *Env Res* 110(3):226–236.

Hartmann, C., Hübnerb, P. and Siegrist, M. 2018. A risk perception gap? Comparing expert, producer and consumer prioritization of food hazard controls. *Food Chem Tox* 116:100–107.

Haas, C.N., Rose, J.B. and Gerba, C.P. 2014. *Quantitative Microbial Risk Assessment*, 2nd edition. Hoboken, NJ: John Wiley & Sons, pp. 427.

Hoffmann, S., Fischbeck, P., Krupnick, A. and McWilliams, M. 2007. Using expert elicitation to link foodborne illnesses in the United States to foods. *J Food Prot* 70(5):1220–1229.

Hoffmann, S. 2010. Ensuring food safety around the globe: The many roles of risk analysis from risk ranking to microbial risk assessment. *Risk Anal* 30(5):711–714.

Jaffee, S., Henson, S., Unnevehr, L., Grace, D. and Cassou, E. 2019. *The Safe Food Imperative: Accelerating Progress in Low- and Middle-Income Countries. Agriculture and Food Series.* Washington, DC: World Bank. doi:10.1596/978-1-4648-1345-0.

Labite, H. and Cummins, E. 2012. A quantitative approach for ranking human health risks from pesticides in Irish groundwater. *Hum Ecol Risk Assess* 18(6):1156–1185.

Langerholc, T., and Lindqvist, R. 2018. Risk ranking of chemical and microbiological hazards in food. *EFSA J* 16(S1):e160813, 9 pp.

Mangen, M.J.J., Batz, M.B., Kasbohrer, A., Hald, T., Morris, J.G., Taylor, M. and Havelaar, A.H. 2010. Integrated approaches for the public health prioritization of foodborne and zoonotic pathogens. *Risk Anal* 30(5):782–797.

Melnyk, L.J., Wang, Z., Li, Z. and Xue, J. 2016. Prioritization of pesticides based on daily dietary exposure potential as determined from the SHEDS model. *Food Chem Toxicol* 96:167–173.

Morgan, M.G., Florig, H.K., DeKay, M.L. and Fischbeck, P. 2000. Categorizing risks for risk ranking. *Risk Anal* 20(1):49–58.

43

Newsome, R., Tran, N., Paoli, G.M., Jaykus, L.A., Tompkin, B., Miliotis, M., Ruthman, T., Hartnett, E., Busta, F.F., Petersen, B., Shank, F., McEntire, J., Hotchkiss, J., Wagner, M. and Schaffner, D.W. 2009. Development of a risk-ranking framework to evaluate potential high-threat microorganisms, toxins, and chemicals in food. *J Food Sci* 74(2):R39–R45.

Nougadère, A., Reninger, J.-C., Volatier, J.-L. and Leblanc, J.-C. 2011. Chronic dietary risk characterization for pesticide residues: A ranking and scoring method integrating agricultural uses and food contamination data. *Food Chem Toxicol* 49(7):1484–1510.

Papadopoulos, A., Sioen, I., Cubadda, F., Ozer, H., Oktay Basegmez, H.I., Turrini, A. Lopez Esteban, M.T., Fernandez San Juan, P.M., Sokolíc-Mihalak, P., Jurkovic, M., De Henauw, S., Aureli, F., Vin, K. and Sirot, V. 2015. TDS exposure project: Application of the analytic hierarchy process for the prioritization of substances to be analyzed in a total diet study. *Food Chem Tox* 76:46–53.

Pires, S.M., Vieira, A.R., Hald, T. and Cole, D. 2014. Source attribution of human salmonellosis: An overview of methods and estimates. *Foodborne Pathog Dis* 11(9):667–676.

Richardson, L.C., Bazaco, M.C., Parker, C.C., Dewey-Mattia, D., Golden, N., Jones, K., Klontz, K., Travis, C., Kufel, J.Z. and Cole, D. 2017. An updated scheme for categorizing foods implicated in foodborne disease outbreaks: A tri-agency collaboration. *Foodborne Pathog Dis* 14(12):701–710.

Robertson, L.J., Sehgal, R. and Goyal, K. 2015. An Indian multicriteria-based risk ranking of foodborne parasites. *Food Res Int* 77:315–319.

Ross, T. and Sumner, J. 2002. A simple, spreadsheet-based, food safety risk assessment tool. *Int J Food Microbiol* 77(1–2):39–53.

Rozell, D.J. 2015. A cautionary note on qualitative risk ranking of homeland security threats. *Homeland Sc Aff* 11(3):1–5.

Ruzante, J.M., Davidson, V.J., Caswell, J., Fazil, A., Cranfield, J.A.L., Henson, S.J., Anders, S.M., Schmidt, C. and Farber, J.M. 2010. A multifactorial risk prioritization framework for foodborne pathogens. *Risk Anal* 30(5):724–742.

Shaffner, D.W. 2008. *Microbial Risk Analysis of Foods. Emerging Issues in Food Safety*, Ed. Doyle, M.P. Washington, DC: ASM Press, pp. 263.

Siegrist, M., Hübner, P. and Hartmann, C. 2018. Risk prioritization in the food domain using deliberative and survey methods: Differences between experts and laypeople. *Risk Anal* 38(3):504–524.

Speybroeck, N., Devleesschauwer, B., Depoorter, R., Dewulf, J., Berkvens, D., Van Huffel, X. and Saegerman, C. 2015. Needs and expectations regarding risk ranking in the food chain: A pilot survey amongst decision makers and stakeholders. *Food Control* 54:135–143.

Stella, P., Cerf, O., Hugas, M., Koutsoumanis, K.P., Nguyen-The, C., Sofos, J.N., Valero, A. and Zwietering, M.H. 2013. Ranking the microbiological safety of foods: A new tool and its application to composite products. *Trends Food Sci Tech* 33(2):124–138.

Stroheker, T., Scholz, G. and Mazzatorta, P. 2017. A new global scientific tool for the assessment and prioritization of chemical hazards in food raw materials. *Food Control* 79:218–226.

Tomuzia, K., Menrath, A., Frentzel, H., Filter, M., Weiser, A.A., Bräunig, J., Buschulte, A. and Appel, B. 2013. Development of a comparative risk ranking system for agents posing a bioterrorism threat to human or animal populations. *Biosec Bioterr Biodef Strat Pract Sci* 11(1)(Supplement 1):S3–S16.

Tsaboula, A., Papadakis, E.-N., Vryzas, Z., Kotopoulou, A., Kintzikoglou, K. and Papadopoulou-Mourkidou, E. 2016. Environmental and human risk hierarchy of pesticides: A prioritization method, based on monitoring, hazard assessment and environmental fate. *Environ Int* 91:78–93.

Van Asselt, E.D., van der Spiegel, M., Noordam, M.Y., Pikkemaat, M.G. and van der Fels-Klerx, H.J. 2013. Risk ranking of chemical hazards in food-A case study on antibiotics in the Netherlands. *Food Res Int* 54(2):1636–1642.

Van Asselt, E.D., Banach, J.L. and van der Fels-Klerx, H.J. 2018a. Prioritization of chemical hazards in spices and herbs for European monitoring programs. *Food Control* 83:7–17.

Van Asselt, E.D., Noordam, M.Y., Pikkemaat, M.G. and Dorgelo, F.O. 2018b. Risk-based monitoring of chemical substances in food: Prioritization by decision trees. *Food Control* 93:112–120.

Van der Fels-Klerx, H.J., Adamse, P., de Jong, J., Hoogenboom, R., de Nijs, M. and Bikker, P. 2017. A model for risk-based monitoring of contaminants in feed ingredients. *Food Control* 72:211–218.

Van der Fels-Klerx, H.J., Van Asselt, E.D., Raley, M., Poulsen, M., Korsgaard, H., Bredsdorff, L., Nauta, M., D'agostino, M., Coles, D., Marvin, H.J.P. and Frewer, L.J. 2018. Critical review of methods for risk ranking of food-related hazards, based on risks for human health. *Crit Rev Food Sci Nutr* 58(2):178–193.

Veterinary Residues Committee. 2015. Developing a systematic approach to ranking residues of veterinary medicines. *Vet Rec* 177(23):590.

Vose, D. 2008. *Risk Analysis: A Quantitative Guide*, 3rd edition. Hoboken, NJ: John Wiley & Sons, pp. 735.

Vromman, V., Maghuin-Rogister, G., Vleminckx, C., Saegerman, G., Pussemier, L. and Huyghebaert, A. 2014. Risk ranking priority of carcinogenic and/or genotoxic environmental contaminants in food in Belgium. *Food Add Cont Part A* 31(5):872–888.

Webster, K., Jardine, C., Cash, S.B., McMullen, L.M. and Lynn, M. 2010. Risk ranking: Investigating expert and public differences in evaluating food safety hazards. *J Food Prot* 73(10):1875–1885.

WHO. 2015. WHO estimates of the global burden of foodborne diseases: Foodborne disease burden epidemiology reference group 2007–2015. https://www.who.int/foodsafety/publications/foodborne_disease/fergreport/en/ (accessed November 10, 2018).

3

Risk Metrics
Quantifying the Impact of Adverse Health Effects

Brecht Devleesschauwer, Sara M. Pires, Barbara B. Kowalcyk,
Robert L. Scharff, Arie H. Havelaar, and Niko Speybroeck

Contents

47

3.1 INTRODUCTION: FROM REACTIVE TO RISK-BASED FOOD SAFETY SYSTEMS

Due to the complexity and changing nature of the food supply, ensuring its safety has been identified as a *wicked problem* – i.e., a problem that arises in complex and interdependent systems and that is difficult or impossible to solve because of incomplete, contradictory, changing, or incomprehensible requirements (Institute of Medicine 2012). Indeed, the food system is multi-faceted, with a large number of stakeholders having diverse interests. The international food production and distribution systems play a major role in the global economy, with significant impacts on population health, income, employment, rural and urban economies, and the environment. Historically, the approach to ensuring food safety has been reactive – responding to crises as they occur – rather than preventive (Koutsoumanis and Aspridou 2016). Globally, many countries lack the infrastructure needed to meet international food safety standards, which in turn, impacts trade and local access to safe food. To address the food challenges of the 21st century, this paradigm is slowly shifting to an integrated, multi-disciplinary, systems-based approach that is informed by the best available science and focuses on prevention. At the same time, there is an increasing need to utilize limited resources so that they effectively address the most important issues and provide the greatest benefits to the most people. As outlined in Chapter 1, risk analysis, which consists of risk assessment, risk management, and risk communication (Figure 3.1),

RISK ASSESSMENT Science-based	RISK MANAGEMENT Policy-based
1. Hazard identification 2. Exposure assessment 3. Dose-response 4. Risk characterization	1. Risk evaluation 2. Option assessment 3. Option implementation 4. Monitoring and review

RISK COMMUNICATION
Interactive exchange of information and opinions

Figure 3.1   Components of risk analysis.

provides a framework for supporting decision making; it is internationally accepted as the best approach to food safety (Food and Agriculture Organization of the United Nations 2006).

A risk-based food safety system is one that uses "a systematic means by which to facilitate decision-making to reduce public health risk in light of limited resources and additional factors that may be considered" (Havelaar et al. 2007; National Research Council 2010). Central to the risk-based framework (Figure 3.2) is an understanding of the risks and burden of disease (i.e., the impact of a disease in the population). Understanding the burden can be complemented by quantifying, ranking, and attributing

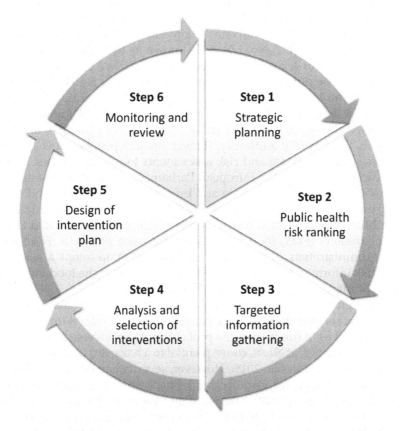

Figure 3.2 Framework for a risk-based food safety system. (Adapted from National Research Council, *Enhancing Food Safety: The Role of the Food and Drug Administration*, The National Academies Press, Washington, DC, 2010.)

the risks to the responsible sources. From there, public health goals can be established – such as the United States Healthy People 2020 goals or the United Nations' Millennium Development Goals – and potential prevention and control interventions can be determined. An evaluation of potential interventions and/or policies allows the determination of their ability to positively impact public health at a reasonable cost in a fair manner. After having identified potential prevention and control strategies, priorities need to be set and resources allocated to those that will have the biggest public health impact. Finally, the effectiveness of the efforts in meeting public health goals and objectives must be quantified.

Risk-based food safety systems are steadily but increasingly being implemented to replace the historical reactive food safety systems. In 2002, the Council of the European Union and the European Parliament adopted Regulation (EC) No. 178/2002, known as the General Food Law of 2002 (http://eur-lex.europa.eu/legal-content/EN/ALL/?uri=CELEX: 32002R0178). One of the key principles of the food law is that "measures adopted by the Member States and the Community governing food and feed should generally be based on risk analysis". The Regulation further created the Rapid Alert System for Food and Feed (RASFF) and the European Food Safety Authority (EFSA), an independent agency that provides scientific advice and risk assessments to relevant bodies in the European Commission, the European Parliament, and Member States. In 2010, recognizing the importance and infrastructure needs around food safety, the United States Congress passed the Food Safety Modernization Act of 2010 (FSMA), the first comprehensive reform of the Food and Drug Administration's (FDA) food safety oversight since 1938 (U.S. Food and Drug Administration 2017). FSMA mandates FDA to adopt a science-based, risk-informed approach to food safety and holds the food industry more accountable for producing safe products.

Although the concept of "risk" is fundamental to these systems, its definition is not entirely standardized. According to the Codex Alimentarius, risk is defined as "a function of the probability of an adverse health effect and the severity of that effect, consequential to a hazard(s) in food" (Codex Alimentarius Commission 2018). However, severity can be viewed in different ways – it may, for instance, be defined as the health or economic impact of the adverse health effects. Many risk assessments do not take severity into account and restrict themselves to the probability of illness (e.g., Bollaerts et al. 2009) or even more distant, the probability of infection (e.g., Hamilton et al. 2006). In toxicology, for instance, the definition of risk by the International Programme on Chemical Safety (2004) is often used, i.e., "the probability of an adverse effect in an organism, system, or (sub)population caused under specified circumstances by exposure to an agent", which thus does not take severity into account.

When quantifying both probability and severity, the function to combine both dimensions can take many different shapes – ranging from a qualitative integration to intricate weighted averages. A risk matrix is, for instance, one of the most basic ways of combining probability and severity (Figure 3.3). Here, both dimensions are expressed in a semi-quantitative way, and risk is expressed as a combination of the two scores – e.g., ranging from low-low to high-high. Despite their intuitive nature, they

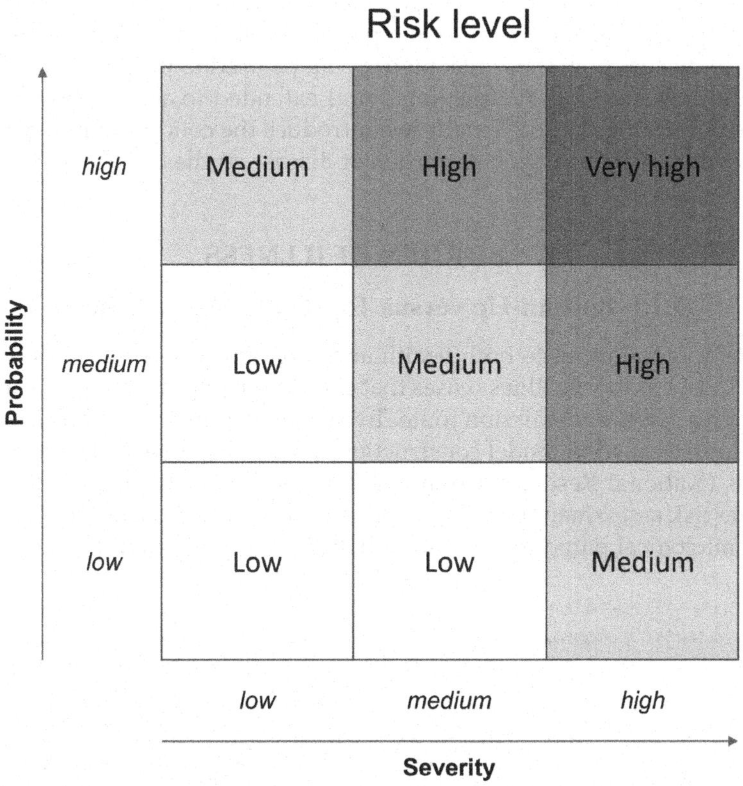

Risk level

Probability	low	medium	high
high	Medium	High	Very high
medium	Low	Medium	High
low	Low	Low	Medium

Severity

Figure 3.3 Risk matrix integrating probability and severity of adverse health effects in a semi-quantitative way to obtain an overall risk level.

pose several limitations for risk-based decision making (Cox, 2008). Their semi-quantitative nature and poor resolution make them less suitable for quantitative risk assessments and risk rankings (see also Chapter 2). Furthermore, the categorization of probability and severity, as well as the definition of the integrated risk levels, is in essence subjective – e.g., what is "medium" risk to one person might well be considered "low" or even "high" risk to another person.

In this chapter, we will focus on the more advanced risk metrics that quantify the health impact or economic impact of foodborne disease, building on Devleesschauwer et al. (2018). Fundamental to these metrics is the burden of illness, i.e., the quantification of the number of foodborne illness cases. We will illustrate each method's utility, data requirements, and output by developing an example on the burden of salmonellosis. We also discuss how these risk metrics can be used to perform risk ranking and how they can be integrated and extended to accommodate further risk ranking criteria. Finally, we introduce the concept of risk-benefit assessment as an extension of burden of disease studies.

3.2 BURDEN OF ILLNESS

3.2.1 Bottom-Up versus Top-Down Approaches

A fundamental input to both health and economic impact metrics is the number of foodborne illness cases that exist in the population or that arise through a given transmission route. Two general approaches, based on the data sources used in model construction, are used to assess the burden of illness (National Research Council 2010): a bottom-up approach following the risk assessment paradigm and a top-down approach following an epidemiological paradigm (Figure 3.4). In theory, both the top-down and

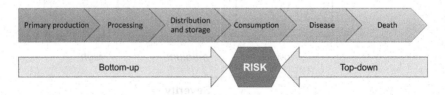

Figure 3.4 Bottom-up and top-down approaches for assessing risk. (Adapted from the EFSA Panel on Biological Hazards (BIOHAZ), *EFSA J.*, 10, 2724, 2012 and Devleesschauwer, B. et al., in *Food Safety Economics: Incentives for a Safer Food Supply*, 2018.)

bottom-up approaches should result in similar estimates for likelihood and severity; in reality, significant data gaps and biases and uncertainty in the metrics make that unlikely (Bouwknegt et al. 2014). The approach selected will probably depend on the risks under consideration and available data. For example, epidemiologic data are typically less specific for assessing risks of exposure to specific food products, such as a particular brand of raw milk cheese, making the bottom-up approach more appealing. Alternatively, epidemiological data are typically more reliable to estimate the total incidence of disease due to a foodborne pathogen, such as campylobacteriosis, making top-down more appealing. EFSA has proposed a strategy to integrate top-down and bottom-up approaches in a Scientific Opinion about risk ranking (EFSA Panel on Biological Hazards (BIOHAZ) 2015).

The bottom-up approach, which derives estimates using the classic risk assessment paradigm that assesses risk using exposure and dose–response information, has been the subject of previous chapters. The focus of this section will therefore be on the top-down approach.

The top-down approach uses information on human disease gathered from public health surveillance and other epidemiological systems to estimate risk at the point of consumption. This can be accomplished according to two main models.

The first approach starts from pathogen-specific surveillance data collected through national surveillance systems. These data typically provide an underestimation of the true burden of illness because of under-ascertainment (i.e., not all patients seek healthcare) and underreporting (i.e., not all healthcare seeking cases will be diagnosed and reported to the national surveillance system). To estimate the true burden of illness, it is therefore necessary to reconstruct the surveillance pyramid. This allows the quantification of multiplication factors, which need to be multiplied with the number of reported cases to obtain an estimate of the true number of cases. For instance, in the United States, Scallan et al. (2011) used data from the Foodborne Diseases Active Surveillance Network to estimate the true number of cases of several foodborne infections in the population. For salmonellosis, the multiplier for under-diagnosis was defined as 29.3 (i.e., $1/29.3 = 3.4\%$ of cases seek healthcare), while the multiplier for underreporting was set at 1.0 (i.e., no underreporting). By applying these multiplication factors to the number of 41,930 reported cases, the true number of cases could be estimated at $41,930 \times 29.3 \times 1.0 = 1,228,549$. Of these cases, 11% were assumed to be travel-related, resulting in an estimated $1,228,549 \times (1 - 0.11) = 1,093,409$ domestically acquired salmonellosis cases.

The other approach starts from burden of illness envelopes, e.g., the total number of diarrhea cases in the population, and attributes these to specific foodborne hazards using population attributable fractions, which may be derived from surveillance, cohort, or cross-sectional studies. For instance, Pires et al. (2015) estimated in a meta-analysis that 2.2% of all diarrhea cases in children <5 worldwide were attributable to salmonellosis. Multiplying this etiological fraction with the total number of diarrhea cases worldwide thus yielded an estimate of the global number of salmonellosis-associated diarrhea cases in children <5.

In the next step, the estimated incidence of a foodborne illness can be attributed to specific transmission routes using results from source attribution studies (Pires et al., 2009). For instance, in an expert elicitation study conducted by Hald et al. (2016), 73% of salmonellosis cases in the AMR A region (World Health Organization [WHO] Region of the Americas – A; including the United States, Canada, and Cuba) were estimated to be foodborne; Hoffmann et al. (2017) further calculated that of those estimated to be foodborne, 22% were attributed to eggs, 22% to poultry meat, and 12% to pork. Continuing our example, this would result in an estimate of 1,093,409 × 0.73 = 798,188 salmonellosis cases obtained through foodborne transmission in the United States, of which 175,601 could be attributed to eggs, 175,601 to poultry meat, and 95,783 to pork.

3.2.2 Outcome Trees

The impact of foodborne disease is defined by the adverse health effects (health states) associated with the concerned foodborne hazard. An important step is, therefore, to design disease outcome trees (or disease models) to define potential outcomes associated with consuming food contaminated with a specific pathogen. A deterministic example using salmonellosis is illustrated in Figure 3.5 based on Centers for Disease Control and Prevention estimates of probabilities associated with healthcare seeking, hospitalization, and death (Scallan et al. 2011) and an estimate of the probability of reactive arthritis resulting from the acute infection (Keithlin et al. 2015). The probabilities associated with eight outcomes (A to G) can be assessed using this tree. For example, the probability of recovering fully without seeking care is 60% for an individual with salmonellosis $(Pr(A) = p_1 * p_2)$. Alternatively, the probability of being hospitalized and acquiring reactive arthritis as a sequela is only 0.1% $(Pr(F) = p_2 * p_6 * p_{10})$.

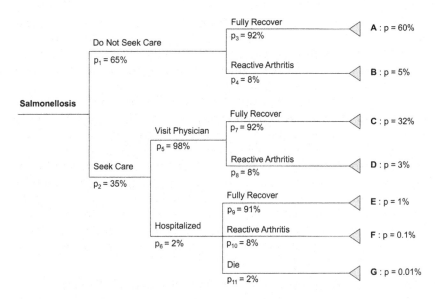

Figure 3.5 Disease outcome tree for *Salmonella* spp. (Based on Scallan, E. et al., *Emerg. Infect. Dis.*, 17, 7–15, 2011 and Keithlin, J. et al., *Epidemiol. Infect.*, 143, 1333–1351, 2015 and adapted from Devleesschauwer, B. et al., in *Food Safety Economics: Incentives for a Safer Food Supply*, 2018.)

The tree in Figure 3.5 makes it clear that the sequela of reactive arthritis is assumed to be equally likely under all severity levels. Whether this is correct or whether the science has just not yet been able to discern these differences is unclear. It can further be noted that this tree is relatively simple, even for salmonellosis. Other outcomes, such as whether the sick person provides a stool sample, is prescribed pharmaceuticals, misses work, or utilizes home healthcare services could be included to develop a more complete decision tree. In cost-of-illness studies (see Section 3.4), these additions to the model may be warranted. Finally, for the sake of simplicity, uncertainty is not expressed in this tree, though the inclusion of uncertainty intervals and scenario and sensitivity analyses is a key characteristic of quality studies.

The choice of values used often relies on the judgments of the modeler. For example, in Figure 3.5, an estimate (8%) from a recent meta-analysis was used for the likelihood of reactive arthritis that focused on diagnoses made by specialists (Keithlin et al. 2015). However, the same meta-analysis also provided an estimate of 6% for all studies, an estimate of

12% for those that had follow-ups within 90 days of the acute illness, and an estimate of 0.2% for those studies involving more than 10,000 persons who had had salmonellosis. Given the wide range of estimates available, modelers are forced to make judgments about which estimates to use and whether to include other estimates in scenario analyses.

Similarly, the choice to limit sequelae to reactive arthritis in this example was likely an oversimplification. Increasing complexity should not be an aim in itself, but many other sequelae have been associated with salmonellosis, including irritable bowel syndrome, inflammatory bowel disease, Crohn's disease, ulcerative colitis, Guillain–Barré syndrome, Miller Fisher syndrome, and hemolytic uremic syndrome. Generally speaking, health outcomes should be included in burden of illness estimates if causation between the acute illness and the outcomes can be sufficiently established. There are both empirical and theoretical criteria for demonstrating causation, including the Bradford Hill criteria (Hill 1965).

3.2.3 From Burden of Illness to Burden of Disease

Burden of disease refers to the total impact of a disease, including health, social, and financial impacts, on society (population burden) and on the individual affected (individual burden). Burden of illness, which provides a picture of the population-level occurrence of foodborne disease, must therefore be extended to account for the severity of the adverse health effects (Batz et al. 2012; Devleesschauwer et al. 2015; Mangen et al. 2010). In the next sections, we will introduce burden of disease metrics that account for the health impacts of the adverse health effects (e.g., reduced quality of life and shorter lifespans) and the economic impacts of the adverse health effects (e.g., increased medical costs, decreased worker productivity, and lower incomes).

3.3 HEALTH IMPACT METRICS

As stated previously, measuring the health impact of foodborne diseases based on the number of prevalent or incident cases, or the number of deaths, does not provide a complete picture of the impact of foodborne diseases on human health. Indeed, these measures quantify the impact of either morbidity or mortality, thus prohibiting a comparative ranking of highly morbid but not necessarily fatal diseases (e.g., mild to moderate diarrhea) and diseases with a high case fatality (e.g., perinatal listeriosis).

On the other hand, they only quantify occurrence of illness or death, thus treating each illness case, or each fatal case, alike. Foodborne diseases may, however, differ in clinical impact and duration of the concerned symptoms, such that the severity of different illness cases may differ. Likewise, fatal cases occurring at different ages will result in different numbers of potential life years lost, such that the impact of deaths may differ across cases.

To overcome the limitations of these simple measures, various summary measures of population health (SMPHs) have been developed as an additional source of information for measuring disease burden. The wide range of proposed SMPHs use time as a general unit of measure; they can further be divided into two broad families: health gaps (i.e., time not lived in good health) and health expectancies (i.e., time lived in good health) (Devleesschauwer et al. 2014a). The most powerful SMPHs allow information on mortality and non-fatal health outcomes to be combined, which requires weighting the time lived with disease or disability according to the health experienced or lost. Currently, the most important SMPH for quantifying burden of disease is the disability-adjusted life year (DALY).

The DALY belongs to the family of health gap measures and is currently the most widely used SMPH in epidemiological research. DALYs find their origin in the Global Burden of Disease (GBD) studies and are officially adopted by the WHO for reporting on health information (Murray et al. 2012; World Health Organization 2018). DALYs measure the health gap from a life lived in perfect health and quantify this health gap as the number of healthy life years lost due to morbidity and mortality. A disease burden of 100 DALYs would thus imply a total loss of 100 healthy life years irrespective of how these healthy life years were lost. Diseases, hazards, or risk factors accounting for more DALYs thus have a higher public health impact.

DALYs extend the notion of mortality gaps to include time lived in health states worse than ideal health (Devleesschauwer et al. 2014b). Specifically, they are the sum of years of life lost due to premature mortality (YLLs) and years lived with disability, adjusted for severity (YLDs):

$$DALY = YLL + YLD$$

YLLs are the product of the number of deaths (M) and the residual life expectancy (RLE) at the age of death:

$$YLL = M * RLE$$

Two approaches exist for defining YLDs. Following an incidence perspective, YLDs are defined as the product of the number of incident cases (N), the duration until remission or death (D), and the disability weight (DW), which reflects the reduction in health-related quality of life on a scale from zero (full health) to one (death):

$$YLD_{inc} = N * D * DW$$

The incidence perspective assigns all health outcomes, including those in future years, to the initial event (e.g., *Campylobacter* infection). This approach, therefore, reflects the future burden of disease resulting from current events.

An alternative formula for calculating YLDs follows a prevalence perspective and defines YLDs as the product of the number of prevalent cases (P) with the disability weight (Murray et al. 2012):

$$YLD_{prev} = P * DW$$

In this prevalence perspective, the health status of a population is assessed at a specific point in time, and prevalent diseases are attributed to events that happened in the past. This approach thus reflects the current burden of disease resulting from previous events. Although both perspectives are valid, the incidence perspective is more sensitive to current epidemiological trends (Murray 1994), including the effects of intervention measures, and therefore often preferred for assessing the burden of foodborne diseases (Devleesschauwer et al. 2015). Furthermore, as YLLs by definition follow an incidence perspective, the use of prevalence YLDs would lead to a "hybrid" DALY measure (Schroeder 2012). Nonetheless, it should also be noted that the incidence and prevalence perspectives would yield identical results when disease epidemiology and population demographics are in a steady state, which can be assumed for acute (i.e., short-term) conditions.

Figure 3.6 presents a theoretical example of calculating DALYs, following the incidence perspective. An individual is born in perfect health. At the age of 20, a given event (e.g., foodborne disease) leads to a decrease of quality of life by 20% (i.e., DW = 0.20), and thereafter the person lives in this new health state for another 50 years, at which point he/she dies prematurely. The burden associated with this disease for this individual (total DALYs) is calculated by summing up the years lived with disability (YLD) with the years of life lost due to premature death (YLL).

The recommended approach for quantifying the health impact of foodborne diseases is the *hazard-based* DALY calculation approach

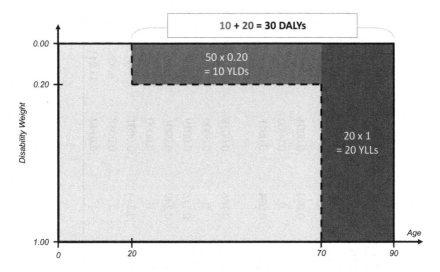

Figure 3.6 Visual example of the disability-adjusted life year metric. DALY: disability-adjusted life year; YLD: years lived with disability; YLL: years of life lost.

(Devleesschauwer et al. 2014c). This approach defines the burden of a specific foodborne disease as that resulting from all health states, i.e., acute symptoms, chronic sequelae, and death, that are causally related to the concerned hazard, and which may become manifest at different time scales or have different severity levels (Mangen et al. 2013). The starting point for quantifying DALYs is therefore the construction of a disease model or outcome tree (Devleesschauwer et al. 2014c).

Table 3.1 demonstrates how DALYs can be computed for salmonellosis following the model introduced in Figure 3.5. In addition to the two sequelae explicitly modeled, i.e., reactive arthritis and death, the index disease, i.e., acute gastroenteritis, is assumed to occur in all salmonellosis patients. We further assume that cases who do not seek care present with mild diarrhea, that cases who only visit a physician present with moderate diarrhea, and that hospitalized cases present with severe diarrhea. Following Haagsma et al. (2008), we assume durations for mild, moderate, and severe diarrhea of 3, 10, and 14 days, respectively. Disability weights are adopted from Salomon et al. (2015), i.e., 0.074, 0.188, and 0.247 for mild, moderate, and severe diarrhea, respectively. For reactive arthritis, we assume a duration of 0.61 years (Mangen et al., 2018) and disability weights of 0.034, 0.116, and 0.344 for mild, moderate,

Table 3.1 Simplified Calculation of Disability-Adjusted Life Years Associated with Salmonellosis

Health State	Transition Probability (%)	Probability (%)	Duration (years)	Disability Weight	DALY per Case	DALY in Population
Salmonellosis						
→ *Do not seek care*	65	65	0.008	0.074	0.0004	316
→ → Recover fully	92	60		0	0.0000	0
→ → → Reactive arthritis	8	5	0.610	0.034	0.0011	861
→ *Seek care*	35					
→ → Visit physician	98	34	0.027	0.188	0.0018	1,410
→ → → Recover fully	92	32		0	0.0000	0
→ → → → Reactive arthritis	8	3	0.610	0.116	0.0019	1,550
→ → → Hospitalized	2	1	0.038	0.247	0.0001	53
→ → → → Recover fully	90	1		0	0.0000	0
→ → → → → Reactive arthritis	8	0.1	0.610	0.344	0.0001	94
→ → → → Die	2	0.01	62	1	0.0087	6,928
Sum of all health states					**0.0140**	**11,211**

and severe cases, respectively (Salomon et al., 2015). For deaths, because we have no knowledge of the age of the cases (and thus of deaths), we assume an average age at death of 30 years, and apply the WHO standard life expectancy table to obtain a residual life expectancy of 62. This results in an average DALY per salmonellosis case of 0.014, or an average loss of approximately 5 days in good health (i.e., $0.014 \times 365 = 5$). This can now be multiplied with burden of illness estimate to obtain the total number of DALYs in the population or due to a given transmission route. Continuing our burden of illness example, the 798,188 estimated foodborne salmonellosis cases in the United States would have resulted in $798,188 \times 0.014 = 11,211$ DALYs.

In addition to the DALY metric, other SMPHs allow quantification of the burden of disease. Among the many alternatives, the most widely used is the quality-adjusted life year (QALY). The QALY belongs to the family of health expectancies and is a standard tool in health economic evaluations and cost–utility analyses in particular. QALYs are healthy life years, obtained by weighting life years according to utility weights, or simply, QALY weights, which reflect individual preferences for time spent in different health states. A number of methods are used to elicit QALY weights, including the standard gamble, time tradeoff, and visual analog scale (Torrance 1986). Common to all methods is their use of a scale that measures health as being between 0 (death) and 1 (perfect health). The use of QALYs across multiple pathogens was made possible by the development of standardized QALY weights associated with multiple dimensions of well-being, which has allowed the generation of condition-specific QALY estimates without costly studies focused specifically on each pathogen in question (though expert opinion is needed to assign QALY weights in this case). For instance, in the EQ-5D multi-attribute utility scale developed by the EuroQoL group, five dimensions of well-being are included (hence the acronym): mobility, self-care, usual activities, pain/discomfort, and anxiety/depression (Herdman et al. 2011).

The health impact of foodborne diseases may be quantified as QALY losses – i.e., the utility losses associated with foodborne disease, which include both disability losses and pain and suffering losses. The measurement of QALY losses must account for the typical sufferer's initial QALY state, which is generally less than 1. Ideally, the initial state should be based on that of the typical person who gets a foodborne disease (who is older or younger and typically more immunocompromised than the average person), though the average population QALY level is frequently used (Batz et al. 2014; Minor et al. 2015; Scharff 2015).

3.4 ECONOMIC IMPACT METRICS

3.4.1 Costs Associated with Foodborne Disease

Decisions in a risk-based food safety system are driven by more than just public health impacts. Risk managers must also consider economic, social, and political factors in the decision-making process. Therefore, it is important to understand the costs associated with foodborne disease: the individual who becomes sick from consuming tainted food, the retailer who sells the contaminated product, the food producer who allows contamination, and the government agencies that monitor, investigate, and regulate all incur costs from foodborne diseases. Table 3.2, an adapted version of the taxonomy originally developed by the USDA Economic Research Service (Roberts 1989), illustrates these costs. Understanding each of these costs is important in a risk-based food safety system, though most efforts to measure economic cost have focused on household costs.

The **household** incurs costs whether or not an individual in the household has been made ill by their consumption of food. Specifically, consumers who are aware of risks associated with foods may face costs if they engage in self-protective efforts. For example, a consumer may choose to buy pasteurized products, avoid risky foods that he/she likes, or cook foods until any potential pathogens are destroyed (at the expense of taste). Each of these measures has a cost to the consumer, either monetary or through lost utility (well-being).

In the presence of illness, the costs include medical costs, productivity losses (to both sick persons and caregivers), pain and suffering losses, and mortality losses. Medical costs include costs for hospitalizations, physician services (both in-patient and out-patient), and drugs used (both prescription and over the counter). Ancillary medical services, such as tests

Table 3.2   Costs Associated with Foodborne Disease (Devleesschauwer et al., 2018)

Households	Food Industry	Public Health Sector
Self-protection	Product recalls	Disease surveillance
Medical care	Plant closing/clean-up	Outbreak response
Productivity losses	Product liability	Recall assistance
Caregiver costs	Reputation costs	Inspection/clean-up
Quality of life	Regulatory compliance	Regulatory enactment
Mortality		Regulatory enforcement

of stool samples and urgent care/emergency room costs, are also included in this category.

Productivity losses occur when an individual is unable to perform productive tasks due to illness (either their own or that of someone they must care for, such as a child). Often, this is measured as the costs of absenteeism from paid work. However, some researchers have chosen to value the time of all ill persons at the average wage in the United States regardless of their work status or age (i.e., the human capital approach, in which the average wage is used as a proxy for the opportunity cost of the individual for time spent ill rather than engaging in his/her normal activities). There are also likely to be reductions in productivity for those who go to work sick, though these losses are likely to be significantly less than those for persons who stay home. Lost household production is also a cost of foodborne disease.

In some economic assessments, a monetary value is assigned to the intangible costs, i.e., the quality of life losses, associated with foodborne disease (quantified as DALYs or QALY losses). The physical discomfort or pain associated with foodborne disease is one way an individual's quality of life is affected, but it is not the only way. Inability to engage in pleasurable activities (or reduced pleasure from those activities) also is an economic cost from foodborne disease. For example, if an individual with a mild case of illness decides not to go with friends to a concert because they do not want to deal with the consequences from a diarrheal illness, their utility is reduced by an amount equal to the value they would have gotten from going to the concert while healthy minus the value they actually got from staying home with the illness. Quality of life losses may also be borne by friends and family who must endure seeing a loved one suffering. These may be quite high, especially when a parent is caring for a very ill child. Similar costs are borne when an individual dies due to a foodborne disease. In some instances, such as when chronic sequelae occur, other household cost categories, such as professional home health-care assistance, may be appropriate to include.

A number of costs accrue to **industry** as a result of foodborne diseases. First, if it is determined that its product has the potential to make people sick, a firm may decide or be instructed to institute a recall of the product. Costs associated with this effort include lost product sales equal to the market value of the recalled product and the cost of collecting and disposing of the product. If the recalled product has been in contact with processing and/or holding facilities, these facilities must conduct a thorough cleaning process, often entailing a lengthy closure of the operation. Next, if anyone was made ill due to the contaminated product, the firm responsible may be exposed to litigation and its attendant costs. Also, media coverage of outbreaks, litigation, and product recalls can have an effect on the reputation/value of the brand. Retailers and wholesalers may also suffer from costs associated with collecting and disposing of recalled product as well as suffering from potential reputation costs if their customers perceive them as sourcing from unscrupulous suppliers (Ribera et al. 2012; Hussain et al. 2013; Pozo and Schroeder 2016). Finally, if the problem is not an isolated one, and there are intervention measures that could remedy the problem, government may respond with costly regulation.

The **public health sector** also incurs costs as a result of foodborne disease. The various surveillance systems that track contamination and illnesses are costly to maintain, and detection and investigation of outbreaks incur additional costs. When recalls are initiated, government personnel are involved, whether or not illnesses have occurred. Inspections and assistance with cleaning up contaminated facilities are also activities funded by government. Finally, the promulgation of regulation involves costs, as do enforcement activities associated with the regulation.

3.4.2 Methods Used to Estimate Costs

As for health impact, different approaches exist for estimating the economic burden associated with foodborne disease. The most commonly applied approach measures the *cost-of-illness* from a societal perspective. Stated or revealed preference methods that generate *willingness to pay/accept* measures are used in some cases to supplement or substitute cost-of-illness studies. DALYs or QALY losses are in some cases monetized for use in cost-of-illness studies, thereby allowing a monetization of the intangible costs. Industry costs are often estimated using case studies, looking at effects on individual companies and industries as a result of food safety incidents. Attempts have been made to estimate recall and litigation costs, but these measures are generally very crude.

3.4.2.1 Cost-of-Illness
The cost-of-illness method is the most widely used approach among economists. The goal of this method is to calculate costs separately and aggregate them for presentation as a single cost number. The following equation illustrates a simple cost-of-illness formula for household costs:

$$Cost_i = DH_i + DNH_i + Productivity_i + QoL_i + Death_i$$

The cost for individual i from a foodborne illness is the sum of direct healthcare (DH), direct non-healthcare (DNH), productivity, quality of life (QoL), and death-related costs. Total costs are defined as the sum of individual costs ($\sum_{i=1}^{n} Cost_i$) and are often used by policymakers as a measure of problem scope, to be used to set agency priorities. Cost per case measures are typically used in regulatory analyses as a means of demonstrating the economic value of an intervention. Cost per case is total costs divided by the number (n) made ill by the pathogen ($\sum_{i=1}^{n} \frac{Cost_i}{n}$). Cost per case is multiplied by number of cases averted (or expected to be averted) by a given intervention to determine intervention effectiveness. The primary focus on household costs means that costs to industry and public health entities are often undervalued.

Direct healthcare and non-healthcare costs are typically evaluated by matching the outcomes of a disease outcome tree (cf. Section 3.2.2) with the utilization of specific resources and the corresponding unit costs. Direct healthcare costs are related to the resources provided by the healthcare sector, such as healthcare provider consultations, diagnosis, medication, and hospitalization. Direct non-healthcare costs (also called *patient costs*) are related to the resources used for healthcare that are not borne by the healthcare system, such as over-the-counter medications and other patient co-payments, and travel expenses to visit a healthcare provider. Unit costs of healthcare resources may be derived from hospital and physician services databases. For example, in 2013, the United States National Inpatient Sample had data on 6455 discharges with a primary diagnosis related to infection with *Salmonella*. The average cost was $9531 for an average of 5.1 days in the hospital.

Earlier efforts to estimate cost-of-illness often relied on interviews or surveys of those who had been sickened in an outbreak. In this case, individuals reported what they (or their insurance companies) had spent on physician services, medication, hospital costs, and other costs. The primary problem with this approach is that the results may not be generalizable to the broader population outside the outbreak area. That said, when other values are not available, estimates from outbreak reports can be useful.

Productivity losses theoretically include lost work in both the paid and household sectors. Where work is compensated, costs include both wages and other compensation for the time away from work. Uncompensated work, or household production (Becker 1965), may also be lost, but it is unclear how much of this an ill person is able to do. A number of studies have looked at lost wages for persons who are ill (Buzby and Roberts 2009; Hoffmann et al. 2012, 2015; Scharff 2012, 2015). The most accurate of these have taken into account both wages and benefits using estimates for cost of compensation, rather than only wages. Also, some studies have included work loss due to caregiving for children (e.g., Scharff 2015). The availability of surveillance data for some pathogens allows the generation of age profiles for those made ill, which can be used to better predict work status and child care needs.

The inclusion of **quality of life losses** in cost-of-illness analyses is controversial. Likewise, losses from death due to foodborne disease imply not just a loss of productivity from premature mortality but also a loss of utility for the deceased. Originally, no cost-of-illness studies attempted to quantify pain and suffering. In the 1990s, however, the FDA began using a monetized QALY estimate for the value of lost QoL, as suggested by Mauskopf and French (1991). Some have argued that the monetization of QALYs is not appropriate because it requires the imposition of a number of restrictive assumptions (Hammit and Haninger 2007), while others have argued that the QALY is the best measure of welfare loss available (Adler 2006). Furthermore, there is no consensus on how to monetize QoL losses. The two main strategies apply either a measure of the value for a statistical life year or the prevailing willingness to pay threshold for gaining a healthy life year. However, the different possible measures often differ in orders of magnitude. In the United States, for instance, measures of the value for a statistical life ranging from $1.6 million to $15.7 million have been proposed (Hoffmann et al., 2015). Furthermore, and irrespective of the measure used, QALY losses in advanced economies will receive a higher monetary value than QALY losses in developing economies, which could raise ethical concerns.

Despite the inclusion of many cost categories in cost-of-illness studies, some are not accounted for, and others can be best seen as rough estimates. For example, the exclusion of self-protective actions and often, QoL losses leads to estimates that are likely to be underestimates of true cost. Some have therefore suggested that the cost-of-illness approach leads to point estimates that give a false sense of certainty. Though uncertainty intervals and sensitivity analyses are increasingly included in these analyses, these typically do not completely account for the structural deficiencies of the approach. Other approaches, most notably the willingness to pay approach, have been suggested as alternatives to the cost-of-illness approach.

3.4.2.2 Willingness to Pay

Foodborne disease cost-of-illness estimates have been criticized for being too limited, i.e., for not including all losses to an ill individual. An alternative is to assess the *willingness to pay* to avoid foodborne disease. Theoretically, this is the most complete measure of utility loss for the affected individual because the individual is allowed to take into account all losses in making their assessment. Currently, the principal method used to elicit willingness to pay for foodborne disease is based on dichotomous choice experiments. Here, individuals are asked to choose between two price/risk combinations for a given food product, where each person chooses between lower-risk/higher-price and higher-risk/lower-price options (Nayga et al. 2006; Teisl and Roe 2010; Haninger and Hammitt 2011). Experimenters vary the price/risk combinations across individuals and in some cases, provide individuals with follow-up price/risk choices to more precisely assess willingness to pay measures. These experiments are more likely to yield meaningful responses when the exchanges are not hypothetical, are conducted in realistic settings, and communicate risks in a meaningful way.

Despite the theoretical appeal of willingness to pay measures, holistic willingness to pay measures have not been used in policy settings for food safety. One reason for this is the cost of conducting these experiments. Second, willingness to pay studies do not include external costs (e.g., costs to one's workplace from absenteeism, the costs to the insurance pool for claims made, and costs to family members for caregiving), and thus are also incomplete. Ideally, these costs would have to be assessed and added in. Perhaps most importantly, the values generated using these methods are not perceived as being plausible by some. This is because willingness to pay estimates are routinely an order of magnitude higher than cost-of-illness estimates and are less sensitive to risk, duration, or consequences than would be expected. For example, Hammitt and Haninger (2007) found that people were implicitly willing to pay $8300 to avoid 1 day described as follows: *"You will have an upset stomach and will feel tired, but these symptoms will not prevent you from going to work or from doing most of your regular activities."* At the same time, the authors found that people were not willing to pay significantly more to avoid 3 days with the same symptoms, and willingness to pay increased less than proportionally with risk. This may be because biases such as the part-whole problem or yea-saying are at work. As a result, the linear extrapolation of individuals' willingness to pay to reduce risk from a single meal or product in an experimental setting to a general willingness to pay measure is likely to overestimate the value of the risk.

3.5 RISK RANKING

In a time of increasing (recognized) threats and decreasing financial resources, there is a growing need to rationally allocate available means (Speybroeck et al., 2015). Consequently, risk ranking is increasingly used within the food safety risk analysis framework (Stella et al. 2013). The aim of these exercises is to prioritize for decision making certain hazards, hazard–commodity pairs, or exposure routes for a given hazard, based on their perceived importance. Figure 3.7, for instance, shows the ranking of the global health impact of 31 foodborne hazards according to the results of WHO Foodborne Disease Burden Epidemiology Reference Group (Havelaar et al. 2015). However, as different stakeholders have their own prioritization objectives and beliefs, the outcome of such exercises is necessarily context-dependent, and no unique or intrinsically correct ranking of risks exists.

It is also increasingly recognized that the use of a single criterion to rank risks may be insufficient, as diseases vary greatly in incidence, clinical manifestations, control measures, transmission potential, and socio-economic impact in animals and humans. *Trichinella* spp., for instance, have a near negligible human health impact in Europe, while their economic impact remains important due to continuous animal/food monitoring

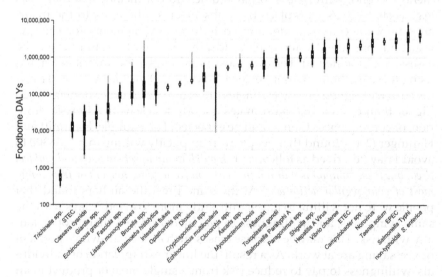

Figure 3.7 Ranking of the global health impact of 31 foodborne hazards. (Adapted from Havelaar, A.H. et al., *PLOS Med.*, 12, e1001923, 2015.)

and trade implications (Devleesschauwer et al. 2015b). It may therefore be required to base the ranking of risks on multiple criteria (Mangen et al. 2010). To quantify the disease burden of various foodborne hazards in the Netherlands in 2011, Mangen et al. (2015) generated both DALY and cost-of-illness estimates at both the population and the individual level. The combination of these different criteria led to four different rankings, with some hazards, most notably *Toxoplasma gondii,* scoring high on multiple criteria.

Ideally, however, a risk ranking exercise should result in a single ranking taking into account multiple criteria. Some authors summed DALYs and economic impact estimates by assuming 1 DALY to correspond to an economic loss equal to the per capita gross national product (Torgerson et al. 2008). This approach belongs to the family of *multi-criteria decision analysis* (MCDA) methods. In MCDA, an overall importance measure is constructed based on different criteria, which are assigned weights reflecting their perceived contribution (Cardoen et al. 2009; Havelaar et al. 2010; FAO/WHO 2014; Bouwknegt et al. 2018). As these weights imply a normative choice, the definition of weights should reflect social preferences or expert opinion. The selection of criteria to be scored typically depends on expert opinion. The scoring of the criteria – be it quantitative, semi-quantitative, or qualitative – can be based on existing data or on expert elicitation. Despite its subjective appearance, MCDA provides a transparent and consistent framework for ranking risks (Anderson et al. 2011). It also allows criteria to be included for which no quantifications are available or possible. In the joint Food and Agriculture Organization of the United Nations and World Health Organization (FAO/WHO) multi-criteria-based ranking of foodborne parasites, trade relevance and impacts on economically vulnerable communities were included in addition to criteria related to health impact. Figure 3.8 shows the outcome of this exercise, confirming the importance of *Taenia solium* at a global level.

A more extensive discussion of risk ranking as a risk management tool to identify food safety priorities at a national level is provided in Chapter 2.

3.6 RISK-BENEFIT ASSESSMENT

While risk assessment, burden of disease, and risk ranking approaches are well-established tools to prioritize food safety problems and allocate resources, they do not always inform public health decisions focused on

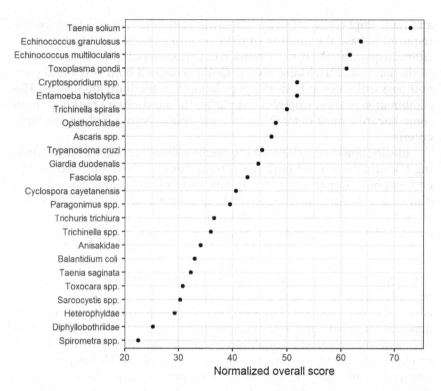

Figure 3.8 Multicriteria-based global ranking of foodborne parasites according to FAO/WHO (2014). (Adapted from Devleesschauwer, B. et al., *Food Waterborne Parasitol.*, 8–9, 1–13, 2017.)

foods in a comprehensive way. As an example, food safety strategies or consumer advice that aims at protecting the population from a given risk (e.g., methylmercury in fish) may not take into account the beneficial effects of that food for individuals' health (such as those of fatty acids, also present in fish). Risk-benefit assessment of foods (RBA) is a relatively new tool that was developed to address such questions. It builds on methodologies typically used in risk assessment and burden of disease studies and allows an integrated evaluation of the negative and positive health effects of a food compound, food, or overall dietary pattern. The application of health metrics is a fundamental piece of such integration, and depending on the question and choice of methodology, these can be single-outcome metrics (e.g., disease incidence or mortality); integrated (or summary) health (e.g., DALYs or QALYs); and economically oriented measures such as willingness to pay (Fransen et al. 2010).

Since its initial developments in the area of food, which were largely achieved in the context of international projects (e.g., Hoekstra et al. 2012; Hart et al. 2013; Verhagen et al. 2012; Pires et al. 2019), substantial progress has been made in the field, including the development of probabilistic RBAs (Thomsen et al., 2019). Chapter 4 describes RBA methodologies, applications, and major challenges in detail.

3.7 CONCLUSIONS

Modern food safety systems are increasingly risk-based. Central to the risk-informed framework are risk analysis and disease burden estimates, providing the foundation for decision making and allocation of resources. Data gathered on the burden of foodborne disease provides important information for risk assessment and subsequently, scientifically grounded risk reduction strategies. Although many risk assessment studies still use basic risk metrics such as the probability of illness, there is increasing attention to more advanced and comprehensive risk metrics, such as summary measures of population health that quantify the intangible costs of foodborne disease, and monetary metrics that quantify the costs to households, industry, and the public sector. A key use of these metrics is to support priority setting within the food safety system. As the literature on the development and use of comprehensive risk metrics evolves, the efficiency of decisions made will improve, and stakeholders will be better protected and served.

ACRONYMS

AMR   WHO Region of the Americas
DALY   disability-adjusted life year
DH   direct healthcare
DNH   direct non-healthcare
EFSA   European Food Safety Authority
FDA   Food and Drug Administration
FSMA   Food Safety Modernization Act
GBD   Global Burden of Disease
MCDA   multi-criteria decision analysis

QALY	quality-adjusted life year
QoL	quality of life
RASFF	Rapid Alert System for Food and Feed
RBA	risk-benefit assessment
SMPH	summary measure of population health
WHO	World Health Organization
YLD	years lived with disability
YLL	years of life lost

REFERENCES

Adler MD. QALY's and policy evaluation: A new perspective. *Yale J Health Policy Law Eth* 2006;6(1):1–92.

Anderson M, Jaykus LA, Beaulieu S, Dennis S. Pathogen-produce pair attribution risk ranking tool to prioritize fresh produce commodity and pathogen combinations for further evaluation (P3ARRT). *Food Control* 2011;22(12):1865–72.

Batz MB, Hoffmann S, Morris JG. Ranking the disease burden of 14 pathogens in food sources in the United States using attribution data from outbreak investigations and expert elicitation. *J Food Prot* 2012;75(7):1278–91.

Batz M, Hoffmann S, Morris JG Jr. Disease-outcome trees, EQ-5D scores, and estimated annual losses of quality-adjusted life years (QALYs) for 14 foodborne pathogens in the United States. *Foodborne Pathog Dis* 2014;11(5):395–402.

Becker GS. A theory of the allocation of time. *Econ J (London)* 1965;75(299):493–517.

Bollaerts KE, Messens W, Delhalle L, Aerts M, Van der Stede Y, Dewulf J, Quoilin S, Maes D, Mintiens K, Grijspeerdt K. Development of a quantitative microbial risk assessment for human salmonellosis through household consumption of fresh minced pork meat in Belgium. *Risk Anal* 2009;29(6):820–40.

Bouwknegt M, Knol AB, van der Sluijs JP, Evers EG. Uncertainty of population risk estimates for pathogens based on QMRA or epidemiology: A case study of *Campylobacter* in the Netherlands. *Risk Anal* 2014;34(5):847–64.

Bouwknegt M, Devleesschauwer B, Graham H, Robertson LJ, van der Giessen JW. The Euro-FBP Workshop Participants. Prioritisation of food-borne parasites in Europe, 2016. *EURO Surveill* 2018;23(9):pii=17-00161.

Buzby JC, Roberts T. The economics of enteric infections: Human foodborne disease costs. *Gastroenterology* 2009;136(6):1851–62.

Cardoen S, Van Huffel X, Berkvens D, Quoilin S, Ducoffre G, Saegerman C, Speybroeck N, Imberechts H, Herman L, Ducatelle R, Dierick K. Evidence-based semiquantitative methodology for prioritization of foodborne zoonoses. *Foodborne Pathog Dis* 2009;6(9):1083–96.

Codex Alimentarius Commission. *Codex Alimentarius Commission Procedural Manual.* 26th ed. Joint FAO/WHO Food Standards Programme; 2018. http://www.fao.org/documents/card/en/c/I8608EN/. Accessed 15 March 2019.

Cox LA Jr. What's wrong with risk matrices? *Risk Anal* 2008;28(2):497–512.

Devleesschauwer B, Maertens de Noordhout C, Smit GSA, Duchateau L, Dorny P, Stein C, et al. Quantifying burden of disease to support public health policy in Belgium: Opportunities and constraints. *BMC Public Health* 2014a;14:1196.

Devleesschauwer B, Havelaar AH, Maertens de Noordhout C, Haagsma JA, Praet N, Dorny P, et al. Calculating disability-adjusted life years to quantify burden of disease. *Int J Public Health* 2014b;59(3):565–9.

Devleesschauwer B, Havelaar AH, Maertens de Noordhout C, Haagsma JA, Praet N, Dorny P, et al. Daly calculation in practice: A stepwise approach. *Int J Public Health* 2014c;59(3):571–4.

Devleesschauwer B, Haagsma JA, Angulo FJ, Bellinger DC, Cole D, Döpfer D, et al. Methodological framework for World Health Organization estimates of the global burden of foodborne disease. *PLOS ONE* 2015;10(12):e0142498.

Devleesschauwer B, Bouwknegt M, Dorny P, Gabriël S, Havelaar AH, Quoilin S, Robertson LJ, Speybroeck N, Torgerson PR, van der Giessen JWB, Trevisan C. Risk ranking of foodborne parasites: State of the art. *Food Waterborne Parasitol* 2017;8–9:1–13.

Devleesschauwer B, Scharff RL, Kowalcyk BB, Havelaar AH. Burden and risk assessment of foodborne disease. In: Roberts T (ed) *Food Safety Economics: Incentives for a Safer Food Supply*; Cham: Springer. 2018, 83–106.

EFSA Panel on Biological Hazards (BIOHAZ). Scientific opinion on the development of a risk ranking framework on biological hazards. *EFSA J* 2012;10:2724.

EFSA Panel on Biological Hazards (BIOHAZ). Scientific opinion on the development of a risk ranking toolbox for the EFSA BIOHAZ Panel. *EFSA J* 2015;13:3939.

Food and Agriculture Organization of the United Nations. Food safety risk analysis: A guide for national food safety authorities (Food and nutrition paper 87) 2006. ftp://ftp.fao.org/docrep/fao/009/a0822e/a0822e.pdf. Accessed 15 March 2019.

Food and Agriculture Organization of the United Nations and World Health Organization. *Multicriteria-Based Ranking for Risk Management of Food-Borne Parasites. Report of a Joint FAO/WHO Expert Meeting*, 3–7 September 2012. 2014. http://www.fao.org/3/a-i3649e.pdf. Accessed 15 March 2019.

Fransen H, de Jong N, Hendriksen M, Mengelers M, Castenmiller J, Hoekstra J, van Leeuwen R, Verhagen H. A tiered approach for risk-benefit assessment of foods. *Risk Anal* 2010;30(5):808–16.

Haagsma JA, Havelaar AH, Janssen BM, Bonsel GJ. Disability adjusted life years and minimal disease: Application of a preference-based relevance criterion to rank enteric pathogens. *Popul Health Metr* 2008;6:7.

Hald T, Aspinall W, Devleesschauwer B, Cooke R, Corrigan T, Havelaar AH, Gibb HJ, Torgerson PR, Kirk MD, Angulo FJ, Lake RJ, Speybroeck N, Hoffmann S. World Health Organization estimates of the relative contributions of food to the burden of disease due to selected foodborne hazards: A structured expert elicitation. *PLOS ONE* 2016;11(1):e0145839.

Hamilton AJ, Stagnitti F, Premier R, Boland AM, Hale G. Quantitative microbial risk assessment models for consumption of raw vegetables irrigated with reclaimed water. *Appl Environ Microbiol* 2006;72(5):3284–90.

Hammitt JK, Haninger K. Willingness to pay for food safety: Sensitivity to duration and severity of illness. *Am J Agr Econ* 2007;89(5):1170–5.

Haninger K, Hammitt JK. Diminishing willingness to pay per quality-adjusted life year: Valuing acute foodborne illness. *Risk Anal* 2011;31(9):1363–80.

Hart A, Hoekstra J, Owen H, Kennedy M, Zeilmaker MJ, de Jong N, Gunnlaugsdottir H. Qalibra: A general model for food risk-benefit assessment that quantifies variability and uncertainty. *Food Chem Toxicol* 2013;54:4–17.

Havelaar AH, Braunig J, Christiansen K, Cornu M, Hald T, Mangen MJ, et al. Towards an integrated approach in supporting microbiological food safety decisions. *Zoonoses Public Health* 2007;54(3–4):103–17.

Havelaar AH, van Rosse F, Bucura C, Toetenel MA, Haagsma JA, Kurowicka D, Heesterbeek JH, Speybroeck N, Langelaar MF, van der Giessen JW, Cooke RM, Braks MA. Prioritizing emerging zoonoses in the Netherlands. *PLOS ONE* 2010;5(11):e13965.

Havelaar AH, Kirk MD, Torgerson PR, Gibb HJ, Hald T, Lake RJ, et al. World Health Organization global estimates and regional comparisons of the burden of foodborne disease in 2010. *PLOS Med* 2015;12(12):e1001923.

Herdman M, Gudex C, Lloyd A, Janssen M, Kind P, Parkin D, Bonsel G, Badia X. Development and preliminary testing of the new five-level version of EQ-5D (EQ-5D-5L). *Qual Life Res* 2011;20(10):1727–36.

Hill AB. The environment and disease: Association or causation? *Proc R Soc Med* 1965;58:295.

Hoekstra J, Hart A, Boobis A, Claupein E, Cockburn A, Hunt A, Knudsen I, Richardson D, Schilter B, Schütte K, Torgerson PR, Verhagen H, Watzl B, Chiodini A. BRAFO tiered approach for benefit-risk assessment of foods. *Food Chem Toxicol* 2012;50(Suppl 4):S684–98.

Hoffmann S, Batz MB, Morris JG Jr. Annual cost of illness and quality-adjusted life year losses in the United States due to 14 foodborne pathogens. *J Food Prot* 2012;75(7):1292–302.

Hoffmann S, Maculloch B, Batz M. *Economic Burden of Major Foodborne Illnesses Acquired in the United States*. EIB-140. Washington, DC: U.S. Department of Agriculture, Economic Research Service; 2015.

Hoffmann S, Devleesschauwer B, Aspinall W, Cooke R, Corrigan T, Havelaar A, Angulo F, Gibb H, Kirk M, Lake R, Speybroeck N, Torgerson P, Hald T. Attribution of global foodborne disease to specific foods: Findings from a World Health Organization structured expert elicitation. *PLOS ONE* 2017;12(9):e0183641.

Hussain MA, Dawson CO. Economic impact of food safety outbreaks on food businesses. *Foods* 2013;2(4):585–9.

International Programme on Chemical Safety. IPCS risk assessment terminology. *Harmonization Project* Document No. 1. 2004. http://www.inchem.org/doc uments/harmproj/harmproj/harmproj1.pdf. Accessed 15 March 2019.

Institute of Medicine. *Improving Food Safety Through a One Health Approach.* Washington, DC: The National Academies Press; 2012.

Keithlin J, Sargeant JM, Thomas MK, Fazil A. Systematic review and meta-analysis of the proportion of non-typhoidal *Salmonella* cases that develop chronic sequelae. *Epidemiol Infect* 2015;143(7):1333–51.

Koutsoumanis KP, Aspridou Z. Moving towards a risk-based food safety management. *Curr Opin Food Sci* 2016;12:36–41.

Mangen MJ, Batz MB, Käsbohrer A, Hald T, Morris JG, Taylor M, et al. Integrated approaches for the public health prioritization of foodborne and zoonotic pathogens. *Risk Anal* 2010;30(5):782–97.

Mangen MJ, Plass D, Havelaar AH, Gibbons CL, Cassini A, Mühlberger N, et al.; BCoDE Consortium. The pathogen- and incidence-based Daly approach: An appropriate [corrected] methodology for estimating the burden of infectious diseases. *PLOS ONE* 2013;8(11):e79740.

Mangen MJ, Bouwknegt M, Friesema IH, Haagsma JA, Kortbeek LM, Tariq L, Wilson M, van Pelt W, Havelaar AH. Cost-of-illness and disease burden of food-related pathogens in the Netherlands, 2011. *Int J Food Microbiol* 2015;196:84–93.

Mangen MJ, Friesema IHM, Pijnacker R, Mughini Gras L, van Pelt W. Disease burden of food-related pathogens in the Netherlands; 2017. RIVM Letter Report 2018-0037. https://www.rivm.nl/bibliotheek/rapporten/2018-0037.pdf. Accessed 15 March 2019.

Mauskopf JA, French MT. Estimating the value of avoiding morbidity and mortality from foodborne illnesses. *Risk Anal* 1991;11(4):619–31.

Minor T, Lasher A, Klontz K, Brown B, Nardinelli C, Zorn D. The per case and total annual costs of foodborne illness in the United States. *Risk Anal* 2015;35(6):1125–39.

Murray CJ. Quantifying the burden of disease: The technical basis for disability-adjusted life years. *Bull World Health Organ* 1994;72(3):429–45.

Murray CJ, Ezzati M, Flaxman AD, Lim S, Lozano R, Michaud C, et al. GBD 2010: Design definitions and metrics. *Lancet* 2012;380(9859):2063–6.

National Research Council. *Enhancing Food Safety: The Role of the Food and Drug Administration.* Washington, DC: The National Academies Press; 2010.

Nayga RM, Woodward R, Aiew W. Willingness to pay for reduced risk of foodborne illness: A nonhypothetical field experiment. *Can J Agr Econ* 2006;54(4):461–75.

Pires SM, Evers EG, van Pelt W, Ayers T, Scallan E, Angulo FJ, Havelaar A, Hald T. Med-Vet-Net Workpackage 28 Working Group. Attributing the human disease burden of foodborne infections to specific sources. *Foodborne Pathog Dis* 2009;6(4):417–24.

Pires SM, Fischer-Walker CL, Lanata CF, Devleesschauwer B, Hall AJ, Kirk MD, Duarte AS, Black RE, Angulo FJ. Aetiology-specific estimates of the global and regional incidence and mortality of diarrhoeal diseases commonly transmitted through food. *PLOS ONE* 2015;10(12):e0142927.

Pires SM, Boué G, Boobis A, Eneroth H, Hoekstra J, Membré JM, Persson IM, Poulsen M, Ruzante J, van Klaveren J, Thomsen ST, Nauta MJ. Risk benefit assessment of foods: Key findings from an international workshop. *Food Res Int* 2019;116:859–69.

Pozo VF, Schroeder TC. Evaluating the costs of meat and poultry recalls to food firms using stock returns. *Food Pol* 2016;59:66–77.

Ribera LA, Plama MA, Paggi M, Knutson M, Masabni JG, Anciso J. Economic analysis of food safety compliance costs and foodborne illness outbreaks in the United States. *HortTechnology* 2012;22(2):150–6.

Roberts T. Human illness costs of foodborne bacteria. *Am J Agr Econ* 1989;71(2):468–74.

Salomon JA, Haagsma JA, Davis A, de Noordhout CM, Polinder S, Havelaar AH, Cassini A, Devleesschauwer B, Kretzschmar M, Speybroeck N, Murray CJ, Vos T. Disability weights for the Global Burden of Disease 2013 study. *Lancet Glob Health* 2015;3(11):e712–23.

Scallan E, Hoekstra RM, Angulo FJ, Tauxe RV, Widdowson M-A, Roy SL, et al. Foodborne illness acquired in the United States—Major pathogens. *Emerg Infect Dis* 2011;17(1):7–15.

Scharff RL. Economic burden from health losses due to foodborne illness in the United States. *J Food Prot* 2012;75(1):123–31.

Scharff RL. State estimates for the annual cost of foodborne illness. *J Food Prot* 2015;78(6):1064–71.

Schroeder SA. Incidence, prevalence, and hybrid approaches to calculating disability-adjusted life years. *Popul Health Metr* 2012;10(1):19.

Speybroeck N, Devleesschauwer B, Depoorter P, Dewulf J, Berkvens D, Van Huffel X, Saegerman C. Needs and expectations regarding risk ranking in the food chain: A pilot survey amongst decision makers and stakeholders. *Food Control* 2015;54:135–43.

Stella P, Cerf O, Hugas M, Koutsoumanis KP, Nguyen-The C, Sofos JN, Valero A, Zwietering MH. Ranking the microbiological safety of foods: A new tool and its application to composite products. *Trends Food Sci Technol* 2013;33(2):124–38.

Teisl MF, Roe BE. Consumer willingness-to-pay to reduce the probability of retail foodborne pathogen contamination. *Food Pol* 2010;35(6):521–30.

Thomsen ST, de Boer W, Pires SM, Devleesschauwer B, Fagt S, Andersen R, Poulsen M, van der Voet H. A probabilistic approach for risk-benefit assessment of food substitutions: A case study on substituting meat by fish. *Food Chem Toxicol* 2019;126:79–96.

Torgerson PR, Schweiger A, Deplazes P, Pohar M, Reichen J, Ammann RW, Tarr PE, Halkik N, Müllhaupt B. Alveolar echinococcosis: From a deadly disease to a well-controlled infection. Relative survival and economic analysis in Switzerland over the last 35 years. *J Hepatol* 2008;49(1):72–7.

Torrance GW. Measurement of health state utilities for economic appraisal: A review. *J Health Econ* 1986;5(1):1–30.

U.S. Food and Drug Administration. FDA food safety modernization act (FSMA); 2017. https://www.fda.gov/Food/GuidanceRegulation/FSMA/default.htm. Accessed 15 March 2019.

Verhagen H, Tijhuis MJ, Gunnlaugsdóttir H, Kalogeras N, Leino O, Luteijn JM, Magnússon SH, Odekerken G, Pohjola MV, Tuomisto JT, Ueland Ø, White BC, Holm F. State of the art in benefit-risk analysis: Introduction. *Food Chem Toxicol* 2012;50(1):2–4.

World Health Organization. WHO methods and data sources for global burden of disease estimates 2000–2016. Global Health Estimates Technical Paper WHO/ HIS/IER/GHE/2018.4; 2018. https://www.who.int/healthinfo/global_bu rden_disease/GlobalDALY_method_2000_2016.pdf. Accessed 15 March 2019.

4

Risk–Benefit Assessment of Foods

Maarten J. Nauta, Lea Sletting Jakobsen, Maria Persson, and Sofie Theresa Thomsen

Contents

4.1 INTRODUCTION

4.1.1 The Need for Risk–Benefit Assessment

Consumption of food is an essential and natural part of life. Much effort is therefore made to ensure that sufficient healthy food is available for the growing world population and that the food on the market is safe for consumption. One of the tools to ensure food safety is risk assessment, as prescribed in the World Trade Organization agreement on the application of sanitary and phytosanitary measures (FAO/WHO 2006). By definition, risk assessment of foods is targeted at the potential adverse health effects due to the intake of chemical or biological agents. In the past decades, an increasing number of risk assessments have been performed to support food safety risk managers in both national and international regulatory bodies and the food industry. These risk assessments have been performed for a large number of hazards and food products, and as a consequence, many potential food hazards have been identified and brought to public attention.

A side effect of this progress and focus on food safety is increased media attention to food risks, which in turn may lead to a public perception that many food products are unsafe or unhealthy. For example, fruit and vegetables may be contaminated with microbial pathogens, which can lead to gastroenteritis (Da Silva Felício et al. 2015), and red and processed meat may be carcinogenic (Bouvard et al. 2015; WCRF/AICR 2007). The perception that food is unhealthy may be reinforced by nutritional risks, such as whole-fat dairy products containing saturated fat, which increases the low-density lipoprotein cholesterol in the blood, and bread containing salt, which increases the risk of hypertension (WHO 2003). In the end, the focus on risk may be confusing, and as a consequence, it may create uncertainty and anxiety among consumers (Ward et al. 2012). The perception that food is very often unhealthy is of course not correct, as food is primarily a basic requirement providing nutrients for a healthy life. Most food products contain nutrients that have beneficial effects on human health. For example, fruit and vegetables are important sources of fibre, vitamins, minerals and bioactive compounds and have been found to decrease the risk of cardiovascular disease (CVD) and cancer (WCRF/AICR 2007; WHO 2003); red meat is an important source of, for example, B vitamins, iron, zinc and selenium; dairy products are an important source of calcium; and bread may contain wholegrains, which have been found to decrease the risk of CVD and some types of cancer (WCRF/AICR 2007;

WHO 2003). If consumers and regulatory bodies focus too much on the risks, the perception may be that these risks outweigh the health benefits associated with the consumption of food. In the end, the nutritional benefits of dietary choices may have a higher impact on public health than the microbiological and chemical hazards in food. Therefore, a more integrated assessment that evaluates both beneficial and adverse health effects is needed: risk–benefit assessment (RBA) of foods.

RBA can be applied for different reasons and for different stakeholders (Boué et al. 2015). One obvious reason is the application of RBA to provide the public and the authorities with comparative information on the health impact of dietary choices. A well-known example is whether or not it is advisable to eat fish, which has been the topic of most published RBAs until now (Boué et al. 2015). RBAs have shown that despite the health risks associated with methylmercury (MeHg), dioxins and *Listeria monocytogenes*, the overall health impact of the recommended fish consumption for the general populations is beneficial due to benefits from n-3 fatty acids and vitamin D. This finding may lead to consumer advice to eat more fish. Another possible reason for performing RBA is the evaluation of the health impact of a proposed intervention by either the authorities or the food industry. Examples include an RBA for a proposed fortification programme or for a change in the production process of a specific food product (Havelaar et al. 2000; Hoekstra et al. 2008).

4.1.2 The Risk–Benefit Assessment Approach and the Role of the Risk–Benefit Question

RBAs of foods have been performed for more than a decade. After the first ground-breaking studies (e.g. Havelaar et al. 2000; Renwick et al. 2004), methodologies for RBA have been developed and case studies have been performed predominantly in European projects, such as BRAFO (Benefit-Risk Analysis of FOods; Boobis et al. 2013). Also, the EFSA Scientific Committee (2010) published the scientific opinion "Guidance on human health risk-benefit assessment of foods". These activities resulted in an agreement on the approach towards RBA and a framework for performing RBA: the tiered approach (Fransen et al. 2010; Hoekstra et al. 2012, see Section 4.3.1).

RBA of foods is a multidisciplinary field that combines research within nutrition, epidemiology, toxicology and microbiology, each of which has its own concepts and traditions. As a consequence of this

81

multidisciplinarity, a variety of definitions has been applied for some of the central concepts (Nauta et al. 2018). For this chapter, we have chosen to apply the definitions given in Box 4.1, which we considered most practical for our purpose.

BOX 4.1   DEFINITIONS AS APPLIED IN THIS CHAPTER

Food component: A nutrient, chemical, microorganism or other constituent present in food. The component is the agent traditionally considered in toxicological and microbiological risk assessment. In nutrition, it is a nutrient or food compound, which can usually be found in different food products.

Adverse health effect: "A change in the morphology, physiology, growth, development, reproduction or life span of a human or (sub) population, that results in an impairment of functional capacity, an impairment of the capacity to compensate for additional stress, or an increase in susceptibility to other influences" (FAO/WHO 2006). This implies that an adverse health effect may be a disease, such as cancer or diarrhoea, but also a more specific health outcome such as liver cancer, infertility or salmonellosis, which reduces quality of life or causes loss of life.

Beneficial health effect: A health effect with the opposite health impact to an adverse health effect. It increases quality of life, lowers or prevents the reduction in quality of life, or prevents or reduces the probability of loss of life.
   Note that not only food components are associated with health effects, but also foods and diets.

Hazard: A biological, chemical or physical agent in, or condition of, food with the potential to cause an adverse health effect (Joint FAO/ WHO Food Standards Programme 2015).
   Note that this definition, which is common in microbiological risk assessment, differs from the one given by IPCS (2004) and EFSA Scientific Committee (2010): "inherent property of an agent or situation having the potential to cause adverse effects when an organism, system, or (sub)population is exposed to that agent". In our definition, the hazard is the agent, not its ability to cause an adverse health effect (Nauta et al. 2018).

Beneficial component: A biological, chemical or physical agent in, or condition of, food with the potential to cause a beneficial health effect.

Note that in these definitions, only a food component can be characterized as hazard and/or beneficial component; a food or a diet cannot.

Risk: A function of the probability of an adverse health effect and the severity of that effect, consequential to a hazard(s) in food (Joint FAO/WHO Food Standards Programme 2015).

Benefit: A function of the probability of a beneficial health effect and the consequences of that effect due to a beneficial component(s) in food.

Note that in RBA, the hazard(s) and beneficial component(s) in food need not be specifically identified: the adverse or beneficial health effect can be found directly from an association between the food and the health effect without knowing which specific component is responsible for this effect.

Health impact: The combined health consequences of the risks and benefits studied in one or more intake scenarios, expressed either qualitatively or quantitatively, by one common health metric or a combination of metrics.

As proposed by others (Hoekstra et al. 2008), the process of an RBA is similar to the process of a traditional risk assessment (Chapter 1, this volume).

Initially, in consultation with the risk–benefit manager, the risk–benefit question (RBQ) is defined, describing the purpose, scope and limitations of the assessment. As part of this step, the population(s) of interest and the intake scenarios to be investigated are defined. A reference intake scenario (often the current food intake scenario in the population of interest) is compared with one or more alternative scenarios. The alternative scenarios serve to assess the health impacts of a change in consumption and may, for example, describe a worst-case scenario, a proposed fortification programme or a scenario based on a previously defined recommended intake.

After the RBQ has been defined, the RBA consists of five steps, the first four of which follow the steps of a traditional food safety risk assessment but are applied to all beneficial and adverse health effects that are identified. In the fifth step, the characterized risks and benefits are integrated to evaluate the overall health impact, for example by using a composite health metric that allows a comparative assessment of risk(s) and benefit(s). This is illustrated in Figure 4.1.

The first step is the identification of the potential health effects, which may be associated with food components, foods and/or diet considered in the RBA. The health effects are included in or excluded from the RBA based on the weight and quality of the scientific evidence. The second step is the characterization of the health effects, which implies that the relation between exposure to a food component or a food and the associated health impact is described, preferably by a dose–response relationship that quantifies the relation between them. In the third step, the food intake or the exposure to the component for the reference scenario and each alternative scenario are described by either single point estimates or probability distributions that describe the variability within the population. This requires human intake data as well as data or model predictions on concentrations of components in the food products considered. The fourth step is the characterization of risks and benefits, where the information on dose–response relations and the exposure assessment are integrated. The outcome of this step is the difference in the incidence of

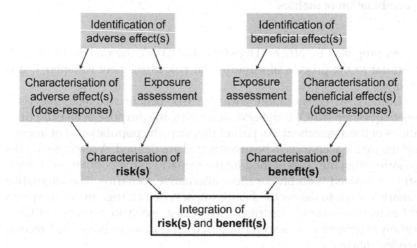

Figure 4.1 The steps taken in risk–benefit assessment of foods. The approach uses the elements of food safety risk assessment for both adverse and beneficial health effects, and combines them in the last step. (Adapted from Tijhuis, M.J. et al., *Food Chem. Toxicol.*, 50, 5–25, 2012.)

the evaluated health effect between the reference and the alternative scenario. In the fifth and final step, the characterized risks and benefits are combined. If they are expressed in a common health metric, the final step allows a comparison of different risks and benefits on a common scale.

The experience with RBA so far has made clear that RBQs are diverse. Together with the potential lack of knowledge in different steps of the RBA, this complicates the development of a general, practical and fit-for-all-purposes methodological framework for RBA. It may therefore be helpful to categorize RBQs and provide an overview of the methodological approaches that fit best into the different RBQ categories. In that way, together with the available data, the RBQ may give guidance on the most suitable methodological approach to be taken in the RBA.

In this chapter, we provide such guidance. First, we present an overview of research questions to clarify the place and relevance of RBA within food safety and nutrition (Section 4.2). Next, we explain how the differentiation of RBQs in qualitative and quantitative questions (Section 4.3) and a categorization based on the levels of aggregation (Section 4.4) impact the approach that has to be taken in the RBA. These levels of aggregation are (1) the food component (such as a micronutrient, for example folic acid); (2) the food or category of foods (such as fish); or (3) the diet (such as a diet where part of the meat is substituted by fish) (Hoekstra et al. 2008). Lastly, we describe the main challenges in RBA and outline some of the future directions that RBA can take (Section 4.5).

4.2 RISK–BENEFIT ASSESSMENT IN FOOD SAFETY AND NUTRITION

RBA fits within the large domains of research in food safety and nutrition, which have a focus on the human health impact of food consumption. Within these domains, a broad range of research questions can be asked, which may demand different methodological approaches. In Table 4.1, we present a classification of research questions in food safety and nutrition, based on the number of food components, foods and health effects considered in the research question. The purpose of the table is to illustrate the diversity of potential research questions and the types of study that are commonly used to answer them as well as the position of RBQs among these questions. Note that the table is not exhaustive in its examples and does not include all possible research questions and types of studies that can be performed to answer them.

Table 4.1 Overview of Potential Research Questions and Types of Study in Food Safety and Nutrition, Based on the Number of Food Components, Foods and Health Effects Considered in the Research Question, with Examples

Food*	Component*	Effect*	Typical Research Questions**	Examples			
				Food	Component	Effect***	Type of Study****
1	NA	1	Does eating f affect e?	Processed meat		Colon cancer (adv)	Etiological study
NA	1	1	What is the threshold intake/DR of c?		PAHs	Colon cancer (adv)	Dose response / Risk assessment
>1	NA	1	Is eating f1 better than f2 with regards to e?	White rice, brown rice		Diabetes (adv)	Etiological study
NA	>1	1	What is the health impact of c1 compared to c2?		Salmonella, Norovirus	Gastroenteritis (adv)	Dose response / Risk assessment
1	1	1	What is the annual incidence of e due to f/c in Denmark?	Chicken meat	Salmonella	Gastroenteritis (adv)	Risk assessment
1	>1	1	What is the annual incidence of e due to f/cs in Denmark?	Processed meat	PAHs, HCAs	Colon cancer (adv)	Risk assessment

(*Continued*)

Table 4.1 (Continued) Overview of Potential Research Questions and Types of Study in Food Safety and Nutrition, Based on the Number of Food Components, Foods and Health Effects Considered in the Research Question, with Examples

Food*	Component*	Effect*	Typical Research Questions**	Examples			
				Food	Component	Effect***	Type of Study****
>1	1	1	What is the annual incidence of e due to fs/c in Denmark?	White rice, brown rice	Arsenic	Cancer (adv)	Risk assessment
>1	>1	1	What is the annual incidence of e due to fs/cs in Denmark?	Vegetables, chicken meat	*Salmonella, Norovirus*	Gastroenteritis (adv)	Risk assessment
1	NA	>1	What are the health effects associated to eating f? Is f healthy overall? How healthy (DALY)?	Processed meat		Colon cancer (adv), anaemia (ben)	Etiological study, RBA
NA	1	>1	Should we advise the intake of supplements with c?		Folic acid	Masking vit B12 deficiency (adv), neural tube development (ben)	RBA
			What is the overall health impact of intake of c?				

(Continued)

87

Table 4.1 (Continued) Overview of Potential Research Questions and Types of Study in Food Safety and Nutrition, Based on the Number of Food Components, Foods and Health Effects Considered in the Research Question, with Examples

				Examples			
Food*	Component*	Effect*	Typical Research Questions**	Food	Component	Effect***	Type of Study****
>1	NA	>1	Should we advise people to eat more f2 instead of f1? What is the overall health impact of eating fs?	White rice, brown rice		Diabetes (adv), cancer (ben)	RBA
NA	>1	>1	What are the health effects of intake of cs? What are DRs?		Arsenic, fiber	== Cancer (adv), == diabetes (ben)	RBA
1	1	>1	Should we fortify f with c in Denmark? If we do, what is the expected health impact in Denmark?	bread	Folic acid	Masking vit B12 deficiency (adv), neural tube development (ben)	RBA
1	>1	>1	Is it healthy to eat more f than we do now? If we do, what is the expected health impact in Denmark?	fish	Methylmercury, vitamin D	== Infant IQ (adv), == osteoporosis (ben)	RBA

(Continued)

Table 4.1 (Continued) Overview of Potential Research Questions and Types of Study in Food Safety and Nutrition, Based on the Number of Food Components, Foods and Health Effects Considered in the Research Question, with Examples

Food*	Component*	Effect*	Typical Research Questions**	Examples			Type of Study****
				Food	Component	Effect***	
>1	1	>1	What are the overall health effects of intake of c from fs?	Vegetables, supplements, white rice, brown rice	Folic acid, fiber, arsenic	Masking vit B12 deficiency (adv), neural tube development (ben), == cancer (adv), == diabetes (ben)	RBA
			What is the expected health impact of c from fs?				
>1	>1	>1	Should we advise people to eat more f2 instead of f1?				RBA
			If we do, what is the expected health impact in Denmark?				

*1 = one, >1 = more than one, NA = none included in the study. For >1 health effect, these are opposite effects.

**Examples of typical research questions; f= food, c= component, e = effect; s indicates plural, number (1,2) indicates first and second mentioned; / indicates "and/or"; DR = dose response.

***adv = adverse effect, ben = beneficial effect; == indicates a link with the component in the adjacent column; opposite effect may be opposite effect within or between persons.

****A typical type of study on the indicated combination of food(s), components(s) and effect(s); not an exhaustive list.

From the table, it can be observed that:

- Studies linking the intake of one or more foods to one or more health effects, without specific consideration of the food components involved, are typically etiological studies within human nutrition (i.e. observational epidemiological studies, randomized control trials, etc.).
- Studies involving food components and one health effect are often risk (or benefit) assessment studies. If there is no specific food defined, these studies focus on the characterization of the adverse or beneficial effects, or dose–response.
- Studies can only be classified as RBAs when both adverse and beneficial health effects are included. If only one food component is included in an RBA, it should be a component that is associated with both adverse and beneficial health effects.
- Studies involving more than one component without including a specific food are uncommon in risk assessment and RBA. Several hazards may be compared in risk ranking studies, but to our knowledge, this is uncommon when opposite health effects are considered.

In anticipation of Section 4.3 on qualitative and quantitative RBA, note that the table includes two examples of research questions for RBA studies, first a qualitative one and then a quantitative one. Here, qualitative questions refer to issues such as whether there is sufficient evidence for an adverse or beneficial effect, whether an intervention is beneficial or not, or whether the risk is larger than the benefit. Quantitative questions typically address a quantitative measure for the health impact; for example, how much the benefit exceeds the risk, or how much more it exceeds the risk after an intervention. (Hence, when referring to "health impact" in Table 4.1, we refer to a quantified impact.) The uncommon case where two components are compared without considering a specific food (see last point in the preceding list) is an exception, as a question related to the overall quantitative health impact is not considered to be realistic. In addition, the list of typical research questions shows that in many cases, and especially for RBA, both qualitative and quantitative assessments of health effects can be required.

In anticipation of Section 4.4 on RBA at different levels of integration, note that among the RBA examples listed in the table, those involving one food component are typically RBAs at the level of the food component, and those involving one food are RBAs at the level of the food. An

RBA with one food and one food component falls into both categories, and the classification will depend on the main interest of the risk–benefit manager. RBAs may involve more than two foods, food components and health effects, and each different health effect may be associated with one or more of these foods and food components. This gives a large number of possible combinations, far from all of which are included in the table. As will be described in Section 4.4, RBAs of a whole diet typically are of this type and include several foods, components and health effects.

4.3 QUALITATIVE AND QUANTITATIVE RISK–BENEFIT ASSESSMENT APPROACHES

4.3.1 Identification of the Overall Health Impact: The Tiered Approach

A fundamental framework for RBA of foods was published as part of the BRAFO project – a European Commission funded project that aimed at developing a framework for comparing risks and benefits of food and food components on a common scale (Boobis et al. 2013; Hoekstra et al. 2012). Also known as "the BRAFO tiered approach", this framework investigates the risks and benefits associated with a theoretical dietary intervention following the steps described in Section 4.1.2. As illustrated in Figure 4.2, the RBQ is investigated in four "tiers". Tiers 1 and 2 investigate the RBQ in a qualitative and/or semi-quantitative way, respectively. Tier 1 investigates whether both adverse and beneficial health effects are associated with consumption. If this is not the case, and all identified health effects point in the same direction, the assessment stops, because the overall health impact is obvious; otherwise, the risk–benefit assessor should proceed to Tier 2. In Tier 2, risks and benefits are compared qualitatively or semi-quantitatively. If either risks or benefits clearly outweigh the other in terms of disease incidence, severity, duration and mortality of the associated health effects, or if exposure evaluated under constraints of, for example, health-based guidance values is clearly either safe or unsafe for the population, the assessment stops, as the overall health impact is clear. If risks or benefits do not clearly outweigh the other, the assessment shifts to a quantitative comparison in Tiers 3 and 4. In Tier 3, risks and benefits are quantified in terms of a common health metric using a deterministic approach. If uncertainty around the estimate is small, this approach may be sufficient to reach a conclusion about the difference in the overall

91

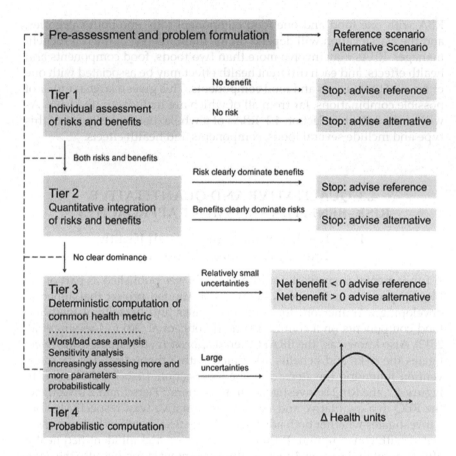

Figure 4.2 Tiered approach with four tiers that indicate when a more advanced RBA is required to decide that the benefits are larger than the risks or vice versa. (Adapted from Hoekstra, J. et al., *Food Chem. Toxicol.*, 50(Supplement 4), S684–S698, 2012.)

health impact of the reference and the alternative scenario. Otherwise, the assessment proceeds to Tier 4, where probabilistic methods are applied in order to quantify the uncertainty of the estimate.

The basis of the tiered approach is a qualitative RBQ on the overall balance of risks and benefits. Two or more intake scenarios are compared in terms of their expected health impact in order to judge which is most beneficial for the population considered. This can support consumers in making a healthy choice and may be the basis for advice to the public or an intervention such as fortification. The advantages of such a qualitative RBQ are that it often reflects the most important question of the risk–benefit manager and that it may be relatively easy to provide a comprehensible answer (especially in Tiers 1 and 2). However, it can also be incomplete, as we may need to know *how large* the benefits and risks are expected to be so that they can be weighed against other aspects than health, such as costs, sustainability and personal food preference. In that case, the RBQ may need to be quantitative from the start, which may require another framework than that outlined in the tiered approach.

4.3.2 Quantification of the Overall Health Impact

In Tiers 3 and 4 of the BRAFO tiered approach (Figure 4.2), the RBA is performed quantitatively, usually because a qualitative approach did not result in an adequate answer to the posed question in Tiers 1 and 2. However, as stated earlier, the RBQ may also initially be framed in a way that calls for a quantitative assessment, so that Tiers 1 and 2 are not relevant. Overall, in a quantitative RBA, the impact of the risks and benefits on human health is quantified and if possible, integrated and summarized. An important challenge of a quantitative approach is that it requires data and models to quantitatively assess the health impact, including a dose–response relation. Often, only threshold values such as a reference dose, margin of exposure, lower threshold intake or upper intake level are available as indicators of the presence of an adverse health impact. These threshold values are, however, not sufficient to quantify health impact in terms of disease incidence or mortality.

Even though a quantitative assessment may be considered data and time consuming, the resulting assessment may be very useful for stakeholders. Quantitative assessments can, for example, be used by risk–benefit managers to measure the impact of different intervention strategies and identify optimum solutions for specific populations. Also, a quantitative assessment can provide crucial information for consumers who want to balance the anticipated health impact of their dietary choices with, for example, economic aspects and personal preferences. Through scenario analyses, a quantitative RBA enables an investigation of which parameters have the highest impact on the final results. A probabilistic assessment allows investigation of the parameters that are contributing most to the overall uncertainty of the assessment through sensitivity or uncertainty analyses, which can provide information on which data sources should be improved to increase the certainty of the estimates.

In a quantitative RBA, the health impacts of the risks and benefits are preferably expressed on the same scale of measurement, as this facilitates the comparison of health impacts. Here, the choice for the scale of measurement is not trivial. The diseases associated with risks and benefits have several dimensions, and their impact on human health may be characterized in various ways, including the disease incidence and mortality in the population of interest. The RBA of fish could, for example, be assessed in terms of how many children are born with a lower IQ compared with how many cases of CVD are prevented. If the number of children born with a lower IQ is much lower than the number of cases of CVD prevented, then the result of the assessment would be that the health impact of increasing fish intake is positive. However, low IQ and CVD are two very different health outcomes, and if the RBQ aims to quantify the health impact in a population, other dimensions of the diseases, including their severity and duration, should be taken into account (Hoekstra et al. 2013). For this purpose, several composite health metrics have been developed, which allow different health effects, beneficial and adverse, to be balanced on the same scale.

The composite health metric most commonly used in RBAs is the disability-adjusted life year (DALY) (Chapter 3), which is also applied in burden of disease studies (Havelaar et al. 2015; Jakobsen et al. 2016; Oberoi et al. 2014). The calculation of DALYs may be challenging, but as user-friendly tools have become available (e.g. the DALY calculator; Devleesschauwer et al. 2014 and the BcoDE toolkit; European Centre for Disease Prevention and Control, 2015), and estimates for the DALY per case for different health effects and food-associated hazards are increasingly available, it may be expected that the use of DALYs will be increasingly facilitated in the future.

In some situations, it may not be necessary or possible to integrate risks and benefits by a composite metric like the DALY. If the risks and benefits assessed affect the same health endpoint, then any measurement of that health outcome could be the common scale of measurement. An example is an RBA on docosahexaenoic acid (DHA) and MeHg in fish, which both affect the IQ of unborn children, but in a positive and negative direction, respectively (Zeilmaker et al. 2013). Here, the overall impact on IQ is an appropriate common metric. If risks and benefits of an RBA cannot be measured using common metrics, other methods are available for comparison, such as multi-criteria decision analysis (MCDA) (Ruzante et al. 2010, 2017).

Finally, the RBA could investigate the economic impact of a given dietary intervention, that is, the money lost or saved, in a society in terms of healthcare costs, work absence, etc. (Mangen et al. 2015). For these estimates, the RBA approach is followed until the fifth step (see Figure 4.1), and the economic analysis is done once the health impact in terms of disease incidence and mortality has been assessed.

4.4 RISK–BENEFIT ASSESSMENT AT DIFFERENT LEVELS OF AGGREGATION

4.4.1 Risk–Benefit Assessment for Food Components, Foods and Diets

In this section, we compare RBQs that are targeted at different levels of aggregation, as illustrated in Figure 4.3. Some examples are given with a schematic representation that frames the RBQ to clarify the difference

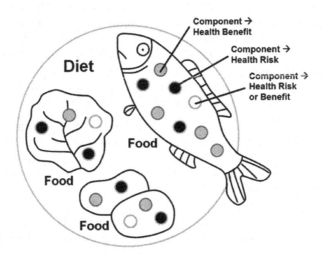

Figure 4.3 RBQs can be defined at three levels of aggregation: 1) the food component, 2) the food and 3) the diet. Each food on the plate consists of various components, which may be beneficial (grey dots), adverse (black dots) or both (white dots) to human health. An RBA on the food component level will assess the impact of a single component on health and requires that the component is a "white dot" associated with both beneficial and adverse health effects. An RBA on the food level will assess the impact of a food on health either by considering the whole food (i.e. fish) or the individual hazards and beneficial components in the food (i.e. MeHg and polyunsaturated fatty acids in fish). An RBA on the diet level assesses the overall impact of increasing and decreasing the consumption of several foods in the diet.

between questions posed at the different levels. In addition, for each level, some of the methods used in RBA are presented, and the specific challenges are discussed. Overall, this can provide guidance on the approaches that can be taken when an RBQ is asked at one of the indicated levels of aggregation.

4.4.2 Food Component Risk–Benefit Assessment

In traditional food safety risk assessment, the food component is the hazardous agent considered, and it always has an assumed negative health effect. However, in RBA, it can also be a beneficial component or a component that is associated with both adverse and beneficial health effects. If such a food component associated with opposite health effects is central in the RBQ, the RBA is performed at the level of a food component. Food components associated with opposite health effects are usually micronutrients, which are either naturally present or added to different food products, or can be part of a food supplement. As microbiological pathogens and chemical contaminants are not associated with beneficial health effects, food component RBA will usually fall within the research discipline of nutrition.

There are different possibilities for the nature and occurrence of the beneficial and adverse effects in a food component RBA:

1) The single food component is associated with opposite health effects. Nitrate is an example. Upon ingestion, it may be converted into other compounds, some of which are known to induce vasodilation and thus prevent hypertension, and others that may cause cancer (Habermeyer et al. 2015).

2) The effects of the food component have different impacts in different population groups. For example, DHA has a beneficial impact on the developing foetus during pregnancy (EFSA NDA Panel 2014) and also a preventive effect on the risk of CVD in the general population (Mozaffarian and Rimm 2006).

3) Different adverse effects occur at doses of the food component that are too low or too high. The benefit occurs around dietary reference values such as the average requirement, recommended intake and/or adequate intake. This is often referred to as the *dual risk phenomenon* in nutrition, which implies that health risks for nutrient intake occur both at doses that are too low and at doses that are too high (Nauta et al. 2018; Renwick et al. 2004). This type of food component RBA has, for example, been studied by (Berjia et al. 2014; Bruins et al. 2015).

In a food component RBA, various types of RBQ can be asked. The first type of question is obviously whether, and to what extent, the overall health effect of the intake of the component is beneficial or detrimental when comparing the different health effects associated with the food component. The folic acid example in Box 4.2 falls in this category of questions. Such an RBQ is particularly relevant when regulatory bodies consider fortification of food products or advice on food supplements, either for the general population or for specific population groups. In addition, this type of question may be relevant for other micronutrients, such as vitamin D (bone mineralization versus bone resorption and hypercalcaemia (Selby et al. 1995), and iron (iron-deficiency anaemia versus oxidative damage due to iron; The Scientific Advisory Committee on Nutrition 2010).

A second type of question at the food component level is the range of intake that is beneficial or the optimum intake of a specific food component. The question of the assessment of the optimal intake level can be studied by application of methods used in RBA, as, for example, shown by Berjia et al (2014) for vitamin D. In this RBA, studies providing dose–response relations for beneficial and adverse health effects were compared and combined by the use of a common health metric to find the optimum intake level. This type of approach could be useful for many food components for which the dual risk phenomenon is relevant.

The example on folic acid in Box 4.2 shows the potential for performing an RBA at the level of the food component, for example to evaluate the risks and benefits of a fortification programme. It also shows its complexity, as the health effect of potentially beneficial food components may depend on the population group considered (e.g. prevention of neural tube defect in the foetus). As for (micro-)nutrients in general, due to the dual risk phenomenon, too low an intake may lead to deficiencies, but too high an intake may have toxic effects. If the food components can be found in different foods in the diet (folic acid can, for example, be found in leafy greens, whole grain products, dairy and meat), the individual intakes will vary largely within the population. This can, for example, complicate a decision about fortification, as it may not be desirable that the health benefit to one population group (which moves from too low to an adequate level) comes at the expense of another (which moves from an adequate to too high a level).

A major challenge for food component RBA is the strength of scientific evidence for the identified adverse and beneficial health effects and the dose–response relations. Food components studied in these RBAs are primarily (micro-)nutrients, for which the health effects generally have to be derived from human observational data, as animal experimental data, if available, may be inconclusive. In the folic acid RBA performed by Hoekstra et al (2008), for example, the majority of the evidence came from human randomized trials, while results of animal studies were only found to provide some additional evidence for the potential increase in risk of acquiring colorectal cancer.

BOX 4.2   EXAMPLE OF A FOOD COMPONENT RBA – FORTIFICATION WITH FOLIC ACID (HOEKSTRA ET AL. 2008)

RBQ: What will be the overall health impact of mandatory fortification of bread with folic acid in the Dutch population in comparison to the current situation?

RBA: The first step of the assessment was to identify the adverse and beneficial health effects associated with folic acid. After considering the scientific evidence, it was decided to compare the beneficial effect of folic acid in terms of reduced risk of neural tube defects in newborns and megaloblastic anaemia and colorectal cancer in adults, with the adverse effect of folic acid in masking vitamin B12 deficiency leading to progression of irreversible neurological symptoms and increased risk of colorectal cancer in adults at very high consumption levels (Figure 4.4). In the second step, the dose–response relations were obtained. Mandatory fortification would thus affect sub-populations differentially: the beneficial effects are mainly in foetal development and persons with low intake; the adverse effects are mainly in elderly, strict vegetarians and persons with an existing (pre)neoplastic lesion. Therefore, in the next step, the exposure to folic acid in the reference scenario (current exposure) and in the defined fortification scenarios (i.e. four scenarios representing the fortification of bread with four different concentrations of folic acid) was computed for each of the sub-populations. In the fourth step, the exposure for the sub-population was combined with dose–response relationships for the health effects relevant for each sub-population to estimate the prevention and/or additional incidence of disease. Finally, the adverse and beneficial effects were expressed in DALYs.

The results showed that the health impact of fortifying bread with folic acid in the overall population was beneficial in the three lowest fortification scenarios; however, in the highest fortification

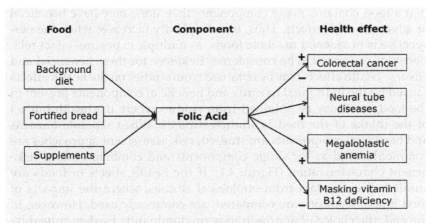

Figure 4.4 Example of the foods, component and health effects considered in a food component RBA for folic acid, as in Hoekstra et al (2008). The food component is found to be associated with both positive (+) and negative (–) health effects. Several foods may contain folic acid; those specifically addressed in the RBA are fortified bread and supplements.

scenario, the adverse health effects outweighed the beneficial ones. As important challenges identified in this RBA, Hoekstra et al. (2008) mention the data demands, the required assumptions and the uncertainty. The evidence of the health effects varies from "probable" to "convincing". Scenarios with and without inclusion of the "probable" health effects can be run to explore the potential importance of the effect. The fact that different risks and benefits apply to different population groups has to be clearly communicated to the risk–benefit managers, who have to decide on weighing these risks and benefits.

Note that in this case, the qualitative question about the overall health impact was answered by using a quantitative metric, because the overall health impact could not be characterized by qualitative arguments only.

4.4.3 Food Risk–Benefit Assessment

As for single food components, consumption of a specific food may be associated with both health risks and benefits. The main difference is that a food contains many components that alone may have beneficial or adverse health effects. Thus, the complexity increases when the perspective is broadened to whole foods, as multiple exposure–effect relationships may need to be considered. Evidence for these beneficial and adverse health effects may be obtained from studies on the health effects caused by the individual hazards and beneficial components present in the food, or it may be obtained from studies where the health impact of the intake of the food is studied directly. When separate hazards and beneficial components are studied, risk assessment approaches are commonly used as for single components and combined into a risk–benefit characterization (Figure 4.1). If the health effects of foods are studied directly, data from etiological studies, where the impacts of food intake scenarios are compared, are commonly used. However, in the end, the choice of approach may predominantly be determined by the available data.

RBA of foods can be used to provide a scientific basis for food recommendations and policies, food manufacturing, food processing and food fortification. In most cases, the RBQ relates to the overall health impact of changes in food consumption, for example due to dietary recommendations or emerging food trends in a population. As fish is the most studied food product in RBA (Boué et al. 2015), we illustrate RBA at the level of food with an example on fish RBA (see Box 4.3 and Figure 4.5). In this example, most health effects are linked to components, but one health effect is obtained from observational studies on fish intake (i.e. at the food level) and has not been associated with individual components.

Human observational studies and intervention studies may describe the association and causation, respectively, between consumption of a food and a given health effect. Meanwhile, for ethical reasons, such studies are rarely available for chemical contaminants in food, and experimental animal models may be needed to investigate the association between exposure to a chemical and the probability of a health effect. However, there are exceptions to this, and the association between exposure to some chemicals and effects in humans has been investigated in observational studies of high exposure populations (e.g. MeHg exposure and neurodevelopmental outcomes in Faroese children; Grandjean et al. 1997, 1998). Similarly, randomized controlled trials have also investigated the effects of fish oil supplements on cardiovascular outcomes (Mozaffarian et al. 2005).

BOX 4.3   EXAMPLE OF FOOD RBA – RBA OF FISH (THOMSEN ET AL. 2018)

RBQ: What will be the overall health impact, in terms of DALYs, of increasing the current fish consumption in the Danish population to the recommended level?

RBA: In the RBA by Thomsen et al. (2018), the health impact of substituting red and processed meat with fish in a Danish diet was quantified. The RBA built on an RBA of increasing fish consumption to the recommended level (i.e. on the food level of aggregation) and was extended to also include a corresponding decrease in the consumption of red and processed meat, which changes it to an RBA performed at the diet level of aggregation. Here, we will address the part of the RBA considering fish consumption, while Box 4.4 will address the integrated RBA of the substitution and consider the decrease in red and processed meat consumption.

As an initial step, the scenarios to be compared in the assessment were defined. The reference scenario was defined as the current observed fish consumption in the Danish adult population (≥15 years), based on data from the Danish National Survey of Diet and Physical Activity (Pedersen et al. 2015). The reference scenario was

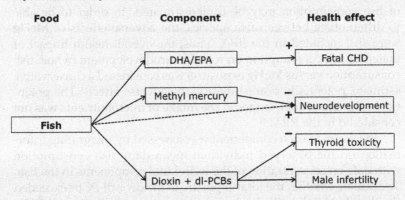

Figure 4.5 Example of the food, components and health effects considered in a food RBA for fish, as in Thomsen et al. (submitted). The food and its components are found to be associated with both beneficial (+) and adverse (−) health effects; the health impact estimates are obtained either from studies on the whole food (dashed line) or from studies on different food components separately.

compared with four alternative scenarios in which the fish consumption was increased to 350 g of fish/week, as recommended by Danish food-based dietary guidelines (Tetens et al. 2013). The four alternative scenarios investigated the health impact of the 350 g of fish/week being either a mix of lean and fatty fish, only fatty fish, only lean fish, or only tuna – a large predatory fish with high concentrations of MeHg. Based on scientific evidence, the authors considered three groups of compounds: DHA and another n-3 fatty acid, eicosapentaenoic acid (EPA), which are associated with decreased risk of fatal CHD in adults; MeHg, associated with adverse effects on foetal neurodevelopment; and dioxin + dioxin-like-polychlorinated biphenyls (dl-PCBs), associated with infertility of male offspring and hypothyroidism. Finally, although individual components in fish (DHA and iodine) have been associated with beneficial effects on neurodevelopment, the beneficial effect of whole fish consumption on neurodevelopment cannot be explained only by the sum of these beneficial compounds (EFSA NDA Panel 2014). Therefore, whole fish consumption was linked to beneficial effects on foetal neurodevelopment in the RBA. As the beneficial effect of fish consumption on neurodevelopment was not adjusted for exposure to MeHg, the authors noted that the estimate of the beneficial effect of fish consumption may be underestimated. In order to be able to differentiate between fish species, the adverse effects of MeHg were still included in the RBA. Thus, the overall health impact of maternal fish consumption on foetal neurodevelopment (whole fish consumption versus MeHg exposure) was considered a conservative estimate, potentially overestimating the adverse effects. The potential health impact of changes in the intake of micronutrients was not considered in the RBA.

Exposure to each component was assessed by integrating information on the observed individual mean daily fish consumption with data on concentration of the individual components in the fish. To evaluate whether the total exposure to dioxin + dl-PCBs exceeded the threshold value, the background exposure to dioxin + dl-PCBs from foods other than fish was also estimated and included in the assessment. On the other hand, fish and seafood were assumed to be the only sources of MeHg, DHA and EPA.

From the risk and benefit characterization, an increase or decrease in disease incidence or mortality due to the changes in fish consumption was estimated; for example, a change in CHD mortality due to a change in the intake of DHA and EPA. This information on disease incidence and mortality was combined with information on the severity and duration of a disease to estimate the disease burden attributable to the change in fish consumption for each health effect in terms of DALYs. By taking the difference between the sum of DALYs in the alternative scenarios and the reference scenario, the overall health impact of increasing fish consumption to the recommended level was estimated.

The results showed an overall health gain from consuming 350 g/week of either a mix, only fatty or only lean fish compared with the current fish consumption (approx. 220 g of a mix of lean and fatty fish/week). Up to approximately 135 DALYs/100,000 adults per year could be prevented if the adult Danish population consumed 350 g of fish/week, with consumption of a mix of lean and fatty or only fatty fish being the most beneficial. Meanwhile, consuming 350 g of tuna/week was found to cause a health loss of approximately 125 DALYs/100,000 compared with the current fish consumption (not taking reduction in meat consumption into consideration). This health loss was mainly attributed to a decreased intake of DHA and EPA causing an increase in fatal CHD incidence, and an increased exposure to MeHg causing an increased incidence in mentally compromised newborns, compared with the current consumption. As for the example on folic acid (Box 4.2), it is important to differentiate sub-populations from each other in terms of health loss in the risk–benefit communication. In this RBA, particularly women who plan to become pregnant constituted a sensitive sub-population, as did individuals with a high risk of CHD.

Note that in this case, the RBQ was quantitative from the start. The aim was not only to provide an answer on whether or not increasing fish consumption was beneficial overall; the RBQ also sought an estimate of the magnitude of the effect.

4.4.4 Diet Risk–Benefit Assessment

Until now, most RBAs have been performed for single foods and food components. In general, this fits well with specific RBQs that have been posed for these foods or food components, where intake scenarios with modified intake of the food or food components were proposed. However, a change in consumption of one food will often result in a change in consumption of other foods as well, and this change in consumption of other foods may be important for the overall health impact of the dietary change. For example, the overall health impact of increasing the consumption of fish depends on the food(s) substituted in the diet and the health impact of these other food(s) (Boobis et al. 2013). This is the challenge of substitution, an important element of diet RBA.

Diets consist of a variety of foods containing an even larger variety of components that are associated with different health effects. Considering a diet in RBA therefore requires more data and more assumptions, and consequently, the uncertainty of the final health impact estimate may be considerable. To evaluate the health impact of changes in overall diets, a central assumption that needs to be made concerns the way in which foods are substituted. A simple assumption would be that all individuals substitute in an equal manner and that only very few foods are substituted. However, it is questionable whether such an assumption holds, and it may not be obvious to what extent it affects the results of the RBA. Further, with an increasing number of foods included in the RBA, the number of health effects to consider and thus, the complexity of the RBA may also increase.

Substitution can be modelled based on weight, amount (e.g. servings) or energy intake. Substitution by amount or servings may be more feasible to communicate than an isocaloric substitution (i.e. a substitution scenario that ensures that the overall calorie intake of the reference and the scenario is the same), but it may lead to a change in energy intake, which may introduce additional health effects to consider in the RBA due to potential weight loss or gain.

One approach in diet RBA is to investigate substitutions that are recommended by, for example, national dietary guidelines. That is, the substitution may not necessarily reflect the choice of the average population but rather, reflect the advice of the authorities. Alternatively, an approach could be to investigate the differences in the risk–benefit balance when a food is substituted by different foods. Box 4.4 provides an example of substituting red and processed meat with fish – a substitution that is supported by the national dietary guidelines in Denmark (Tetens et al. 2013). It illustrates some of the challenges associated with diet RBA and some potential solutions.

BOX 4.4 EXAMPLE OF DIET RBA – SUBSTITUTION OF RED AND PROCESSED MEAT WITH FISH (THOMSEN ET AL. 2018)

RBQ: What will be the overall health impact, in terms of DALYs, of increasing the current fish consumption in the Danish population to the recommended level?

RBA: An increase in fish consumption, as in the example in Box 4.3, is likely to cause a decrease in consumption of other foods; for example, other sources of protein, such as red and processed meat. In line with this, various national food-based dietary guidelines recommend increasing intake of fish and limiting consumption of red and processed meat (to less than 500 g/week) (Kromhout et al. 2016; Nasjonalt råd for ernæring 2011; Tetens et al. 2013; U.S. Department of Health and Human Services and U.S. Department of Agriculture 2015). Thus, food-based dietary guidelines implicitly recommend substituting red and processed meat with fish. In this example, we build on the example given in Box 4.3 based on the RBA by Thomsen et al. In this study, the health impact of substituting red and processed meat with fish in a Danish diet was quantified. As in Box 4.3, four alternative scenarios were compared with the current Danish consumption of fish and meat. All alternative scenarios investigated the health impact of increasing fish consumption to 350 g/week and differed from one another only in terms of the fish consumed. The increased fish consumption in the alternative scenarios compared with the current fish consumption was compensated by a decrease in consumption of red and processed meat, modelled by applying substitution factors based on portion sizes of fish and meat.

In addition to the health effects associated with fish consumption included for fish (Box 4.3, Figure 4.5), including the substitution of red and processed meat introduces additional health effects to be considered. Based on the available evidence, both red and processed meat were found to be associated with colorectal cancer (WCRF/AICR 2011), and processed meat was additionally found to be associated with increased risk of non-cardia stomach cancer (WCRF/AICR 2016). However, red meat is also a good source of various important micronutrients, such as B vitamins, iron, zinc and selenium (Pedersen et al. 2015), so a reduction in red meat consumption would lead to a decrease in the intake of these micronutrients. For example, a potential decrease in iron intake would be expected to increase the incidence of iron-deficiency anaemia (The Scientific Advisory Committee on Nutrition 2010). Meanwhile, due to challenges in establishing iron status and lack of good dose–response data for iron, this and other micronutrients were not considered in the RBA for either fish or meat.

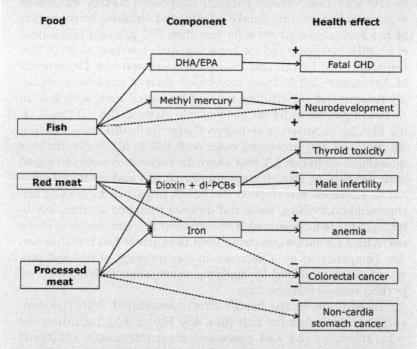

Figure 4.6 Example of the foods, food components and health effects considered in a diet RBA of substitution of fish for meat, as in Thomsen et al. (submitted). The same component may be found in several foods, and the foods and their components are found to be associated with both beneficial (+) and adverse (–) health effects; the health effect estimates are either obtained from studies on the whole food (dashed lines, or from studies on different food components separately.

The results showed that substituting red and processed meat with fish further increased the benefit compared with considering fish consumption alone. Had fish been replacing another food, the picture might have looked different. The additional benefit of decreasing the consumption of red and processed meat caused only a minor decrease in colorectal and stomach cancer incidence and only around an additional 20 DALYs averted/100,000 per year compared with only considering fish in the RBA. Meanwhile, the authors noted that the consumption of red and processed meat was on average not decreased to below the recommended maximum 500 g/week, whereas the guideline on fish (350 g/week) was achieved in the whole study population.

Dioxin and dl-PCBs comprised the only group of components considered in the RBA that was present in both fish and meat. Exposure to dioxin and dl-PCBs did not change much due to the substitution except when all fish consumed was fatty. Still, the impact on the health effects associated with exposure to dioxin and dl-PCBs due to the substitution was negligible (Figure 4.6).

The RBQ for this study was quantitative from the beginning with the aim to quantify the overall health impact of the substitution and not just determining whether or not the substitution was beneficial. Also, only adverse effects were included for red and processed meat, making the risk–benefit balance primarily dependent on the fish consumption (amount and type of fish consumed) rather than the decrease in consumption of red and processed meat. Red and processed meat are important sources of micronutrients like iron, but it was not feasible to include this in this RBA.

4.5 FUTURE PERSPECTIVES AND CHALLENGES OF RISK–BENEFIT ASSESSMENT OF FOODS

The research area of risk–benefit assessment of foods has made significant progress in the past two decades. Methodological frameworks such as the BRAFO tiered approach have been developed, and a significant number of case studies have been performed (Boué et al. 2015). After the finalization of the European projects, RBA has been taken up by a variety of research groups. At the National Food Institute in Denmark, a risk–benefit research group was established in 2015 (Nauta et al. 2018). The group focused its efforts on method development and also organized a European Food Safety Authority (EFSA)-supported workshop on risk–benefit assessments in foods in 2017 (Pires et al. 2019). These publications, as well as Boué (2017), describe future challenges and perspectives in RBA. They stress that major challenges in RBA are 1) the missing data for dose–response relationships for many food components and foods and 2) the conceptual difference in the approaches taken in etiological studies and food safety risk assessments, which use different types of scientific evidence and may be associated with different types of uncertainty. As suggested by these researchers, there is a need for more RBA case studies to evaluate the practical usefulness of the methodological approaches that have been developed and to discuss possible solutions for the identified challenges. An evident challenge to RBA of foods is how it can support risk–benefit management based on the results of an assessment. In many instances, if the overall health impact of an intervention in the general population, such as a fortification programme, is beneficial, adverse health effects may still occur in sub-populations. Based on these challenges, the EFSA-supported workshop concluded that there is a need for increased international collaboration (Pires et al. 2019).

In this chapter, we have shown where RBA fits within food safety and nutrition research, and explained how different RBQs can be categorized in terms of the qualitative or quantitative nature of the RBQ and in terms of the level of aggregation considered. This categorization may be helpful in future RBA case studies. It may, for example, give guidance on the relevance of the use of quantitative metrics in the RBA, which is typically required if a qualitative RBA is insufficient to decide on the balance between risk and benefit (Section 4.3.1) or if the RBQ requires a quantitative assessment from the start (Section 4.3.2). Also, identification of the level of aggregation of interest may be helpful when decisions are made on the inclusion or exclusion of food components and foods. If more foods and/or components are included, the RBA becomes more complete, but the data need is also increased, and most probably, the uncertainties are large.

It is important that the choice of food components and foods included in the assessment depends not only on the RBQ but also on the strength of scientific evidence for the associated health effects and the available data. If some of the potential health effects associated with a food component are less well studied, or the effect, on average, is relatively small, it may be unlikely that statistically significant evidence will be found in etiological studies. Epidemiological analyses of human observational studies, for example, require a large number of participants and a long timeframe before risk factors are identified with sufficient strength of evidence, certainly if they concern the risk for relatively rare diseases. Still, the effect may be real and may affect the result of the RBA. In that case, it may be advisable to do a quantitative "what-if" analysis, where the potential quantitative impact of a health effect is included even if the effect itself is not found to be statistically significant. This can be helpful to assess how important additional research into this health effect could be.

Another important challenge in RBA lies in the discrepancy between risks and benefits, where risks are commonly identified as soon as there is some evidence for a food safety problem, whereas benefits are only acknowledged when the scientific evidence is convincing (Boobis et al. 2013; Nauta et al. 2018). Although this makes perfect sense in food policy, where consumers need to be protected against health hazards and incorrect health claims, it is not helpful for a fair comparison of risks and benefits in an RBA. Also, when a dose–response relation is derived from either human or experimental animal studies, simplifying assumptions are inevitable and increase the uncertainty of the final estimates. These challenges obviously complicate RBA and can only be met when more and higher-quality data are made available and when approaches for uncertainty analysis are worked out in more detail.

The difference in definitions used in different underlying disciplines impacts the terminology in RBA and may therefore be confusing, which causes a challenge when integrating the different disciplines. For example, the adaption of the CODEX definition of hazard (Box 4.1), which refers to the hazard as the agent and not to its potential to cause an effect, means that "hazard identification" gains another meaning as well. First of all, the specific hazard and/or the beneficial counterpart in food (i.e. the agents) do not necessarily need to be identified for RBA; only the health effects associated with the whole food are needed. Second, others have defined the beneficial counterpart of "hazard" as "benefit" (EFSA Scientific Committee 2010), potentially causing confusion, with "benefit" also being the term used for the beneficial counterpart of "risk". Thus, in order to be consistent and reduce potential confusion, we chose to refer to "identification of adverse effect(s)" instead of "hazard identification" on the risk side, and to "identification of beneficial effect(s)" on the benefit side (Tijhuis et al. 2012; see Figure 4.1).

In addition to RBA, other methods are available to study the health risks and benefits associated with food intake and to support decision making in food safety and nutrition. An interesting approach is the use of mathematical optimization techniques to characterize beneficial dietary changes. In this approach, a set of requirements based on lower and upper intake levels for different food components and foods is defined, creating a so-called "feasible region" of intakes that fulfil these requirements. Next, the technique analyses how consumers currently fall into the feasible region and what is needed to get there. Due to their multidimensional property, mathematical optimization methods are suitable for addressing the complexity of data on food consumption, concentrations, recommendations for nutrients and contaminants, etc. (Barre et al. 2016; Maillot et al. 2009, 2010; Persson et al. 2018; Sirot et al. 2012). A commonly used application in the area of optimization models in food research is the modelling of dietary recommendations that deviate as little as possible from the observed consumption while meeting the pre-defined constraints on nutrients and contaminants. The argument for this approach is that recommendations that differ least from current intakes are the most relevant and achievable for consumers. The complexity of the method is flexible; the number of foods in question can be one or more, and even recommendations for whole diets can be generated with optimization techniques. Other applications can be to minimize the exposure to a contaminant and at the same time maximize the exposure to a nutrient. Furthermore, the optimization approach can be expanded to include other food-related issues, such as sustainability and economy (Darmon et al. 2002; Horgan et al. 2016; Kramer et al. 2017; Maillot et al. 2017; van Dooren et al. 2015; Persson et al. 2019).

RBA of foods is a relevant research area, combining food safety and nutrition, and can provide important decision support for policy making in food safety and nutrition. Due to its multidisciplinary nature, it allows the use of a more holistic approach, which may prevent confusing or even conflicting dietary recommendations and food safety guidelines from being set. Still, the area is challenging and may cover a large number of research questions with increasing complexity. The categorization of questions as qualitative or quantitative RBQs and on the basis of the level of aggregation, as proposed in this chapter, can be a step forward for the development of a generic harmonized framework for risk–benefit assessments of foods, which will facilitate its application in the future.

ACKNOWLEDGEMENT

The preparation of this book chapter was funded through the Metrix project by the Ministry for Environment and Food in Denmark. We thank Marianne Uhre Jakobsen, Sara Pires and Morten Poulsen for helpful comments on an earlier version of this chapter.

REFERENCES

Barre T, Vieux F, Perignon M, Cravedi J-P, Amiot M-J, Micard V, et al. 2016. Reaching nutritional adequacy does not necessarily increase exposure to food contaminants: Evidence from a whole-diet modeling approach. *J. Nutr.* 146(10):2149–2157; doi:10.3945/jn.116.234294.

Berjia FL, Hoekstra J, Verhagen H, Poulsen M, Andersen R, Nauta M. 2014. Finding the optimum scenario in risk-benefit assessment: An example on vitamin D. *Eur. J. Nutr. Food Saf.* 4(4):558–576.

Boobis A, Chiodini A, Hoekstra J, Lagiou P, Przyrembel H, Schlatter J, et al. 2013. Critical appraisal of the assessment of benefits and risks for foods, "BRAFO Consensus Working Group." *Food Chem. Toxicol.* 55:659–675; doi:10.1016/j. fct.2012.10.028.

Boué G. 2017. *Public Health Risk-Benefit Assessment in Foods. Methodological Development with Application to Infant Milk-Based Diet.* PhD Thesis. ONIRIS, INRA, France.

Boué G, Guillou S, Antignac J-P, Bizec B, Membré J-M. 2015. Public health risk-benefit assessment associated with food consumption–A review. *Eur. J. Nutr. Food Saf.* 5:32–58; doi:10.9734/EJNFS/2015/12285.

Bouvard V, Loomis D, Guyton KZ, Grosse Y, Ghissassi FEl, Benbrahim-Tallaa L, et al. 2015. Carcinogenicity of consumption of red and processed meat. *Lancet. Oncol.* 16(16):1599–1600; doi:10.1016/S1470-2045(15)00444-1.

Bruins MJ, Mugambi G, Verkaik-Kloosterman J, Hoekstra J, Kraemer K, Osendarp S, et al. 2015. Addressing the risk of inadequate and excessive micronutrient intakes: Traditional versus new approaches to setting adequate and safe micronutrient levels in foods. *Food Nutr. Res.* 59:1–10; doi:10.3402/fnr. v59.26020.

Da Silva Felício MT, Hald T, Liebana E, Allende A, Hugas M, Nguyen-The C, et al. 2015. Risk ranking of pathogens in ready-to-eat unprocessed foods of non-animal origin (FoNAO) in the EU: Initial evaluation using outbreak data (2007–2011). *Int. J. Food Microbiol.* 195:9–19; doi:10.1016/j. ijfoodmicro.2014.11.005.

Darmon N, Ferguson EL, Briend A. 2002. A cost constraint alone has adverse effects on food selection and nutrient density: An analysis of human diets by linear programming. *J. Nutr.* 132(12):3764–3771.

Devleesschauwer B, McDonald S, Haagsma J, Praet N, Havelaar A, Speybroeck N. 2014. DALY: The DALY Calculator - A GUI for stochastic DALY calculation in R. R package version 1.3.0. Available: http://cran.rproject.org/package=DALY (accessed 1 June 2018).

EFSA NDA Panel. 2014. Scientific opinion on health benefits of seafood (fish and shellfish) consumption in relation to health risks associated with exposure to methylmercury. *EFSA J.* 12(7):3761 [80 pp.]; doi:10.2903/j.efsa.2014.3761.

EFSA Scientific Committee. 2010. Guidance on human health risk-benefit assessment of foods EFSA Scientific Committee. *EFSA J.* 8:1673; doi:10.2093/j.efsa.2010.1673.

European Centre for Disease Prevention and Control. 2015. *ECDC BCOoDE Toolkit [Software Application].*

FAO/WHO. 2006. Food safety risk analysis. A guide for national authorities. *FAO Food Nutr. Pap.* 87.

Fransen H, De Jong N, Hendriksen M, Mengelers M, Castenmiller J, Hoekstra J, et al. 2010. A tiered approach for risk-benefit assessment of foods. *Risk Anal.* 30(5):808–816; doi:10.1111/j.1539-6924.2009.01350.x.

Grandjean P, Weihe P, White RF, Debes F. 1998. Cognitive performance of children prenatally exposed to "safe" levels of methylmercury. *Environ. Res.* 77(2):165–172.

Grandjean P, Weihe P, White RF, Debes F, Araki S, Yokoyama K, et al. 1997. Cognitive deficit in 7-year-old children with prenatal exposure to methylmercury. *Neurotoxicol. Teratol.* 19(6):417–428; doi:10.1016/S0892-0362(97)00097-4.

Habermeyer M, Roth A, Guth S, Diel P, Engel K-H, Epe B, et al. 2015. Nitrate and nitrite in the diet: How to assess their benefit and risk for human health. *Mol. Nutr. Food Res.* 59(1):106–128; doi:10.1002/mnfr.201400286.

Havelaar AH, De Hollander AEM, Teunis PFM, Evers EG, Van Kranen HJ, Versteegh JFM, et al. 2000. Balancing the risks and benefits of drinking water disinfection: Disability adjusted life-years on the scale. *Environ. Health Perspect.* 108(4):315–321; doi:10.1289/ehp.00108315.

Havelaar AH, Kirk MD, Torgerson PR, Gibb HJ, Hald T, Lake RJ, et al. 2015. World Health Organization global estimates and regional comparisons of the burden of foodborne disease in 2010. *PLOS Med.* 12(12): e1001923; doi:10.1371/journal.pmed.1001923.

Hoekstra J, Hart A, Boobis A, Claupein E, Cockburn A, Hunt A, et al. 2012. BRAFO tiered approach for benefit-risk assessment of foods. *Food Chem. Toxicol.* 50(Supplement 4):S684–S698; doi:10.1016/j.fct.2010.05.049.

Hoekstra J, Hart A, Owen H, Zeilmaker M, Bokkers B, Thorgilsson B, et al. 2013. Fish, contaminants and human health: Quantifying and weighing benefits and risks. *Food Chem. Toxicol.* 54:18–29; doi:10.1016/j.fct.2012.01.013.

Hoekstra J, Verkaik-Kloosterman J, Rompelberg C, van Kranen H, Zeilmaker M, Verhagen H, et al. 2008. Integrated risk–benefit analyses: Method development with folic acid as example. *Food Chem. Toxicol.* 46(3):893–909; doi:10.1016/j.fct.2007.10.015.

Horgan GW, Perrin A, Whybrow S, Macdiarmid JI. 2016. Achieving dietary recommendations and reducing greenhouse gas emissions: Modelling diets to minimise the change from current intakes. *Int. J. Behav. Nutr. Phys. Act.* 13:1–11; doi:10.1186/s12966-016-0370-1.

IPCS. 2004. *IPCS Harmonization Project Document no. 1. IPCS Risk Assessment Terminology.*

Jakobsen LS, Granby K, Knudsen VK, Nauta M, Pires SM, Poulsen M. 2016. Burden of disease of dietary exposure to acrylamide in Denmark. *Food Chem. Toxicol.* 90: 151–159; doi:10.1016/j.fct.2016.01.021.

Joint FAO/WHO Food Standards Programme. 2015. Codex Alimentarius Commission: Procedural Manual. Available: http://www.fao.org/3/a-i5079e.pdf. Accessed 19 June 2020

Kramer GF, Tyszler M, Veer PV, Blonk H. 2017. Decreasing the overall environmental impact of the Dutch diet: How to find healthy and sustainable diets with limited changes. *Public Health Nutr.* 20(9):1699–1709; doi:10.1017/S1368980017000349.

Kromhout D, Spaaij CJK, de Goede J, Weggemans RM. 2016. The 2015 Dutch food-based dietary guidelines. *Eur. J. Clin. Nutr.* 70: 869–878; doi:10.1038/ejcn.2016.52.

Maillot M, Vieux F, Amiot MJ, Darmon N. 2010. Individual diet modeling translates nutrient recommendations into realistic and individual-specific food choices. *Am. J. Clin. Nutr.* 91(2):421–430; doi:10.3945/ajcn.2009.28426.

Maillot M, Vieux F, Delaere F, Lluch A, Darmon N. 2017. Dietary changes needed to reach nutritional adequacy without increasing diet cost according to income: An analysis among French adults. *PLOS ONE* 12(3):e0174679; doi:10.1371/journal.pone.0174679.

Maillot M, Vieux F, Ferguson E, Volatier J-L, Amiot MJ, Darmon N. 2009. To meet nutrient recommendations, most French adults need to expand their habitual food repertoire. *J. Nutr.* 139(9):1721–1727; doi:10.3945/jn.109.107318.

Mangen M-JJ, Bouwknegt M, Friesema IHM, Haagsma JA, Kortbeek LM, Tariq L, et al. 2015. Cost-of-illness and disease burden of food-related pathogens in the Netherlands, 2011. *Int. J. Food Microbiol.* 196:84–93; doi:10.1016/j.ijfoodmicro.2014.11.022.

Mozaffarian D, Geelen A, Brouwer IA, Geleijnse JM, Zock PL, Katan MB. 2005. Effect of fish oil on heart rate in humans: A meta-analysis of randomized controlled trials. *Circulation* 112(13):1945–1952; doi:10.1161/CIRCULATIONAHA.105.556886.

Mozaffarian D, Rimm EB. 2006. Fish intake, contaminants, and human health: Evaluating the risks and the benefits. *JAMA* 296(15):1885–1900; doi:10.1001/jama.296.15.1885.

Nasjonalt råd for ernæring. 2011. *Kostråd for å fremme folkehelsen og forebygge kroniske sykdommer. Metodologi og vitenskapelig kunnskapsgrunnlag [Food Based Dietary Guidelines in Promotion of Public Health and Prevention of Chronic Diseases. Methods and Scientific Evidence].*

Nauta MJ, Andersen R, Pilegaard K, Pires SM, Ravn-Haren G, Tetens I, et al. 2018. Meeting the challenges in the development of risk-benefit assessment of foods. *Trends Food Sci. Technol.* 76: 90–100; doi:10.1016/j.tifs.2018.04.004.

Oberoi S, Barchowsky A, Wu F. 2014. The global burden of disease for skin, lung, and bladder cancer caused by arsenic in food. *Cancer Epidemiol. Biomark. Prev. Prev. Oncol.* 23(7):1187–1194; doi:10.1158/1055-9965.EPI-13-1317.

Pedersen AN, Christensen T, Knudsen J, Matthiessen VK, Rosenlund-Sørensen M, Biltoft-Jensen A, et al. 2015. Danskernes kostvaner 2011–2013. *Hovedresultater [Dietary Habits in Denmark 2011–2013. Main Results]*. National Food Institute, Technical University of Denmark. Søborg, Denmark. 1–209.

Persson M, Fagt S, Pires SM, Poulsen M, Vieux F, Nauta MJ. 2018. Use of mathematical optimization models to derive healthy and safe fish intake. *J. Nutr.* 148(2):275–284; doi:10.1093/jn/nxx010.

Persson, M., Fagt, S. and Nauta, M. 2019. Optimising healthy and safe fish intake recommendations: a trade-off between personal preference and cost. *Br. J. Nutr.* 122:206–219.

Pires SM, Boué G, Boobis A, Eneroth H, Hoekstra J, Persson IM, et al. Risk benefit assessment of foods: key findings from an international workshop. *Food Res. Int.* 116: 859–869.

Renwick AG, Flynn A, Fletcher RJ, Müller DJ, Tuijtelaars S, Verhagen H. 2004. Risk–benefit analysis of micronutrients. *Food Chem. Toxicol.* 42(12):1903–1922; doi:10.1016/j.fct.2004.07.013.

Ruzante JM, Davidson VJ, Caswell J, Fazil A, Cranfield JAL, Henson SJ, et al. 2010. A multifactorial risk prioritization framework for foodborne pathogens. *Risk Anal.* 30(5):724–742; doi:10.1111/j.1539-6924.2009.01278.x.

Ruzante JM, Grieger K, Woodward K, Lambertini E, Kowalcyk B. 2017. The use of multi-criteria decision analysis in food safety risk-benefit assessment. *Food Prot. Trends* 37: 132–139.

Selby PL, Davies M, Marks JS, Mawer EB. 1995. Vitamin D intoxication causes hypercalcaemia by increased bone resorption which responds to pamidronate. *Clin. Endocrinol. (Oxf.)* 43(5):531–536; doi:10.1111/j.1365-2265.1995.tb02916.x.

Sirot V, Leblanc J-C, Margaritis I. 2012. A risk–benefit analysis approach to seafood intake to determine optimal consumption. *Br. J. Nutr.* 107(12):1812–1822; doi:10.1017/S0007114511005010.

Tetens I, Andersen LB, Astrup A, Ulla Holmboe Gondolf, Kjeld H, Marianne UJ, Knudsen VK, et al. 2013. *Evidensgrundlaget for danske råd om kost og fysisk aktivitet [The evidence-base for the Danish guidelines for diet and physical activity]*. Søborg, Denmark: National Food Institute, Technical University of Denmark.

The Scientific Advisory Committee on Nutrition. 2010. *Iron and Health.* London, UK: TSO.

Thomsen ST, Pires SM, Devleesschauwer B, Poulsen M, Fagt S, Ygil KH, et al. 2018. Investigating the risk-benefit balance of substituting red and processed meat with fish in a Danish diet. *Food Chem. Toxicol.* 120:50–63; doi:10.1016/j.fct.2018.06.063.

Tijhuis MJ, De Jong N, Pohjola MV, Gunnlaugsdóttir H, Hendriksen M, Hoekstra J, et al. 2012. State of the art in benefit risk analysis: Food and nutrition. *Food Chem. Toxicol.* 50(1):5–25; doi:10.1016/j.fct.2011.06.010.

U.S. Department of Health and Human Services and U.S. Department of Agriculture. 2015. *2015–2020 Dietary Guidelines for Americans*; doi:10.1097/NT.0b013e31826c50af.

van Dooren C, Tyszler M, Kramer GFH, Aiking H. 2015. Combining low price, low climate impact and high nutritional value in one shopping basket through diet optimization by linear programming. *Sustainability* 7(9):12837–12855; doi:10.3390/su70912837.

Ward PR, Henderson J, Coveney J, Meyer S. 2012. How do South Australian consumers negotiate and respond to information in the media about food and nutrition? *J. Sociol.* 48(1):23–41; doi:10.1177/1440783311407947.

WCRF/AICR. 2007. *Food, Nutrition, Physical Activity, and the Prevention of Cancer: A Global Perspective.*

WCRF/AICR. 2011. *Continuous Update Project Report. Food, Nutrition, Physical Activity, and the Prevention of Colorectal Cancer.*

WCRF/AICR. 2016. *Continuous Update Project Report: Diet, Nutrition, Physical Activity and Stomach Cancer.*

WHO. 2003. Diet, nutrition and the prevention of chronic diseases: Report of a joint WHO/FAO expert consultation. *WHO Tech. Rep. Ser.* 916.

Zeilmaker MJ, Hoekstra J, Van Eijkeren JCH, De Jong N, Hart A, Kennedy M, et al. 2013. Fish consumption during child bearing age: A quantitative risk benefit analysis on neurodevelopment. *Food Chem. Toxicol.* 54:30–34; doi:10.1016/j.fct.2011.10.068.

5

Application of Quantitative Risk Assessment Methods for Food Quality

Fernanda B. Campagnollo, Juliana Lane Paixão dos Santos,
Verônica O. Alvarenga, Rafael D. Chaves, Leonardo do Prado-Silva, and
Anderson S. Sant'Ana

Contents

117

5.1 INTRODUCTION

Risk analysis is a logical, structured, and consistent process that aims to provide information about the risk of introduction, expression, and dissemination of diseases, assessing their economic impact and the consequences for public and animal health (Chapman, Otten, Fazil, Ernst, & Smith, 2016). It is a tool that encompasses three elements with different objectives; however, they are strongly linked: risk assessment, risk management, and risk communication (Buchanan, 2004; Lammerding, 1997; OMS & FAO, 2014).

Quantitative microbiological risk assessment (QMRA) is the scientific framework that characterizes the microbiological risks involved in the food production chain (Chapman et al., 2016; WHO/FAO, 2008). Currently, several models, methodologies, and procedures assist in risk analysis for decision making (Aven & Zio, 2018). The concepts involved in information collection and risk management are based on the current state of knowledge and information available in the literature (Aven & Zio, 2018).

QMRA aims to assess scientifically the risks of occurrence of food spoilage and the likely severity, known or potential, resulting from human exposure to food hazards. It includes four steps: hazard identification; hazard characterization; exposure assessment; and risk characterization (Lammerding, 1997; OMS & FAO, 2014; Vose, 2008).

The QMRA model should not only be used to estimate the probability of microbiological risk exposure but should also provide information that helps to elaborate strategies, interventions, and hazard controls that can be found along the production chain (Barron, Redmond, & Butler, 2002). These models incorporate techniques for examining the variability associated with raw materials, microorganisms, and production processes. Additionally, simulations and sensitivity analyses may predict and project the critical points that can be found along the production chain. Furthermore, mitigating effect measures can be integrated into the model for quantifying the process effectiveness and the responses of the production system (Barron et al., 2002).

Risk analysis allows control points to be identified along the food processing chain, which are essential for making preventive and corrective decisions. Thus, it is possible to evaluate the benefits and costs of each action, thereby enhancing the efficiency of the risk management process (Membré & Boué, 2018).

Although there are several QMRA models developed for pathogenic microorganisms, only a few quantify food spoilage (Snyder & Worobo, 2018). The control of microorganisms associated with food spoilage impacts directly on the food quality. Thus, the development of QMRA models for spoilage microorganisms may be a valuable tool to give support for control and intervention measures to prevent food deterioration along the chain. This chapter explores the concepts used to develop spoilage QMRA models.

5.2 CONCEPTS OF RISK ASSESSMENT FOR MICROBIAL SPOILAGE

Spoilage can be defined as a process or a change that makes a product undesirable or unacceptable for the consumer. The spoilage is the result of the biochemical activity of microbial metabolism that produces alcohols, sulfur compounds, hydrocarbons, fluorescent pigments, organic acids, esters, carbonates, and diamines, which modify food taste, odor, and texture (Abdel-Aziz, Asker, Keera, & Mahmoud, 2016; Nychas & Panagou, 2011; Petruzzi, Corbo, Sinigaglia, & Bevilacqua, 2017). The spoilage can occur at any stage of the chain. Thus, careful monitoring from production to distribution, as well as product storage in retail and household refrigerators, is essential to ensure its safety and quality (Abdel-Aziz et al., 2016; Nychas & Panagou, 2011).

QMRA models developed to evaluate spoilage use the exposure assessment in a broader sense. They consider the contribution of each step during food processing or distribution, which is quantified with the ultimate goal of identifying and evaluating management options to control or reduce that risk (Gougouli & Koutsoumanis, 2017). Furthermore, the variability and uncertainty affecting the microbial response are preponderant factors in estimating the risk of spoilage (Koutsoumanis, 2009). These concepts will be discussed later in this chapter.

The QMRA model structure considers as fundamental parameters hazard definition, product or food matrix, the exposure scenario, and the target population. The definition of risk is based on the likelihood of an adverse effect on an organism, system, or population caused under the circumstances specified by exposure to an agent (Barlow et al., 2015; Vose, 2008). For food QMRA, the term "risk" has been used to describe the likelihood that food is unfit for consumption, whether for reasons of contamination or spoilage (WHO/FAO, 2008).

The first step in developing a food spoilage QMRA model is to determine the sources that may cause variation along the food chain. These sources are related to the production environment, storage, and microbiological and consumer variability (Koutsoumanis, 2009). The next step is hazard identification, which indicates the effects considered to be adverse regardless of the dose required or the specific mechanisms involved in the process.

Thus, hazard characterization provides a quantitative and qualitative estimate of adverse effects, so that the dose–response relationship and mode of action for these effects can be established (Vose, 2008). Exposure assessment involves the assessment of the modes, magnitudes, duration and time of actual or anticipated exposure, and the number and nature of those likely to be exposed (WHO/FAO, 2008). Risk characterization, the final stage of risk assessment, is the estimation of the likelihood of food spoilage as an exposure consequence, taking into account the results of hazard identification, hazard characterization, and exposure assessment (Membré & Boué, 2018).

The risk assessment implementation may be biased (vulnerable) by many uncertainties resulting from information deficiencies or critical gaps (Kouamé-Sina et al., 2012). When this occurs, plausible assumptions are made according to the current state of scientific knowledge, taking into account these uncertainties, so that the assessment can be achieved (Vose, 2008). Therefore, risk assessment is considered a complex mix of currently available data and assumptions based on prevailing scientific data.

5.3 ASSESSMENT OF MICROBIOLOGICAL SPOILAGE OF FOOD AND BEVERAGES: CHEMICAL AND MICROBIOLOGICAL CHANGES

The microbiology of food spoilage has been well characterized, but the challenge is to determine the relationship between microbial composition and the presence of microbial metabolites, resulting in a proper evaluation of microbiologic spoilage (in't Veld, 1996; Remenant, Jaffrès, Dousset, Pilet, & Zagorec, 2015).

In order to manage food quality risks, it is essential to identify which foods, spoilage microorganisms, or situations contribute to chemical or microbiological changes and also to find out the magnitude of the impact caused by these changes. Such information is necessary to make rational decisions about the most effective interventions for reduction of food spoilage. With the development of microbial risk assessment concepts, it was recognized that the focal point for controlling spoilage microorganisms and determining food quality should consider the chemical and microbiological changes together, not only the concentration of a microorganism in the food. Additionally, Remenant et al. (2015) point out that spoilage may be a result of the presence of different microorganism species and not always due to the dominant one, and that there may be a synergistic effect, which can increase the magnitude of spoilage compared with the presence of only one spoilage microorganism.

The establishment of microbiological criteria defines the acceptability of a product, a batch, or a process based on the absence, presence, or a defined number of microorganisms and the number of their toxins or metabolites per unit of mass, volume, area, or batch. Such criteria can be used as tools to assess the safety and quality of foods, and they are generally determined by quantitative risk assessment and feasibility (Barlow et al., 2015). As an example, Rukchon, Nopwinyuwong, Trevanich, Jinkarn, & Suppakul (2014) explain that the quantification of chemical changes in poultry meat can provide information about the level of spoilage. Indicators such as biogenic amines, volatile bases, nucleotide breakdown products, and volatile acidity can be used to assess meat quality and freshness during storage. However, the ideal microbial metabolite to be used as the indicator should follow some principles such as the absence (or at least, presence in a low concentration) in fresh product, should increase over the storage period, and should be produced by the predominant microorganisms and have a good correlation with the product's organoleptic assessment (Lianou, Panagou, & Nychas, 2016).

As crucial as detecting spoilage during storage by the food industry and retailers is the search for a cheap, simple, and accurate measurement device for detecting food spoilage at a consumer level. However, the existing spoilage detectors/indicators are sensitive to specific metabolites present in a specific food product or group of food products and are not validated for general application in every type of food. Also, spoilage indicators are still in a developmental stage at universities or innovative companies (de Jong et al., 2005). Some examples of published work on food spoilage indicators include (1) a colorimetric mixed pH dye-based indicator for real-time monitoring of intermediate-moisture dessert spoilage by measuring the response to CO_2 (Nopwinyuwong, Trevanich, & Suppakul, 2010); (2) a colorimetric sensor array for fish spoilage monitoring in which chemo-sensitive compounds were incorporated in an array for colorimetric detection of typical fish spoilage compounds (trimethylamine, dimethylamine, cadaverine, and putrescine) at room temperature and 4°C (Morsy et al., 2016); (3) an on-package dual-sensor label based on pH indicators (methyl red and bromocresol purple) for real-time monitoring of beef freshness at room and chiller temperatures (Kuswandi & Nurfawaidi, 2017); and (4) the use of an electronic nose, based on a chemical sensor array of six metal oxide semiconductors, to diagnose fungal contamination, to detect high fumonisin content, and to predict fumonisin concentration in maize cultures (Gobbi, Falasconi, Torelli, & Sberveglieri, 2011).

There are different reasons to explain why there has not been a single attempt to quantify spoilage of specific food products: (i) available methodologies are too slow, as they provide retrospective information about spoilage and cannot be applied for online monitoring; and (ii) there are different technologies for food preservation (i.e., vacuum, modified atmospheres, etc.), which influences the choice of a specific methodology and can affect its application. Furthermore, some considerations that make it challenging to choose the ideal metabolite for spoilage assessment should be taken into account: specificity to a given microorganism – if this microorganism is not present or is inhibited by other microorganisms (naturally present or intentionally added to the product), incorrect results may be achieved; specificity to a given substrate – if this substrate is absent or is present in low concentration, the metabolite cannot be produced, and spoilage does not occur; influence of environmental conditions (i.e., pH, temperature, water activity, oxygen tension, etc.) – the microorganism will not produce the metabolite – if environmental conditions are not in favor of such production. Besides, ideal metabolites should not demand complex procedures for their measurement and should not demand too much time and equipment. If the detection of the metabolite is difficult, rapid analytical methods or tools for quantifying their indicators should be chosen. Finally, validation of such methodologies and inspection of authorities for control purposes are also necessary (Lianou et al., 2016).

Classical microbial determination of perishable foods is of limited value for predictive prognoses because foods are sold or consumed earlier than the results of microbiological assays are accessible (in't Veld, 1996). There is interest among the food industry, which includes retailers, consumers' rights organizations, and food safety controlling bodies, in the development of accurate, cost-effective, rapid, reliable, non-invasive, and non-destructive methods or devices to evaluate the real-time freshness of food products (Rukchon et al., 2014). New methodologies and instruments for early quantitative detection of spoilage microorganisms or their metabolites have been developed and include novel analytical approaches based on biosensors, sensor arrays, and spectroscopy techniques in tandem with chemometrics. Examples of such methodologies consist of metagenomics, enzymatic reactor systems, electronic noses (arrays of sensors), potentiometric measurements by electronic tongues, Fourier transform infrared spectroscopy, near-infrared spectroscopy, hyperspectral imaging techniques, and multispectral imaging technologies (Lianou et al., 2016).

5.4 STAGES OF RISK ASSESSMENT FOR MICROBIAL SPOILAGE

According to the principles and guidelines of the Codex Alimentarius Commission (CAC, 1999) for the conduct of a microbiological risk assessment, a scientifically based process should consist of four steps: (i) hazard identification, (ii) hazard characterization, (iii) exposure assessment, and (iv) risk characterization. Also, a microbiological risk assessment should consider the dynamics of microorganism growth, survival, and death in the target food to be evaluated and the complexity of the interaction between this agent and humans who have consumed this food (CAC, 1999). Indeed, the most important implication of a microbial risk assessment is that it allows a deductive assessment of the effect of potential intervention measures along the whole food chain, and the combinations of these potential intervention measures, on public health or food quality (Havelaar, Evers, & Nauta, 2008). Such information allows decision makers in the food industry to compare different interventions and recognize those that can result in effective reduction of microbial spoilage and therefore, food quality improvement. Identification of data gaps and target research with considerable value for food quality can be achieved through microbial risk assessment (Koutsoumanis & Aspridou, 2016).

First, to build a microbial risk assessment, it is necessary to state explicitly the question to be answered and to frame the scope by outlying the objective and the goals in order to solve the question raised satisfactorily (Membré & Boué, 2018). A general framework should be structured in order to create a risk assessment for microbial spoilage, and the general steps include setting the scale of the food system; mixing of ingredients; processing batches; reduction of spoilage microorganisms by thermal treatment; assessing potential for cross-contamination or recontamination; assessing the potential for microbial growth, survival, and death; assessing variability of consumption; and assessing direct and indirect economic impacts (McNab, 1998). Membré & Lambert (2008) illustrated an application of risk assessment at Unilever, where a probabilistic exposure assessment was developed to assess the safety of refrigerated processed foods of extended durability with regard to *Bacillus cereus*. The factors included in this assessment were raw material contamination, heat treatment, bacterial spore heat resistance, injured spore lag time and growth potential, chilled supply-chain market (split into a warehouse, retail, and consumers), and consumers' habits, such as time spent shopping and time before consuming. In the end, a sensitivity analysis was performed to identify the impact of each factor on the output of this risk assessment.

In the step of hazard identification for microbial agents, the purpose is to identify the microorganisms or the microbial toxins of relevance to food, capable of causing adverse health effects (for pathogens) or an inedible product (for spoilage microorganisms), which is a qualitative process. Information regarding the microorganism of concern can be obtained from the scientific literature, from industry or government databases, reports, books, and experiments, or from expert opinion (CAC, 1999; Membré & Boué, 2018). Considering the case of microbial spoilage in a risk assessment, relevant information includes, for example, data in areas such as investigations of the characteristics of microorganisms, the interaction between microorganisms and their environment through the food chain from primary production up to and including consumption, and studies on analogous microorganisms and situations.

The initial activity in microbial risk assessment is primarily concerned with figuring out the primary sources of exposure to the microorganism or finding out which microorganism deserves most attention in a particular food or food group. Microbiological and chemical evidence should also be considered in support of spoilage information. Inferences from experimental and on-site observations are relevant in gaining insight about the nature and behavior of the target microorganism (Lammerding & Fazil, 2000).

The next step, hazard characterization, represents the step that provides a qualitative or quantitative description of a microorganism or its toxin in food, describing the severity and extent of adverse effects related to ingestion of this microorganism (CAC, 1999). When the microbial agent is a pathogenic one, the adverse health effects (a type of illness) are studied, but if the microbial agent is a spoilage one, the effects associated with its presence in the food (a type of product quality fault) are analyzed (Membré & Boué, 2018). There are some essential aspects that need to be considered in hazard characterization: microorganisms are capable of replicating; microorganisms can change depending on their interaction with the environment; genetic material can be transferred between microorganisms; microorganisms can be spread through secondary and tertiary transmission; microorganisms can persist in certain environments, leading to continued contamination; sometimes, low doses of some microorganisms can cause a severe effect; and food attributes may alter microbial growth (CAC, 1999).

The exposure assessment step for microbiological agents includes an assessment of the potential extent of food contamination by a particular microorganism or its toxins (CAC, 1999); it is the estimation of the likely intake of pathogenic microorganisms or the likely level of a spoilage microorganism in the food considering the changes from raw materials to consumption (Membré & Boué, 2018). The exposure assessment phase of a microbial risk assessment is a dynamic process because of potential multiplication and death of microorganisms. Models and assumptions are necessary to translate the available data into quantitative estimates of the amount of microorganism in the product, and the unit of food that is of interest should be specified, i.e., by product weight or batch. Predictive microbial models are beneficial for this description, since they use mathematical expressions to describe how microorganism numbers change with time and how the rate of change is affected by environmental conditions (Lammerding & Fazil, 2000). While primary models describe the microbial concentration over time, secondary models characterize the effect of formulation and process on growth rate and lag phase (Membré & Boué, 2018). The current applications of predictive modeling in an industrial context are broad but can be summarized into product innovation (assessing microbial proliferation, growth limits, or inactivation rate in order to develop new products and processes, reformulate existing products, or determine storage conditions and shelf-life), operational support (supporting food safety and quality decisions related to implementing or running a food manufacturing operation or assessing the impact of process deviations), and incident support (estimating the impact on product quality in the case of problems with products on the market) (Membré & Boué, 2018).

Some aspects must be considered for the design of the exposure assessment, such as the frequency of contamination of the target food by the spoilage microorganism and its level in this food over time. These aspects may be influenced by microorganism characteristics, microbial ecology of the food, initial contamination of raw material (including regional differences and seasonality of production), degree of sanitation and process control, processing methods and preparation practices (cooking, holding, etc.), packaging, distribution, and storage. Also, other factors to be included are cross-contamination by handling and the potential impact of abusive environmental time/temperature relationships (CAC, 1999). Lammerding & Fazil (2000) explain that mixing or blending raw materials or ingredients may result in an increase or a reduction of contamination if highly contaminated and uncontaminated food are mixed. For example, minced meat is produced from different animal carcasses, bulk tank milk comes from different cows, and broken eggs are combined before pasteurization.

The final step, the risk characterization, represents the integration of the hazard identification, hazard characterization, and exposure assessment determinations in order to obtain a risk estimate; it provides a qualitative or quantitative estimate of the probability of occurrence and severity of known or potential adverse health effects in a given population when considering pathogenic microorganisms, or an estimate of the likelihood of food quality changes when considering spoilage microorganisms, including a description of the uncertainties associated with these estimates. It is important to highlight that the output is not provided as a single value but as a range of values with their probability of occurrence and should be interpreted. Variability, uncertainty, and assumptions identified in all previous steps influence the degree of confidence in the final estimation of risk (CAC, 1999; Membré & Boué, 2018). Differentiation between uncertainty and variability is significant and will be explained further in this chapter.

Havelaar et al. (2008) emphasize that the risk assessment model development should take place before data collection, since developing a model is cheaper than collecting data, and this would result in more focused surveillance and monitoring activities. Also, the availability of appropriate data is a key need for any risk assessment, which should be representative of the system under study and sufficiently express the variability of microbial contamination. Quantitative information on the behavior of spoilage microorganisms (microorganism characteristics in combination with environmental conditions) in the food chain is a central point in risk assessment models. Additionally, detailed information on food consumption and the determination of the dynamic nature of microbial contamination at a consumer's house (growth under suboptimal storage temperature, inactivation due to cooking and cross-contamination) should be obtained in order to adequately model the fate of the target microorganism at the end of the food chain.

Koutsoumanis (2009) explains that in most risk assessments published up to now, pathogenic microorganisms are traditionally considered, and microbial spoilage is not taken into account. Also, an erroneous estimation of risk can occur when spoilage is ignored, since the same conditions that lead to critical levels of hazard agents frequently favor spoilage; i.e., abusive temperatures during storage may increase the likelihood of high numbers of pathogens at the time of consumption. However, spoilage also may occur at this time and lower the probability of consuming this product. So, the identification of product quality at the moment of consumption can result in more realistic risk assessments. The same author presented a systematic approach for taking spoilage into account in risk assessment of *Escherichia coli* O157: H7 in ground beef combined with pseudomonads. A positive correlation between the concentrations of *E. coli* and pseudomonads after retail storage was obtained due to the application of abusive storage conditions (high temperature and long storage time), which led to extended growth for both microorganisms. So, it was observed that packages with high numbers of *E. coli* have an increased probability of showing clear spoilage before consumption and that ignoring spoilage in risk assessment could lead to a significant overestimation of risk. Other examples of risk assessment for microbial spoilage will be provided in the last section of this chapter.

5.5 MODELING APPROACHES FOR SPOILAGE RISK ASSESSMENT

5.5.1 Qualitative and Quantitative Exposure Assessment

Qualitative and quantitative methods can exploit the exposure assessment for food spoilage. The main difference between qualitative and quantitative assessments is related to data treatment, which is a descriptive analysis in qualitative assessments and a mathematical evaluation in quantitative assessments. Some points should be discussed before defining the type of assessment that will be conducted, and the main issue is data availability.

When the available data are insufficient for numerical analysis, a qualitative assessment must be conducted in order to determine the risks of the exposure factors. Usually, the results of a risk assessment are divided into low, medium, and high, although there may be cases of very low and very high depending on the situation. Although the data needed to perform a qualitative assessment are relatively more straightforward than for a quantitative assessment, the results may lead to different conclusions (e.g., the threshold of the difference between a low and very low risk). However, through this analytical approach, it is more challenging to predict medium risks than low and high.

Through the quantitative approach, the numerical data can be analyzed, so that the results are more reliable than qualitative assessments. In order to compile all numerical data, the development of a mathematical model is needed to compare and evaluate the factors that affect spoilage risk to foods. The "input" (e.g., time, temperature, or pH) used to perform a mathematical model determines the "output" that result in response to the risk assessment. Most works on risk assessment do not take into account the process of spoilage, and the problem is that the conditions of critical levels of hazard may favor spoilage (Koutsoumanis, 2009; Nauta, Litman, & Barker, 2003). Depending on the type of inputs, quantitative assessments can be separated into two types of analysis: deterministic and stochastic. In the deterministic method, single values such as the average are used to generate the outputs, and in the stochastic method, all available data for each input are used to generate a distribution of the possible results.

Due to the difficulty of solving stochastic models analytically, software (e.g., @RISK®) is usually used in a Monte Carlo simulation to evaluate the model. Being an extension of the "what-if" test, Monte Carlo simulation has been widely used in quantitative spoilage risk assessment in foods such as meats and canned vegetables (Koutsoumanis, 2009; Rigaux, André, Albert, & Carlin, 2014). Once the model is converted into a Monte Carlo simulation, a likelihood distribution is formed. On the other hand, the software evaluates the model according to the distribution of variables by selecting specific values, and each value can be selected according to the probability used to describe the variable. By iteration of the selected values and equations to describe the model, the spoilage risk exposure is calculated (WHO/FAO, 2008).

5.5.1.1 Model Development

The model development is based on the routes of interest by which the target of the assessment can be exposed to the risk (expressed by text, diagrams, or mathematical models). Commonly used, a mathematical model is primarily used to describe the route of the spoilage risk exposure and all factors that may influence the final assessment. The model structure should be developed in such a way as to facilitate the analysis. Each of the independent variables that affect the risk exposure need their inputs and outputs paired correctly. So, with Monte Carlo simulation software, the user's need indicates which variables must be followed.

5.5.2 Model Types and Availability

There are many types of model available to be used as a platform for new models to be developed in the context of food spoilage. Rates, limits, and probability of growth as a function of intrinsic and extrinsic character-istics of the food and environment are some of the possibilities. Besides, there are online platforms, such as PMP (Predictive Microbiology Portal) and ComBase Predictor, which describe the behavior of pathogens and spoilage microorganisms. These tools are handy for industry and aca-demia in order to solve problems quickly and additionally possess a user-friendly and intuitive interface.

5.5.3 Application of Predictive Microbiology within Exposure Assessment

Predictive models of microbiology play an important role in exposure assessment, though two critical points still need to be considered in order to adjust their accuracy. For example, it is challenging for a mathematical model to predict the closeness of all the variables involved in the growth or inactivation of a microorganism. Throughout, the models developed must be submitted to a validation process whereby it is possible to deter-mine to what extent the model created can represent the microbial behav-ior. Conventionally, it is desirable for a mathematical model to have as few parameters as possible so that its outputs have a high reliability (J Baranyi, Ross, Mcmeekin, & Roberts, 1996). As described by Ross (1999), the addi-tion of one variable to the mathematical model represents an increase of approximately ±10–15%. This means that as the number of variables of the model increases, the confidence of the prediction tends to decrease.

5.6 VARIABILITY, UNCERTAINTY, AND SENSITIVITY FOR SPOILAGE RISK ASSESSMENT

Although the words "uncertainty" and "variability" may be used in the same context, their meanings are completely different. Uncertainty, according to the Cambridge Dictionary, can be defined as "a situation in which something is not known, or something that is not known or certain." When we transfer this word to the risk analysis world, we can define it as "the lack of perfect knowledge, information, sampling or measurement errors of a given variate value" (Pouillot, Albert, Cornu, & Denis, 2003; Régis Pouillot & Delignette-Muller, 2010), and the effects of this uncertainty can be minimized by adding further experimental assays. On the other hand, "variability," according to the same dictionary, means "lack of consistency or fixed pattern." Again, when we use it in the microbiological context, it can be defined as the "true heterogeneity of a population irreducible by additional measurements" (Pouillot et al., 2003). It is important to understand the differences between these two concepts for a better understanding in modeling microbial risk assessment (Nauta, 2000).

In model-based optimization, uncertainty is always an inherent problem when we are dealing with living organisms and can originate from external disturbances (Tellen et al., 2015) or unmodeled process variables (Kaern, Elston, Blake, & Collins, 2005). In biological processes, this uncertainty can be present due to inherent biological variability (Nimmegeers, Telen, Logist, & Van Impe, 2016) between genetically identical cells (Kaern et al., 2005). The wrong estimation could result from not taking this uncertainty into account, generating inaccurate models (Nagy & Braatz, 2004). This leads to a challenging optimization problem under uncertainty, which requires a sturdy solution (Nimmegeers et al., 2016).

Variability is due not only to microbiological characteristics but also to these "external disturbances," such as quality of the raw material, variability in the storage times and temperatures, microbial load, and the determination of the intrinsic properties of foods. In order to estimate an accurate shelf life for a given product, the variability of external factors cannot be excluded (Chotyakul, Lamela, & Torres, 2012). However, it is very difficult to achieve a significant reduction in uncertainty when multiple factors are considered in the microbial model used for this estimation. As with the ability or inability to evaluate uncertainty and variability, a source of variation in model parameters, separately, will determine whether a proposed model will be successfully employed (Nauta, 2000; Pouillot et al., 2003). However, more accurate and realistic predictions are difficult to achieve due to the different sources of variability (Zwietering, 2015). The fitness and robustness of strains depend on the physiological state, cell history, genetic and phenotypic variability within a population (i.e., population heterogeneity), and strain variability (den Besten, Aryani, Metselaar, & Zwietering, 2017). Sturdy controls (i.e., variables that can be manipulated throughout the process) ensure that constraints are met and an overall better objective function estimate is guaranteed.

In this scenario, sensitivity analysis can be used in order to identify which of the inputted variables are more sensitive to a specific QMRA. A large number of methods are available to apply sensitivity analysis, although the challenge lies in an accurate separation between variability and uncertainty inputs. The sensitivity analysis underlines that depending on the process characteristics, microbiological variability can become the most important determining factor affecting variability in the final contamination levels (den Besten et al., 2017). A procedure is proposed that focuses on the relationship between risk estimates obtained by Monte Carlo simulation and the location of pseudo-randomly sampled input variables within the uncertainty and variability distributions (Busschaert, Geeraerd, Uyttendaele, & van Impe, 2011).

The majority of papers and established models are designed for pathogenic microorganisms. Limited information on microbiological variability is available for spoilage microorganisms (Aryani, Besten, & Zwietering, 2016). Lactic acid bacteria (LAB) and molds are the most common organisms used in spoilage research. For example, variability was demonstrated in the growth of individual fungal spores at the lowest inoculum possible (Samapundo, Devlieghere, De Meulenaer, & Debevere, 2007) with the same physiological state (Gougouli & Koutsoumanis, 2012).

Variability was described mostly by normal, lognormal, uniform, or Pert distributions (Pujol, Albert, Magras, Brian, & Membré, 2015). However, risk variability distributions and the evaluation of uncertainty associated with this spoilage risk can be obtained through a two-dimensional Monte Carlo simulation separately propagating uncertainty and variability through the model (Mokhtari & Frey, 2005; Régis Pouillot & Delignette-Muller, 2010).

The variation of the primary model parameters can be analyzed based on the propagation of errors intentionally introduced into experimental data that are assumed to be normally distributed (Poschet et al., 2004). However, the variation of the primary model parameters has not been taken into account in determining the uncertainty of the parameters of the secondary model (Giannakourou & Stoforos, 2017).

5.7 EXAMPLES OF RISK ASSESSMENT FOR FOOD SPOILAGE

Quantitative risk assessment for food quality can be applied through two approaches. The first one corresponds to a systemic approach considering spoilage risk in microbial risk assessment of foodborne poisoning. The main idea of this approach is to combine the data and predictive models (kinetic spoilage modeling) of target spoilage microorganisms with those of pathogens (Koutsoumanis, 2009). The second approach is related to the estimation of spoilage risk of a certain food product associated with a target spoilage microorganism. Four examples of risk assessment for food spoilage including both approaches are discussed and presented in this section (Table 5.1).

As an example of the first approach, a study of exposure assessment of *Escherichia coli* O157: H7 and pseudomonads in ground beef ($a_w = 0.990$, pH = 5.7) will be discussed (Koutsoumanis, 2009). The exposure assessment of this study started from the moment when the ground beef was packed. At this stage, the distributions of the initial levels of the specific spoilage microorganism (pseudomonads) and *E. coli* O157: H7 (potential pathogen) were obtained. No correlation was observed between these two levels in the fresh produce, and all packages were included as acceptable (the level

Table 5.1 Risk Assessment Models Applied to Food Spoilage

Microorganisms	Food Product	Risk Model Outputs	Reference
Pseudomonads, *E. coli* O157: H7	Ground beef	The concentration of pseudomonads and *E. coli* O157: H7 at the end of retail storage; % of spoilage packages	Koutsoumanis, 2009
Geobacillus stearothermophilus	Canned green beans	The concentration of *G. stearothermophilus* after sterilization and incubation test; % of green-bean cans with the defect	Rigaux et al., 2014
Geobacillus stearothermophilus	Evaporated milk	Time to spoilage ($10^{7.4}$ CFU/mL), risk categorization	Kakagianni and Koutsoumanis, 2018
Aspergillus niger	Yogurt	The probability distribution of the number of cups with a visible mycelium of *A. niger* at the time of consumption	Gougouli and Koutsoumanis, 2017

of pseudomonads was lower than spoilage levels). Next, data of distributions of retail storage conditions (time and temperature) were described (Cassin, Lammerding, Todd, Ross, & McColl, 1998). The growth of the two microorganisms during retail storage was then calculated, considering no lag phase and a maximum population of 10^8 and 10^{10} colony forming units (CFU)/g for *E. coli* O157: H7 and pseudomonads, respectively. The maximum specific growth rates were estimated from predictive models. Based on these data, the concentrations of the two microorganisms at the end of retail storage were then estimated with the Monte Carlo simulation technique (10,000 iterations). Spoiled packages of ground beef at the end of retail storage were identified by comparing the estimated levels of pseudomonads with the spoilage levels. These were represented by a normal distribution taking into account microbial and sensory analysis of the product (Koutsoumanis, Stamatiou, Skandamis, & Nychas, 2006).

In contrast to the initial levels, a positive correlation was observed between the concentration of *E. coli* and pseudomonads in ground beef packages after retail. The increase in storage temperature and storage time would lead to extended growth for both microorganisms. Consequently, the packages with a high concentration of *E. coli* also had high counts of the spoilage pseudomonads. Also, the initial level of a certain spoilage microorganism will directly affect the estimation of hazard risk. For instance, if this level is very high, the risk will be lower, as this will reduce the shelf life of the food products. Therefore, these results indicate that not considering spoilage in risk assessment studies might result in overestimation of the risk of a pathogen. Moreover, the risk may also be affected by the differences in temperature sensitivity between the pathogen and the spoilage microorganism and by their potential interaction. The incorporation of these factors might result in a more accurate estimation of risk in food safety studies (Koutsoumanis, 2009).

Based on the second approach of risk assessment applied to food spoilage, three examples are presented (Table 5.1).

The first example is a quantitative assessment of the risk of microbial spoilage of canned green beans by *Geobacillus stearothermophilus* (Rigaux et al., 2014). In this study, the risk of spoilage was determined using a non-stability rate of cans submitted to a biological stability test, which is used as a quality indicator. During this test, the canned beans were incubated at 55°C for 7 days, and the spoilage was defined by the growth of *G. stearothermophilus*, i.e., the percentage of positive cans reaching concentrations of 7 log CFU/g. In this study, the modeled unit was defined as one can containing 445 g of green beans filled with 405 g of covering brine (final weight = 850 g). The risk assessment model estimated the changes in *G. stearothermophilus* concentration along a canned green-bean processing chain, including the initial contamination of fresh, unprocessed green beans, cross-contamination during processing, inactivation processes, the probability of survival and growth, and the estimation of pH of the green beans at different stages of processing (Rigaux et al., 2014). The global risk model was represented by a direct acyclic graph (DAG), which shows the conditional dependence between the model parameters. A total of 20 model parameters were classified into four categories: fixed, variable, uncertain, and variable and uncertain. They included concentrations of *G. stearothermophilus*, inactivation parameters (such as D- and Z-values), environmental parameters such as the pH of the green beans during processing, cardinal pH values, as well as processing parameters such as F_0 (sterilization reference value). This information was obtained by employing collected data, expert opinions, literature data, and information based on the food processor. The development of the risk model included five main steps: (i) determination of occurrence of *G. stearothermophilus* through processing; (ii) predictions of changes in *G. stearothermophilus* concentrations during processing through predictive models and non-stability prevalence; (iii) model validation; (iv) determination of the most influential factors by sensitivity analysis; and (v) what-if scenarios to test the influence of microbiological phenomena and options for spoilage risk management (Rigaux et al., 2014). The model predicted that about 92% of the cans containing at least one spore after sterilization would develop spoilage, considering the high growth ability of germinated spores. The prediction of non-stability rate due to *G. stearothermophilus* was 0.5% with 95% confidence interval (CI) = (0.2%–1.2%). After validation with independent data, a specific non-stability rate estimation of 0.9% with 95% CI = (0.5%–1.2%) was obtained. Sensitivity analysis and what-if scenarios were used to identify the highest sources of variability and uncertainty in the final model. The sensitivity analysis indicated that two processes, blanching (cross-contamination) and inactivation due to sterilization, were the highest sources of variability and uncertainty.

Moreover, the analysis showed that the pH of the product during sterilization is also critical. What-if scenarios are handy visual tools and may be used to test the effect of some decisions on the management of spoilage risk considering different scenarios (Membré & Boué, 2018; Rigaux et al., 2014). The risk variability distributions and evaluation of uncertainty associated with this spoilage risk were obtained by using two-dimensional

Monte Carlo simulations taking into account uncertainty and variability separately through the model.

The second example is regarding a study of spoilage risk of evaporated milk by *G. stearothermophilus* (Kakagianni & Koutsoumanis, 2018) (Table 5.1). In this study, the spoilage risk of evaporated milk was mapped in the Mediterranean region based on the effect of storage temperature on *G. stearothermophilus* growth. A total of 115 scenarios were tested simulating the conditions of transportation and storage of evaporated milk in 23 Mediterranean capitals. The spoilage of evaporated milk by *G. stearothermophilus* is characterized by the coagulation of milk caused by acid production due to the metabolic activity of cells (Kakagianni, Gougouli, & Koutsoumanis, 2016). The predicted extent and variability of growth during the shelf life were then used to estimate the risk of spoilage. Unlike the previous example of *G. stearothermophilus* in green beans, the initial bacterial contamination was considered to be 10^1 spores/mL, which is the maximum level according to the dairy industry in Greece. Thus, possible changes to these levels were not taken into account in evaluating the spoilage risk. To predict the growth of *G. stearothermophilus* during the shelf life of the evaporated milk, temperature profiles covering 23 Mediterranean capitals were obtained (over 5 years). A maximum population of $10^{7.4}$ CFU/mL was fixed, corresponding to the maximum population density when change in milk pH, and consequently spoilage, is observed (Kakagianni et al., 2016). The growth prediction was based on the combination of the secondary cardinal model with inflection (Rosso, Lobry, & Flandrois, 1993) and the primary model of Baranyi and Roberts (Baranyi & Roberts, 1994). The prediction of growth under dynamic conditions was based on the time–temperature profile of evaporated milk storage and the estimated "momentary" maximum specific growth. The latter was based on the assumption that after a temperature shift, the growth rate adapts instantaneously to the new temperature environment. The prediction of the total growth of *G. stearothermophilus* in evaporated milk of each region took into account the variability related to the 1 year shelf life temperature profile. The risk was based on time to spoilage, which in this case, was the time required for the *G. stearothermophilus* population increase by 6.4 log CFU/mL. The risk of spoilage was then categorized into four levels according to the increase in the level of the bacteria during shelf life: high risk, moderate risk, low risk, and very low risk. A geographical risk assessment was then obtained on the basis of the ranking of Mediterranean capitals concerning the spoilage risk. The results showed that most of the Mediterranean capitals present a low risk of spoilage (Kakagianni & Koutsoumanis, 2018). The predictions will be ultimately used to evaluate possible adjustments to the expiration date of evaporated milk and as a support decision to the dairy industry in order to reduce the risk of spoilage.

The third example is regarding the application of a risk assessment model to fungal spoilage through a case study of yogurt spoilage by *Aspergillus niger* (Gougouli & Koutsoumanis, 2017). In contrast to risk assessments for food spoilage by bacteria, the risk assessment concept was first adjusted to fungal responses and spoilage risk related to the risk of spoilage due to mycelium appearance (Gougouli & Koutsoumanis, 2017). In this context, a risk spoilage model was developed taking into account different sources of variability: environmental (time and temperature distributions during yogurt chill chain); biological (= ability of fungal spores to form visible mycelia); and consumer, regarding consumers' judgment of the dimensions of a visible mycelium leading to rejection of the product. Initially, some assumptions were considered for the development of the risk model. The first was that packaging was considered to be the only stage of contamination of yogurt, so that no more contamination occurs after packaging and before consumption. The second was that the chill chain was considered to comprise each stage of yogurt storage (from factory to domestic storage). Based on this, a stochastic modeling approach was used, taking into account the sources of variability. A kinetic growth model, required for the development of the spoilage risk model, was first obtained. *A. niger* diameters were then predicted through the kinetic growth model under static and dynamic conditions for each storage stage of the chill chain (Gougouli & Koutsoumanis, 2010, 2012). Distributions characterizing environmental storage and lag time variability were further incorporated into the kinetic model, which was converted from a deterministic to a stochastic approach. A probabilistic growth/no growth model was also introduced into the kinetic growth model in order to define the range of applicability of the kinetic model and at the same time to incorporate the biological variability in the ability of single spores to grow and form visible mycelia at each storage temperature. Finally, Monte Carlo simulations were performed, and probability distributions of fungal diameter predictions at each stage of the chill chain were obtained.

The risk model was then developed based on the output of a fungal growth model combined with consumer variability regarding the visible limit for product rejection. At this stage, spoilage was defined when the estimated fungal diameter (D_i) was equal to or higher than the visible limit, and the model resulted in 1. If the fungal diameter (D_i) was lower than the visible limit, the model resulted in 0. Up to this point, it was assumed that each yogurt cup was contaminated with one *A. niger* spore. Four batches were tested, representing the production of 5000, 10,000, 50,000, and 100,000 yogurt cups, which was similar to the number of iterations. The simulations resulted in probabilities of observing spoiled yogurts in a batch with *n* yogurt cups at the time of consumption.

The final output was obtained by combining the true prevalence of yogurt cups contaminated with *A. niger* spore, estimated from challenge tests, with the probability of visible fungal growth before the end of the shelf life. For instance, the risk assessment output showed that for a batch of 100,000 cups of which 1% were contaminated with *A. niger*, eight cups (95% CI = [5–14]) were estimated to present visual mycelium at the time of consumption (= spoilage). This study presented a basis for risk assessment for fungal spoilage, which may be applied to diverse food products and their potential spoilers. Moreover, the risk model has the value of translating the prevalence of the fungus into the number of spoiled products at the time of consumption as well as being used to support decision making in yogurt production.

5.8 CONCLUDING REMARKS

Information regarding QMRA for food spoilage is still quite limited. QMRA models developed for deterioration need to focus on the probability of food deterioration and not neglect the sources of variability along the food chain. Further studies can fill this gap.

REFERENCES

Abdel-Aziz, S. M., Asker, M. M. S., Keera, A. A., & Mahmoud, M. G. (2016). Microbial food spoilage: Control strategies for shelf life extension. In: *Microbes in Food and Health* (pp. 239–264) N. Garg, S.M. Abdel-Aziz, A. Aeron (Eds.). Cham: Springer International Publishing. doi:10.1007/978-3-319-25277-3_13

Aryani, D. C., Besten, H. M. W. Den, & Zwietering, M. H. (2016). Quantifying variability in growth and thermal inactivation kinetics of *Lactobacillus plantarum*. *Applied and Environmental Microbiology, 82*(16), 4896–4908. doi:10.1128/AEM.00277-16.Editor

Aven, T., & Zio, E. (2018). Quality of risk assessment: Definition and verification. In: *Knowledge in Risk Assessment and Management* (1st ed., pp. 143–164) T. Aven & E. Zio (Eds.). Hoboken, NJ: Jonh Wiley & Sons, Inc.

Baranyi, J., & Roberts, T. A. (1994). A dynamic approach to predicting bacterial growth in food. *International Journal of Food Microbiology, 23*(3–4), 277–294. doi:10.1016/0168-1605(94)90157-0

Baranyi, J., Ross, T., Mcmeekin, T. A., & Roberts, T. A. (1996). Effects of parameterization on the performance of empirical models used in "predictive microbiology." *Food Microbiology, 13*(1), 83–91.

Barlow, S. M., Boobis, A. R., Bridges, J., Cockburn, A., Dekant, W., Hepburn, P., Houben, G. F., König, J., Nauta, M. J., Schuermans, J., & Bánáti, D. (2015). The role of hazard- and risk-based approaches in ensuring food safety. *Trends in Food Science and Technology, 46*(2), 176–188. doi:10.1016/j.tifs.2015.10.007

Barron, U. G., Redmond, G., & Butler, F. (2002). What is microbiological risk assessment? *Irish Microbial Quantitative Risk Assessment Network.* 1–9. Retrieved from: https://pdfs.semanticscholar.org/aeca/9568ad4b23bd1398aaabc2 24cec0a65a4f78.pdf Accessed on April 02, 2019.

Buchanan, R. L. (2004). Principles of risk analysis as applied to microbial food safety concerns. *Mitt. Lebensm. Hyg., 12*(October 2003), 6–12.

Busschaert, P., Geeraerd, A. H., Uyttendaele, M., & van Impe, J. F. (2011). Sensitivity analysis of a two-dimensional quantitative microbiological risk assessment: Keeping variability and uncertainty separated. *Risk Analysis, 31*(8), 1295–1307. doi:10.1111/j.1539-6924.2011.01592.x

Codex Alimentarius Commission (CAC). (1999). Principles for the Establishment and Application of Microbiological Criteria for Foods (CAC/GL 30-1999). Retrieved from http://www.fao.org/docrep/004/y1579e/y1579e05.htm Accessed on April 10, 2019.

Cassin, M. H., Lammerding, A. M., Todd, E. C. D., Ross, W., & McColl, R. S. (1998). Quantitative risk assessment for *Escherichia coli* O157: H7 in ground beef hamburgers. *International Journal of Food Microbiology, 41*(1), 21–44. doi:10.1016/S0168-1605(98)00028-2

Chapman, B., Otten, A., Fazil, A., Ernst, N., & Smith, B. A. (2016). A review of quantitative microbial risk assessment and consumer process models for Campylobacter in broiler chickens. *Microbial Risk Analysis, 2–3,* 3–15. doi:10.1016/j.mran.2016.07.001

Chotyakul, N., Lamela, C. P., & Torres, J. A. (2012). Effect of model parameter variability on the uncertainty of refrigerated microbial shelf-life estimates. *Journal of Food Process Engineering, 35*(6), 829–839. doi:10.1111/j.1745-4530.2010.00631.x

de Jong, A. R., Boumans, H., Slaghek, T., Van Veen, J., Rijk, R., & Van Zandvoort, M. (2005). Active and intelligent packaging for food: Is it the future? *Food Additives and Contaminants, 22*(10), 975–979. doi:10.1080/02652030500336254

den Besten, H. M. W., Aryani, D. C., Metselaar, K. I., & Zwietering, M. H. (2017). Microbial variability in growth and heat resistance of a pathogen and a spoiler: All variabilities are equal, but some are more equal than others. *International Journal of Food Microbiology, 240,* 24–31. doi:10.1016/j.ijfoodmicro.2016.04.025

Giannakourou, M., & Stoforos, N. (2017). A theoretical analysis for assessing the variability of secondary model thermal inactivation kinetic parameters. *Foods*, *6*(1), 7. doi:10.3390/foods6010007

Gobbi, E., Falasconi, M., Torelli, E., & Sberveglieri, G. (2011). Electronic nose predicts high and low fumonisin contamination in maize cultures. *Food Research International*, *44*(4), 992–999. doi:10.1016/j.foodres.2011.02.041

Gougouli, M., & Koutsoumanis, K. P. (2010). Modelling growth of *Penicillium expansum* and *Aspergillus niger* at constant and fluctuating temperature conditions. *International Journal of Food Microbiology*, *140*(2–3), 254–262. doi:10.1016/j.ijfoodmicro.2010.03.021

Gougouli, M., & Koutsoumanis, K. P. (2012). Modeling germination of fungal spores at constant and fluctuating temperature conditions. *International Journal of Food Microbiology*, *152*(3), 153–161. doi:10.1016/j.ijfoodmicro.2011.07.030

Gougouli, M., & Koutsoumanis, K. P. (2017). Risk assessment of fungal spoilage: A case study of *Aspergillus niger* on yogurt. *Food Microbiology*, *65*, 264–273. doi:10.1016/j.fm.2017.03.009

Havelaar, A. H., Evers, E. G., & Nauta, M. J. (2008). Challenges of quantitative microbial risk assessment at EU level. *Trends in Food Science and Technology*, *19*(Suppl. 1), 26–33. doi:10.1016/j.tifs.2008.09.003

in't Veld, J. H. J. H. I. (1996). Microbial and biochemical spoilage of foods: An overview. *International Journal of Food Microbiology*, *33*(1), 1–18. doi:10.1016/0168-1605(96)01139-7

Kaern, M., Elston, T. C., Blake, W. J., & Collins, J. L. (2005). Stochasticity in gene expression: From theories to phenotypes. *Nature Reviews: Genetics*, *6*(6), 451–464.

Kakagianni, M., Gougouli, M., & Koutsoumanis, K. P. (2016). Development and application of *Geobacillus stearothermophilus* growth model for predicting spoilage of evaporated milk. *Food Microbiology*, *57*, 28–35. doi:10.1016/J.FM.2016.01.001

Kakagianni, M., & Koutsoumanis, K. P. (2018). Mapping the risk of evaporated milk spoilage in the Mediterranean region based on the effect of temperature conditions on *Geobacillus stearothermophilus* growth. *Food Research International*, *111*(April), 104–110. doi:10.1016/j.foodres.2018.05.002

Kouamé-Sina, S. M., Makita, K., Costard, S., Grace, D., Dadié, A., Dje, M., & Bonfoh, B. (2012). Hazard identification and exposure assessment for bacterial risk assessment of informally marketed milk in Abidjan, Cote d'Ivoire. *Food and Nutrition Bulletin*, *33*(4), 223–234. doi:10.1177/156482651203300402

Koutsoumanis, K. (2009). Modeling food spoilage in microbial risk assessment. *Journal of Food Protection*, *72*(2), 425–427.

Koutsoumanis, K., Stamatiou, A., Skandamis, P., & Nychas, G. J. (2006). Development of a microbial model for the combined effect of temperature and pH on spoilage of ground meat, and validation of the model under dynamic temperature conditions. *Applied and Environmental Microbiology*, *72*(1), 124–134. doi:10.1128/AEM.72.1.124

Koutsoumanis, K. P., & Aspridou, Z. (2016). Moving towards a risk-based food safety management. *Current Opinion in Food Science, 12*, 36–41. doi:10.1016/j.cofs.2016.06.008

Kuswandi, B., & Nurfawaidi, A. (2017). On-package dual sensors label based on pH indicators for real-time monitoring of beef freshness. *Food Control, 82*, 91–100. doi:10.1016/j.foodcont.2017.06.028

Lammerding, A. M. (1997). An overview of microbial food safety risk assessment. *Journal of Food Protection, 60*(11), 1420–1425.

Lammerding, A. M., & Fazil, A. (2000). Hazard identification and exposure assessment for microbial food safety risk assessment. *International Journal of Food Microbiology, 58*(3), 147–157. doi:10.1016/S0168-1605(00)00269-5

Lianou, A., Panagou, E. Z., & Nychas, G.-J. E. (2016). Microbiological spoilage of foods and beverages. In: *The Stability and Shelf Life of Food* (pp. 3–42). P. Subramaniam (Ed.), Cambridge, UK: Elsevier. doi:10.1016/B978-0-08-100435-7.00001-0

McNab, W. B. (1998). A general framework illustrating an approach to quantitative microbial food safety risk assessment. *Journal of Food Protection, 61*(9), 1216–1228.

Membré, J. M., & Boué, G. (2018). Quantitative microbiological risk assessment in food industry: Theory and practical application. *Food Research International, 106*(November 2017), 1132–1139. doi:10.1016/j.foodres.2017.11.025

Membré, J. M., & Lambert, R. J. W. (2008). Application of predictive modelling techniques in industry: From food design up to risk assessment. *International Journal of Food Microbiology, 128*(1), 10–15. doi:10.1016/j.ijfoodmicro.2008.07.006

Mokhtari, A., & Frey, H. C. (2005). Sensitivity analysis of a two-dimensional probabilistic risk assessment model using analysis of variance. *Risk Analysis, 25*(6), 1511–1529. doi:10.1111/j.1539-6924.2005.00679.x

Morsy, M. K., Zór, K., Kostesha, N., Alstrøm, T. S., Heiskanen, A., El-Tanahi, H., Sharoba, A., Papkovsky, D., Larsen, J., Khalaf, H., Jakobsen, M. H., & Emnéus, J. (2016). Development and validation of a colorimetric sensor array for fish spoilage monitoring. *Food Control, 60*, 346–352. doi:10.1016/j.foodcont.2015.07.038

Nagy, Z. K., & Braatz, R. D. (2004). Open-loop and closed-loop robust optimal control of batch processes using distributional and worst-case analysis. *Journal of Process Control, 14*(4), 411–422.

Nauta, M. J. (2000). Separation of uncertainty and variability in quantitative microbial risk assessment models. *International Journal of Food Microbiology, 57*(1), 9–18.

Nauta, M. J., Litman, S., & Barker, G. C. (2003). A retail and consumer phase model for exposure assessment of *Bacillus cereus*. *International Journal of Food Microbiology, 83*(2), 205–218. doi:10.1016/S0168-1605(02)00374-4

Nimmegeers, P., Telen, D., Logist, F., & Van Impe, J. (2016). Dynamic optimization of biological networks under parametric uncertainty. *BMC Systems Biology, 10*(1), 1–20. doi:10.1186/s12918-016-0328-6

Nopwinyuwong, A., Trevanich, S., & Suppakul, P. (2010). Development of a novel colorimetric indicator label for monitoring freshness of interme-diate-moisture dessert spoilage. *Talanta, 81*(3), 1126–1132. doi:10.1016/j.talanta.2010.02.008

Nychas, G.-J. E., & Panagou, E. (2011). Microbiological spoilage of foods and bever-ages. In: *Food and Beverage Stability and Shelf Life* (pp. 3–28). D. Kilcast and P. Subramaniam (Eds.), Cambridge, UK: Elsevier. doi:10.1533/9780857092540.1.3

Organización Mundial de la Salud (OMS). The *Food and Agriculture Organization* (FAO). *Comisión del Codex Alimentarius.* (2014). Principios y directrices para la aplicacion de la evaluacion de riesgos microbiologicos. 7. CAC/GL 30-1999. Available at: http://www.fao.org/3/Y1579S/y1579s05.htm Accessed March 22 2019.

Petruzzi, L., Corbo, M. R., Sinigaglia, M., & Bevilacqua, A. (2017). Microbial spoil-age of foods: Fundamentals. In: *The Microbiological Quality of Food: Foodborne Spoilers* (pp. 1–21). A. Bevilacqua, M.R. Corbo and M. Sinigaglia (Eds.), Cambridge, UK: Elsevier. doi:10.1016/B978-0-08-100502-6.00002-9

Poschet, F., Bernaerts, K., Geeraerd, A. H., Scheerlinck, N., Nicolai, B. M., & Van Impe, J. F. (2004). Sensitivity analysis of microbial growth parameter dis-tributions with respect to data quality and quantity by using Monte Carlo analysis. *Mathematics and Computers in Simulation, 65*(3), 231–243.

Pouillot, R., Albert, I., Cornu, M., & Denis, J. B. (2003). Estimation of uncertainty and variability in bacterial growth using Bayesian inference. Application to *Listeria monocytogenes. International Journal of Food Microbiology, 81*(2), 87–104. doi:10.1016/S0168-1605(02)00192-7

Pouillot, R., & Delignette-Muller, M. L. (2010). Evaluating variability and uncer-tainty separately in microbial quantitative risk assessment using two R pack-ages. *International Journal of Food Microbiology, 142*(3), 330–340. doi:10.1016/j.ijfoodmicro.2010.07.011

Pujol, L., Albert, I., Magras, C., Brian, N., & Membré, J. (2015). Estimation and eval-uation of management options to control and / or reduce the risk of not com-plying with commercial sterility. *International Journal of Food Microbiology, 213*, 124–129. doi:10.1016/j.ijfoodmicro.2015.05.014

Remenant, B., Jaffrès, E., Dousset, X., Pilet, M. F., & Zagorec, M. (2015). Bacterial spoilers of food: Behavior, fitness and functional properties. *Food Microbiology, 45*(A), 45–53. doi:10.1016/j.fm.2014.03.009

Rigaux, C., André, S., Albert, I., & Carlin, F. (2014). Quantitative assessment of the risk of microbial spoilage in foods. Prediction of non-stability at 55°C caused by *Geobacillus stearothermophilus* in canned green beans. *International Journal of Food Microbiology, 171*, 119–128. doi:10.1016/j.ijfoodmicro.2013.11.014

Ross, T. (1999). *Predictive Food Microbiology Models in the Meat Industry.* Sydney: Meat and Livestock Australia.

Rosso, L., Lobry, J. R., & Flandrois, J. P. (1993). An unexpected correlation between cardinal temperatures of microbial growth highlighted by a new model. *Journal of Theoretical Biology, 162*(4), 447–463. doi:10.1006/JTBI.1993.1099

Rukchon, C., Nopwinyuwong, A., Trevanich, S., Jinkarn, T., & Suppakul, P. (2014). Development of a food spoilage indicator for monitoring freshness of skinless chicken breast. *Talanta, 130,* 547–554. doi:10.1016/j.talanta.2014.07.048

Samapundo, S., Devlieghere, F., De Meulenaer, B., & Debevere, J. (2007). Growth kinetics of cultures from single spores of *Aspergillus flavus* and *Fusarium verticillioides* on yellow dent corn meal. *Food Microbiology, 24*(4), 336–345.

Snyder, A. B., & Worobo, R. W. (2018). The incidence and impact of microbial spoilage in the production of fruit and vegetable juices as reported by juice manufacturers. *Food Control, 85,* 144–150. doi:10.1016/j.foodcont.2017.09.025

Tellen, D., Vallerio, M., Cabianca, L., Houska, B., Van Impe, J., & Logist, F. (2015). Approximate robust optimization of nonlinear systems under parametric uncertainty and process noise. *Journal of Process Control, 33,* 140–154.

Vose, D. (2008). *Risk Analysis: A Quantitative Guide.* Chichester: John Wiley.

WHO/FAO [World Health Organization/Food and Agriculture Organization]. (2008). Exposure assessment of microbiological hazards in food. *Microbiological Risk Assessment Series, 7,* 11–32.

6

Estimating Concentration Distributions
The Effect of Measurement Limits with Small Data

Jukka Ranta

Contents

143

6.1 INTRODUCTION: MODELING THE OBSERVATION

The variation of concentration values (e.g. chemicals or microbes) over inspected units (e.g. food items) is described by a statistical distribution. This is usually some standard distribution function to be fitted either to the original values or some transformed values, such as logarithms. The estimation of the model parameters proceeds depending on what the data actually were. Often, data consist of measurements thought to represent exact concentration values. More generally, data can represent any relevant information of the measurements, not only exact. This can still be useful if we can describe how that information is linked to the model parameters by specifying a probability model for the kind of observations we have. Apart from looking for parameter point estimates, Bayesian statistical inference is based on solving a probability distribution (uncertainty distribution) for the unknown model parameters, conditionally on the observed data. However, data may be incompletely specified with concentrations reported below or above limits, partially missing values, etc. Hence, the model needs to accommodate features of data as far as they are available while acknowledging the limitations in their reporting, format, or other quality.

Statistical estimation from "exact" values is often complemented with values reported to be within known intervals. These are commonly denoted and modeled as *censored* data because the underlying exact value is then "censored" from us. However, care is needed when interpreting the specific features of data. Censored data can occur in many applications. There is a long tradition of censored data analysis in survival data analysis (Cox and Oakes, 1984) where the event times (of deaths, failures, etc.) often either have not yet occurred during the follow-up time or have already occurred at some unknown time before the follow-up even started. Accordingly, they are said to be right or left censored. In addition, interval censored data may occur naturally in many applications and observation schemes. For example, a caries in a tooth is usually noted to have occurred after the previous visit but before the current visit to the dentist. Event times, weights, counts, and concentrations can likewise be censored. Essentially, censoring means directional uncertainty of the exact value being measured. From censored data, we can only know with certainty the limits for the possible, yet unseen, values of the object of measurement.

Microbiological samples, from e.g. food units within a lot, are often subjected to qualitative testing, resulting in reporting either positive or negative detection for the presence of a specific bacterium, although the limit of detection (LOD) itself is also a matter of some uncertainty. Among the detected positives, further quantitative results can be obtained, usually resulting in reports of estimated concentration levels expressed as colony forming units per gram (cfu/g) or per milliliter. Typically, some concentrations among the positives fall below the reported level of quantification (LOQ). This upper limit is then reported, e.g. "<10 cfu/g". In such a case, since the analytical sample portion is a randomly drawn fraction of the food unit, the original unit could by chance still have a concentration slightly above or anything below the LOQ, or it may be truly uncontaminated (Duarte et al., 2015). The objects of estimation are the concentrations in the food units and the distribution among them. It should be noted that a reported cfu-value is itself already an estimate, while the actual raw data would be the observed plate counts or observed patterns from dilution series with known sample volumes and dilution factors (Chik et al., 2018). There are no left censored values in such original raw data observations.

With measurements for chemical compounds, LOD and LOQ can be assigned based on the statistical upper limit from blank sample apparent measurements (LoB) and the measurements from known low-concentration analytes. The limits depend on the chosen error rates for false positives and false negatives (Armbruster and Pry, 2008; Belter et al., 2014). The determination of those limits has been discussed in a guidance (Wenzl et al., 2016). So, the LOD is the lowest concentration that can be reliably distinguished from the blank. Nevertheless, a value below the LOD could still be either from a true blank or from a unit with very low concentration. An LOQ is the lowest concentration that can be both reliably detected and measured with the required precision, although the LOQ may coincide with the LOD. For statistical inference from such data, substitution and imputation methods for values below the LOD, or between the LOD and the LOQ, have been proposed (Office of Pesticide Programs, 2000; Helsel, 2006).

A critical assumption for the subsequent statistical analysis is whether the data are thought to represent positive concentrations, excluding truly zero concentrations. If that can be assumed, then knowing which samples are from the true contaminations and the total statistical sample size allows straightforward estimation of prevalence and a separate estimation of concentrations among the positive contaminations. If this cannot be assumed, a concentration value that was too small to be quantified could be either a true zero or just a small value below (or around) LOQ. Then, the modeling needs to be extended to describe the unknown prevalence of true zeros, e.g. with zero-inflated models (Duarte and Nauta, 2015) or in general, latent variable models for describing the underlying unknown true (binary) status of each measurement. Theoretical true positive values of contamination could then be described by either a distribution for discrete counts or a continuous distribution, depending on the data. Within the model, positivity implies that such a value originated from the concentration distribution even if it happened to be small.

In any case, the reported concentration values are still estimates of the concentrations in the original food units, although these may well be the only data available for statistical modeling. It is unfortunate that more detailed reports of the microbiological data of the pattern of observed outcomes from dilution tubes or plate counts are rarely published. Likewise, statistical analysis of chemical measurements could benefit from more detailed information on measurement errors. Any estimation procedure obviously depends on the quality of the available data. Risk assessments often exploit published data and depend critically on their quality, be they detailed original raw data, coarse summary data, or both.

Higher levels of hierarchical structures are not addressed here, such as samples within batches within producers, but only analytical test samples from units representing a population of similar food units. At first, let us assume that a data set has been obtained providing concentrations representing truly contaminated samples. If these measured concentrations were taken as accurate values, the sample mean and sample standard deviation could be used to describe the distribution of concentrations among the study population of contaminated units. However, it is fairly common for even 95% of the microbiological measurements (even among the known positives) to be below the reported limit of enumeration. Obviously, estimates based only on the few remaining values above the limit would grossly overestimate the mean and induce bias in the estimated standard deviation. Substitution methods, i.e. setting arbitrary default values such as lower bounds, upper bounds, or mid bounds for the censored data, does not remove the bias, as discussed with chemical hazards (Helsel, 2006). With both chemical and microbiological concentrations, the problem is partially similar, although some chemical compounds may be assumed to be so widespread that every sample might be assumed to have at least a small concentration, detectable or not. For some others, it may be appropriate to assume that truly zero concentrations exist with some prevalence, as with microbiological hazards (LaFleur et al., 2011). Therefore, the same modeling approach can be applicable in part to both chemical and microbiological concentrations after a careful case-by-case judgment of the data and assumptions.

6.2 PARAMETER ESTIMATION FROM CENSORED DATA OF KNOWN POSITIVE CONCENTRATIONS

For the example, it is assumed that the data represent sampled concentrations from truly positively contaminated food units. A common approach for concentration distributions is to assume a normal model for the logarithmic ($\log_{10}$) concentrations: $y \sim N(\mu, \sigma^2)$ with mean μ and variance σ^2. The parameters (μ, σ) of the distribution are to be estimated from observed log-concentrations $y_1,...,y_n$, of which m measurements are "accurate" and $n-m$ are censored below a reported log-value c. For bacteria concentrations, *assumed* to represent truly contaminated food units with at least one bacterium, we could also assume a lower concentration limit as $1/V$ for the volume or mass V of the unit. This may not be known or it may vary between units, so that a crude simplification could use the sample volume or weight as a substitute. Setting a number for the lower limit for concentration is artificial, but it can be more realistic than assuming that positive concentrations could be arbitrarily close to zero, i.e. log-value approaching minus infinity. For example, consider an interval $[\log_{10}(1/25), c]$ with the upper bound c and the lower bound based on a sample of 25 grams. When a large proportion of reported values fall below c, the lower limit will constrain the estimation, preventing parameter estimates drifting to infinity. For chemical concentrations, the lower limit can be hard to set, which further complicates the estimation from heavily censored data. The LOD may serve as the lower limit for the positive values below LOQ.

In *maximum likelihood estimation* (MLE), the parameter estimates are the values of μ, σ that maximize the so-called likelihood function of the parameters. This is written as the probability of the observed complete set of data, including both censored and accurate data points. The estimates are thus $\hat{\mu}, \hat{\sigma} = \text{argmax}L(\mu, \sigma; \text{data}) = \text{argmax}\, P(\text{data}|\mu, \sigma)$. The exact form depends on the form of the data, for example:

$$P(\text{data}|\mu, \sigma) = P(n-m \text{ censored values}|\mu, \sigma) \times P(m \text{ accurate values } y_i|\mu, \sigma)$$

$$= \prod_{i=1}^{n-m} P(-\infty < y_i < c|\mu, \sigma) \prod_{i=1}^{m} N(y_i|\mu, \sigma^2)$$

147

$$= \left[\int_{-\infty}^{c} \frac{1}{\sigma\sqrt{2\pi}} \exp\left(-0.5(y-\mu)^2/\sigma^2\right) dy \right]^{n-m} \prod_{i=1}^{m} \frac{1}{\sigma\sqrt{2\pi}} \exp\left(-0.5(y_i-\mu)^2/\sigma^2\right)$$

In order to find the parameter values that maximize this expression, numerical methods are needed, because an analytical solution is not available. Such methods were used e.g. in Lorimer and Kiermeier (2007), Busschaert et al. (2010), and Pouillot and Delignette-Muller (2010) for obtaining parameter estimates. These were based on reported data as face value of the concentrations (cfu/g) as discussed here. However, raw data could represent the actual observed pattern in the positive dilution tubes or plate counts (Pouillot et al., 2013; Duarte et al., 2015), leading to different mathematical expressions based on Poisson distribution. Both situations are discussed in the following sections.

The uncertainty of the maximum likelihood estimates has often been explored in the literature by using bootstrap methods in which the data are repeatedly re-sampled to create a set of possible data sets in order to find an estimate from each. The set of estimates then shows the bootstrap uncertainty, describing how the estimates would have changed had the data been slightly different from what they were. A non-parametric bootstrap re-samples the actual data values, whereas a parametric bootstrap is based on a parametric distribution of data. The approach of simulating what the data might have been and then performing the estimation for each hypothetical data set may seem an unnecessarily convoluted approach. A straightforward method is to directly simulate the uncertainty distribution of parameters, and this is succinctly based on the actually given data via Bayes' theorem.

In *Bayesian estimation*, posterior distribution of the parameters describes the parameter uncertainty. It is a probability distribution of the unknown parameters conditionally on the data we had and the prior distribution of the parameters. This is computed routinely using Monte Carlo simulation methods. If we choose uniform prior over a sufficient parameter range, the density function $P(\mu, \sigma)$ of the prior is constant, and we see that the posterior is exactly proportional to the likelihood function we had. This follows from what is known as the "Bayes theorem" or "Bayes formula": $P(\mu, \sigma | \text{data}) \propto P(\text{data} | \mu, \sigma) P(\mu, \sigma)$. Here, $\propto$ means "proportional to". Essentially, this is derived from the laws of probability: the product rule and conditional probability, treating all unknowns as "random" with a probability distribution. The shape and the maximum point of the likelihood (or posterior density) function can then be graphically inspected as a two-dimensional (2-D) surface plot of $P(\text{data}|\mu,\sigma)$; see Figure 6.1. If all measurements fall below limit c, i.e. $[-\infty, c]$, parameter estimates are not uniquely identifiable, and we can merely conclude that the mean μ is probably below c. How far below it would be hard to tell. With just a few measurements above c, the estimates for both μ, σ can already be roughly identified from the contour plot but with large uncertainty. In this case, there is no clear peak in the likelihood function contours, since the peak nearly flattens out diagonally north-west; see Figure 6.1. Even then, the most likely combinations of μ, σ are visible, so that we can rule out some combinations of mean and standard deviation as unlikely. Also, knowing the value of one parameter can still help to identify the most likely values for the other. This could be exploited if we could draw information for one of the parameters from other data sets (assuming that either μ or σ is common over both sets of data). More structured evidence synthesis could be done using hierarchical Bayesian modeling.

The same estimation methods apply in the same manner to other distribution functions that are suitable for describing positive concentrations. The fit of the distribution to the empirical data could be inspected and the best fitting distribution function chosen. A common concern is how well the distribution can predict rare but high concentration values, which can be important for the risk of disease. With enough data, it might become evident that a single distribution function cannot fit the data

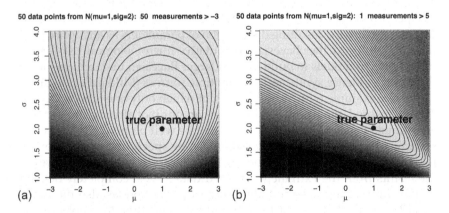

Figure 6.1 Likelihood contours. (a) None of the 50 simulated data points were left censored $<c$. (b) Only one of the 50 simulated data points was not left censored. Black point shows the underlying true parameter values $\mu = 1$, $\sigma = 2$.

properly. This could be the case if the empirical concentration distribution shows multiple peaks that could result e.g. from different food production processes. A finite mixture of distributions could then be a flexible approach, but the number of parameters is increased, and the identifiability of their estimates can be challenging (Burmaster and Wilson, 2000). With very small data, it is harder to reject or accept any particular distribution model, so that this choice itself becomes uncertain.

6.2.1 Bayesian Computation

Bayesian inference is based on the likelihood function exploiting the full data set, together with a specification of prior distribution for parameters, here $P(\mu, \sigma)$. The modeling is then accomplished by computing the posterior distribution, here $P(\mu, \sigma |data)$, which often requires Markov chain Monte Carlo simulations (MCMC). The models are relatively easily constructed with the BUGS language, and the more difficult numerical task of Monte Carlo algorithms can be left to OpenBUGS (Lunn et al., 2013), JAGS (Plummer, 2003), or Stan (Carpenter et al., 2017) with some differences in implementation and techniques. These tools allow the user to focus more on the models rather than the technical simulation algorithms. We are then also not restricted only to uniform priors. Modifications of the basic model are often easy to explore for sensitivity of the results. Comparison with previous bootstrap methods was explored in (Busschaert et al., 2011). The computations can also be done within R software by using e.g. the R2OpenBUGS package to run BUGS in the background. An advantage is that one can use the resulting distribution of parameters as an input (prior) distribution for further Bayesian quantitative microbiological risk assessment (QMRA) models to build a much larger Bayesian hierarchical model exploiting evidence synthesis from multiple sources (Mikkelä et al., 2016).

In our example with 50 data points, of which only one is accurate and 49 are censored, the likelihood function contour plot already showed that an estimate for mean and variance cannot be well pinpointed. The corresponding Bayesian model can be simulated by specifying uniform priors for both parameters over a fixed range of values; see Figure 6.2. However, the results from the extremely censored data almost completely depend on the chosen ranges, since the posterior distribution (as also the likelihood function) flattens out towards infinity, and this does not provide a proper probability distribution unless the boundaries are artificially forced. An improper distribution is not a distribution, because it does not

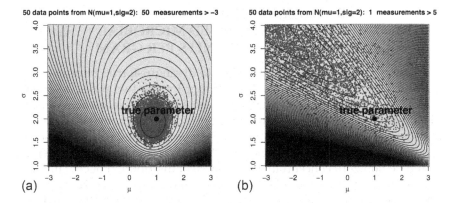

Figure 6.2 Likelihood contours and simulated parameter distribution from Bayesian model with uniform priors. (a) Censoring limit is −3. (b) Censoring limit is 5. True parameter value shown in black. Simulated sample of parameters from posterior distribution in gray dots.

integrate to one. In addition to uniform priors, other objective prior distributions can also be chosen among so-called conjugate distributions with minimal information and without preset boundaries for parameters, and if these are still proper distributions (although vague), the posterior may be theoretically proper to compute, although it may be only barely proper in practice. Moreover, there is always the possibility to use external information, based on expert judgment or literature, for more informative priors. Obviously, the more limited the data, the more effect the priors have. In the limit, all estimation methods fail if the data provide null information. The parameters are then not identifiable in either Bayesian or classical statistics. Although a solution could be forced by more constrained prior distributions relying on expert knowledge, it would not be a solution based on actual data alone.

With the Bayesian approach, if the posterior is proper, the combined uncertainty of all model parameters can be evaluated consistently as a joint distribution of the parameters and visualized for parameter combinations of interest. One can then explore multiparameter correlations and what can be concluded from the data. Also, summaries from the large MCMC sample representing the full posterior distribution can be drawn. There is no need to select any specific point estimation method (MLE, least squares estimate, moment estimate, etc.) when the focus is on the whole distribution as a measure of uncertainty.

6.3 BAYESIAN ESTIMATION FOR TRUE ZEROS WITH REPORTED MICROBIAL COLONY FORMING UNITS PER GRAM

If we assume no qualitative information to distinguish true zeros in data, a mixture model can be applied to account for the two possibilities. A mixture distribution is the weighted sum of two or more component distributions so that the unknown weights, or proportions, of the components also become model parameters to be estimated. In this case, a proportion of true zeros is unknown, while the rest of the measurements would be non-zero concentrations following a distribution of contamination. Some of the reported "<LOQ" values that were truly non-zeros could even have originated from food units that have a concentration slightly larger than the LOQ by chance, so we cannot strictly claim them as censored values below the LOQ. But the purpose of a careful analytic sampling is to ensure representativeness as far as possible, e.g. by avoiding possible clustering effects. It should be unlikely that those reported to be "<LOQ" would originate from concentrations much larger than LOQ by chance. We should, then, mostly be able to be confident that they are still small, if not true zeros, or else the value of the measurements becomes questionable. In the following model, a mixture of zeros and non-zeros is specified, and a censoring limit LOQ is used. Typically, mixture models require more data than simple models. They are more demanding for estimation, since the same data pattern can often be equally explained by several parameter combinations – again, a concern of identifiability. Obviously, the apparent zeros could occur equally well from a population with small prevalence and high concentrations among positives – or vice versa: high prevalence and low concentrations among positives. The reported concentrations are here taken as the only primary data source, providing for each measurement either a reported numerical value or a missing value "NA", which may have been positive, yet "<LOQ", or a true zero.

The model can be constructed by adding a latent binary variable I_i describing the true status underlying each measurement result. $I_i = 1$ denotes a true contamination, and $I_i = 0$ denotes a true zero. It would also be possible to estimate the other parameters without I_i-variables in the model, but these can be used for e.g. also estimating the finite sample true prevalence $\sum I_i/n$. For the detected positive and measured concentrations ($y_i > \text{LOQ}$), obviously $I_i = 1$, assuming no false positives exist. The

152

observable outcome for the ith measurement is denoted as $O_i = (y_i, D_i)$, which consists of the binary variable D_i for detection of a measured quantitative concentration larger than LOQ ($D_i = 1$ for "yes") and the corresponding measured concentration y_i, which is observed only in the case of a detected positive and is unknown ("NA") otherwise. The conditional probability for observation O_i, given the I_i, is

$$P(O_i \mid I_i) = P(y_i, D_i \mid I_i) = P(y_i \mid D_i, I_i)P(D_i \mid I_i)$$

$$= \begin{cases} 1 \times ((1 - I_i) + I_i F(c)) & \text{if } D_i = 0 \\ f(y_i \mid D_i = 1) \times I_i(1 - F(c)) & \text{if } D_i = 1 \end{cases}$$

where $f(y_i \mid D_i = 1)$ is a truncated distribution function and F is cumulative distribution. Because of the condition $D_i = 1$, the Y_i values must follow the part of the distribution that falls above $c = $ LOQ:

$$f(y_i \mid D_i = 1) = \frac{f(y_i \mid \mu, \sigma)}{1 - F(c \mid \mu, \sigma)} \text{ for } y_i > c.$$

The full likelihood with parameters $\mu, \sigma, p, I_1, \ldots, I_n$ is then

$$\prod_{i=1}^{n} P(O_i \mid I_i, \mu, \sigma)P(I_i \mid p).$$

The Bayesian model is completed by setting a prior for the additional parameter, e.g. $p \sim U(0, 1)$. From this model, we can explore the joint probability distribution of all three parameters μ, σ, p together with the individual latent variables I_i that remain uncertain for the nondetects. The unknown concentration values y_i of the true contaminations need not be imputed for the parameter estimation unless one is specifically interested in inferring them. The BUGS code is provided in the Appendix. While the data remain as before, there are now more parameters in the model, which makes the estimations more uncertain than in a simpler model with fewer parameters. Also, the uniform priors for μ, σ have arbitrary boundaries, which particularly under small data, can imply too much probability for high values, which would be considered practically implausible, and a posterior that remains nearly flat (nearly improper distribution) over a huge range of values. Other objective vague priors could be used, e.g. the often used conjugate priors for precision $\tau = 1/\sigma^2 \sim \Gamma(0.01, 0.01)$ and mean $\mu \sim N(0, 1/0.001)$, which are barely proper distributions; see Figure 6.3.

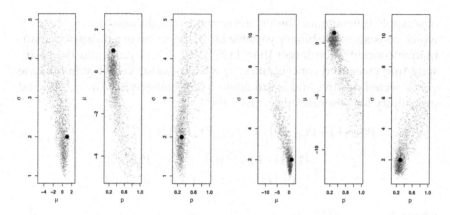

Figure 6.3 Example posterior distributions (samples in gray dots) from the model with nondetects, which could be either true zeros or values below LOQ. The uncertainty about μ, σ is now larger, and there is also uncertainty about prevalence *p*. Underlying true values shown in black. Three frames on the left: using uniform priors on μ, σ, *p*. Three frames on the right: using priors $\mu \sim N(0,1/0.001), 1/\sigma^2 \sim \Gamma(0.01,0.01), p \sim U(0,1)$.

These might still lead to unstable estimation if data are very limited, so caution should be taken. Finally, expert knowledge or literature data from a similar context might be employed for constructing more informative or partially informative priors for some, if not all, parameters. Narrowing down the possible values for one parameter in the prior can decrease the posterior uncertainty of the others too.

6.4 BAYESIAN ESTIMATION FOR TRUE ZEROS WITH REPORTED MICROBIAL PLATE COUNTS

A similar analysis for true zeros can be done if the data consist of the plate counts instead of reported cfu/g-values and their reported LOQs. The conditional model for the counts from a food unit is then a zero-inflated Poisson distribution, which is a mixture distribution of true zeros (with probability $1 - p$) and counts from a truly contaminated food unit with concentration μ_i (with probability *p*). The counts resulting from truly contaminated units (with concentrations μ_i) can obviously also result in zero sample counts, but there is no explicit LOQ involved in the model. Note that, in this example, the observed counts are not modeled with an "unconditional" distribution as in Gonzales-Barron et al. (2010), but instead conditionally on the true status I_i and the possible concentrations μ_i of the contaminated food units. Both I_i and μ_i are variable between units. The conditional distribution for the observed counts x_i is then

$$P(x_i \mid I_i, \mu_i) = \text{Poisson}(I_i 10^{\mu_i} w_i d_i)$$

where I_i is the binary indicator for the true contamination of the ith unit, μ_i is the true $\log_{10}$ concentration of the unit (if contaminated), w_i is the weight of the analyzed sample, and d_i the dilution factor. In turn, $P(I_i \mid p) = \text{Bernoulli}(p)$ and $P(\mu_i \mid \mu, \sigma) = N(\mu, \sigma^2)$ so that the main parameters of interest are μ, σ and p. The posterior distribution of all model parameters is now

$$P(p, \mu, \sigma, I_1, \ldots, I_n, \mu_1, \ldots, \mu_n \mid x_1, \ldots, x_n) \propto \prod_{i=1}^{n} P(x_i \mid I_i, \mu_i) P(I_i \mid p)$$

$$P(\mu_i \mid \mu, \sigma) P(p, \mu, \sigma)$$

The BUGS code for this is in the Appendix, and the results from this model, Figure 6.4, are comparable to the previous model, where concentration values were assumed to be treated as exact data with left censoring. Apart from the smallest counts, which would transform into "concentration <LOQ", the expected set of observed non-zeros in both data sets in the two approaches are similar, after accounting for the dilution factors d and weights w. Comparisons can be made by simulating zero inflated true concentrations from which Poisson distributed observed counts are generated (to create raw count data) and then the derived "cfu-data" based on those counts (to create corresponding concentration data with censored values). The uncertainty of parameter estimates grows likewise with the increasing proportion of observed zeros.

6.5 CONCLUDING REMARKS

- The model choice should reflect the type of available data. More complicated models with more parameters require more data than simple models with few parameters. It is not possible to decide beforehand on the model without considering the data. Although a simple model may be "too simple", "more realistic" models can fail estimations if attempted with insufficient information in the data.

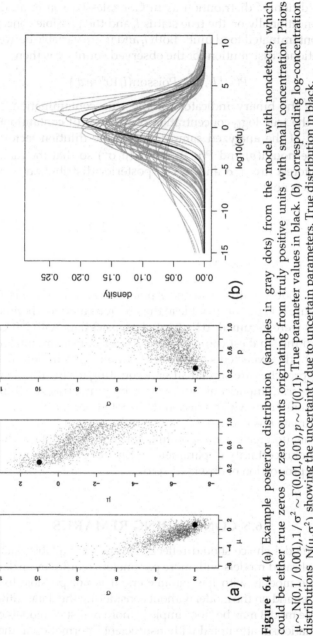

Figure 6.4 (a) Example posterior distribution (samples in gray dots) from the model with nondetects, which could be either true zeros or zero counts originating from truly positive units with small concentration. Priors $\mu \sim N(0.1/0.001), 1/\sigma^2 \sim \Gamma(0.01, 0.01), p \sim U(0,1)$. True parameter values in black. (b) Corresponding log-concentration distributions $N(\mu, \sigma^2)$ showing the uncertainty due to uncertain parameters. True distribution in black.

- With microbial concentrations, more detailed reporting of the outcomes of original plate counts or dilution tube patterns allows more realistic modeling of the Poisson process randomness. Parameters of the concentration distribution can then be estimated from the (uncensored) counts and patterns. Reporting such raw data should be encouraged.

- Most published historical data report only the estimated concentrations and LOQs, in which case zero-inflated mixture models are feasible approaches to account for the uncertainty due to unknown true zeros and true non-zeros.

- With very small data, adding parameters to describe simultaneously the distributions of true zeros and non-zero concentration values may add realism but also increases the uncertainty of the estimates, and the prior distributions should be carefully chosen.

- If all samples are reported left censored "<LOQ", all estimation methods fail, and no useful results for both mean and variance can be drawn from the data. With only a very few accurate measurements above the LOQ, the problem is not much easier, and there is no guaranteed solution with any method. Therefore, it is useful to visually plot the likelihood function around plausible parameter values to see whether any clear peak can be found in the parameter landscape.

- For a parametric model, in non-Bayesian classical estimation, the estimates and their depicted uncertainties depend on the estimation method, e.g. a maximum likelihood estimate, a least squares estimate, or a moment estimate, etc., which can behave differently. The Bayesian approach proposes a generally applicable approach based on the probability distribution of all unknown parameters simultaneously, which is explicitly conditional on the stated data and the priors.

- Images produced from 2-D (and 3-D) posterior distributions of parameter combinations provide informative visual results of the combined uncertainty of the parameters, based on evidence (data+prior). The parameter distribution can be used as an input distribution for further QMRA.

APPENDIX 6: VISUAL EXPLORING
USING R AND OPENBUGS
A.6.1 Censored Data and Likelihood Contour Plots

If standard distribution functions are chosen to model the concentrations, the likelihood function over the 2-D parameter space is easily computed numerically and visualized. The following R-codes give an example with log-normal distribution. The contour plot can be used to visually locate the maximum likelihood estimates (MLE). In Bayesian analysis with uniform prior distributions, these are also the maximum posterior density estimates. To create an artificial data set, the code generates a sample of 50 from $N(1, 2^2)$ distribution and sets a censoring limit at a given value c. The log-likelihood function is evaluated for every combination of parameter values in a grid of 100×100 points.

```
c <- -3 # censoring limit
y <- rnorm(50,mean=1,sd=2) # create data or use existing data
mu.grid <- seq(-3,3,length=100) # define grid for plotting
sig.grid <- seq(1,4,length=100) # define grid for plotting
M <- matrix(NA,100,100) # define matrix for likelihood
    values within grid
for(i in 1:100){
for(j in 1:100){
M[i,j] <- sum(log(dnorm(y[y>c],mean=mu.grid[i],sd=sig.
    grid[j])))+ sum(y<=c)*log(pnorm(c,mu.grid[i],sig.
    grid[j]))
if(is.infinite(M[i,j])){M[i,j]<-NA}
}
}
image(mu.grid,sig.grid,M,col=gray.colors(50), xlab=
    expression(mu),ylab=expression(sigma), main=paste(as.
    character(50),"data points from N(mu=1,sig=2):
    ", as.character(sum(y>c))," measurements >",as.
    character(c)),cex.main=1)
contour(mu.grid,sig.grid,M,add=TRUE,method="simple",
    200,drawlabels=FALSE)
points(1,2,col="black",pch=16,cex=1.5)
text(1*1.05,2*1.08,"true parameter",cex=1.5,font=2,col=
    "black")
```

A.6.2 Bayesian Model with Censored Data

In Bayesian statistics, the uncertainty of the parameter estimates is obtained by evaluating the posterior distribution. For this, the following BUGS code can be used. By using R2OpenBUGS, we can process the data first in R, then call OpenBUGS to simulate a large sample of parameter values from the posterior distribution, and finally summarize/visualize the results in R. In this model, uniform priors were defined for μ, σ, but these could be replaced by other distributions, and/or the model could be extended in other ways.

```
model{ # BUGS model code to be contained in the file
    'cenmodel.txt'
for(i in 1:m){ y[i] ~ dnorm(mu,tau) } # accurate data
    points
for(i in m+1:n){ y[i] ~ dnorm(mu,tau)C(,c) } # censored
    data points
mu ~ dunif(mu.lo,mu.up) # prior for mu
tau <- pow(sig,-2) # conversion from standard deviation to
    precision
sig ~ dunif(0,sig.up) # prior for sigma
}
# Doing Bayesian model within R software, calling OpenBUGS
    in background:
library(R2OpenBUGS)
c <- -3 # set censoring limit
y <- rnorm(50,mean=1,sd=2) # create data randomly, or use
    existing data
m <- sum(y>c) # number of accurate values
n <- length(y) # number of all data values
y <- c(y[y>c],rep(NA,n-m)) # vector of accurate values and
    NA for censored
sig.up <- 5 # upper limit for sigma prior, set as needed
mu.lo <- -5 # lower limit for mu prior, set as needed
mu.up <- 5 # upper limit for mu prior, set as needed

data<-list("y","c","n","m","sig.up","mu.lo","mu.up")
    # specify data
inits<-function(){list(mu=0,sig=1)} # specify initial
    values for MCMC
parameters<-c("mu","sig") # specify which parameter
    simulations are output
```

```
mymodelresult <- bugs(data,inits,parameters,model.file=
    "cenmodel.txt",
n.iter=10500,n.burnin=500,n.chains=1)
attach.bugs(mymodelresult)
plot(mu,sig,col="gray40",pch=16,cex=0.4,xlab=expression
    (mu),ylab=expression(sigma))
dev.new()
par(mfrow=c(2,1))
plot(density(mu),lwd=2,xlab=expression(mu),main="Marginal
    distribution")
plot(density(sig),lwd=2,xlab=expression(sigma),main=
    "Marginal distribution")
```

A.6.3 Bayesian Model with Censored Data and True Zeros

The following BUGS model for true zeros can be run with the same data "y" as previously. However, in this case, initial values are specified for a larger set of parameters, and we can draw results for each of them.

```
model{ # BUGS model code to be contained in the file
    'cenmodel2.txt'
for(i in 1:n){
I[i] ~ dbern(p) # P(true positive)=p
detect[i] ~ dbern(prob[i]) # observed detections
prob[i] <- I[i]*(1-phi((c-mu)/sig)) # probability to
    detect
}
for(i in 1:m){ # observed m detects
detect[i] <- 1
}
for(i in m+1:n){ # observed n-m nondetects
detect[i] <- 0
}
for(i in 1:m){ # values for the known detects
y[i] ~ dnorm(mu,tau)T(c,)
}
falsenegs <- sum(I[m+1:n])/(n-m) # proportion (TRUE+)
    among the apparent zeros
sig <- pow(tau,-0.5)
tau ~ dgamma(0.01,0.01) # prior for precision
mu ~ dnorm(0,0.001) # prior for mean
p ~ dunif(0,1) # prior for prevalence
}
```

```
# R code:
c <- -3 # set censoring limit
y <- rnorm(50,1,2) # generate data
ptrue <- 0.3 # true prevalence set here
n <- length(y)
truepos <- rbinom(n,1,ptrue)
y[which((truepos==0)|(y<c))] <- NA
y <- c(y[(y>=c)&!is.na(y)],rep(NA,sum(is.na(y))))
m <- sum(!is.na(y))
data<-list("y","c","n","m")
inits<-function(){list(mu=0,sig=1,p=0.5,I=rep(1,n))}
parameters<-c("mu","sig","p","I","falsenegs")
mymodelresult <- bugs(data,inits,parameters,model.file=
    "cenmodel2.txt", n.iter=10500,n.burnin=500,n.chains=1)
attach.bugs(mymodelresult)
dev.new()
par(mfrow=c(1,3))
th<-seq(1,10000,by=5)
plot(mu[th],sig[th],col="gray40",pch=16,cex=0.2,
xlab=expression(mu),ylab=expression(sigma),font=2,cex.
    lab=1.3)
points(1,2,col="black",pch=16,cex=2)
plot(p[th],mu[th],col="gray40",pch=16,cex=0.2,
xlab="p",ylab=expression(mu),font=2,cex.lab=1.3)
points(ptrue,1,col="black",pch=16,cex=2)
plot(p[th],sig[th],col="gray40",pch=16,cex=0.2,
xlab="p",ylab=expression(sigma),font=2,cex.lab=1.3)
points(ptrue,2,col="black",pch=16,cex=2)
```

A.6.4 Bayesian Model with Plate Count Data

```
model{ # BUGS model code to be contained in the file
  'cenmodel3.txt'
for(i in 1:n){
I[i] ~ dbern(p) # P(true non-zero)=p
counts[i] ~ dpois(lambda[i]) # observed plate counts
lambda[i] <- I[i]*pow(10,musample[i])*w[i]*dilution[i]
musample[i] ~ dnorm(mu,tau) # log10-concentration of the
    sampled unit
}
falsenegs <- sum(I[m+1:n])/(n-m) # proportion among the
    apparent zeros
sig <- pow(tau,-0.5)
```

161

```
tau ~ dgamma(0.01,0.01) # prior for precision
mu ~ dnorm(0,0.001) # prior for mean
p ~ dunif(0,1) # prior for prevalence
}

# R code:
n<- 50
w <- rep(25,n) # sample weights, e.g. 25 grams
dilution <- rep(0.01,n)
y <- rnorm(n,mean=1,sd=2) # generate concentrations
ptrue <- 0.3
truepos <- rbinom(n,1,ptrue)
counts <- sort(rpois(n,dilution*w*truepos*10^(y)),
    decreasing=TRUE)
m <- sum(counts>0)
data <- list("counts","n","m","w","dilution")
inits<-function(){list(mu=0,tau=1,p=0.5,I=rep(1,n))}
parameters<-c("mu","sig","p","I","lambda","musample",
    "falsenegs")
mymodelresult <- bugs(data,inits,parameters,model.file=
    "cenmodel3.txt", n.iter=10500,n.burnin=500,n.chains=1)
attach.bugs(mymodelresult)
```

REFERENCES

Armbruster DA, Pry T. Limit of blank, limit of detection and limit of quantification. *The Clinical Biochemist Reviews* 29 Suppl 1 (2008), S49–52.

Belter M, Sajnóg A, Barałkiewicz D. Over a century of detection and quantification capabilities in analytical chemistry – Historical overview and trends. *Talanta* 129 (2014), 606–616.

Burmaster DE, Wilson AM. Fitting second-order finite mixture models to data with many censored values using maximum likelihood estimation. *Risk Analysis* 20(2) (2000), 261–271.

Busschaert P, Geeraerd AH, Uyttendaele M, Van Impe JF. Estimating distributions out of qualitative and (semi)quantitative microbiological contamination data for use in risk assessment. *International Journal of Food Microbiology* 138(3) (2010), 260–269.

Busschaert P, Geeraerd AH, Uyttendaele M, Van Impe JF. Hierarchical Bayesian analysis of censored microbiological contamination data for use in risk assessment and mitigation. *Food Microbiology* 28(4) (2011), 712–719.

Carpenter B, Gelman A, Hoffman MD, Lee D, Goodrich B, Betancourt M, Brubaker M, Guo J, Li P, Riddell A. Stan: A probabilistic programming language. *Journal of Statistical Software* 76(1) (2017), 1. doi: 10.18637/jss.v076.i01.

Chik AHS, Schmidt PJ, Emelko MB. Learning something from nothing: The critical importance of rethinking microbial non-detects. *Frontiers in Microbiology* 9 (2018), 2304. doi: 10.3389/fmicb.2018.02304.

Cox DR, Oakes D. *Analysis of Survival Data*. Chapman & Hall, Boca Raton, FL, 1984.

Duarte ASR, Nauta MJ. Impact of microbial count distributions on human health risk estimates. *International Journal of Food Microbiology* 195 (2015), 48–57.

Duarte ASR, Stockmarr A, Nauta MJ. Fitting a distribution to microbial counts: Making sense of zeroes. *International Journal of Food Microbiology* 196 (2015), 40–50.

Gonzales-Barron U, Kerr M, Sheridan JJ, Butler F, Count data distributions and their zero-modified equivalents as a framework for modelling microbial data with a relatively high occurrence of zero counts. *International Journal of Food Microbiology* 136 (2010), 268–277.

LaFleur B, Lee W, Billhiemer D, Lockhart C, Liu J, Merchant N. Statistical methods for assays with limits of detection: Serum bile acid as a differentiator between patients with normal colons, adenomas, and colorectal cancer. *Journal of Carcinogenesis* 10 (2011), 12.

Lorimer MF, Kiermeier A. Analysing microbiological data: Tobit or not Tobit? *International Journal of Food Microbiology* 116(3) (2007), 313–318.

Lunn D, Jackson C, Best N, Thomas A, Spiegelhalter D. *The BUGS Book. A Practical Introduction to Bayesian Analysis.* Chapman & Hall/CRC, Boca Raton, FL, 2013.

Helsel DR. Fabricating data: How substituting values for nondetects can ruin results, and what can be done about it. *Chemosphere* 65(11) (2006), 2434–2439.

Mikkelä A, Ranta J, González M, Hakkinen M, Tuominen P. Campylobacter QMRA: A Bayesian estimation of prevalence and concentration in retail foods under clustering and heavy censoring. *Risk Analysis* 36(11) (2016), 2065–2080.

Office of Pesticide Programs. Assigning values to nondetected/non-quantified pesticide residues in human health food exposure assessments. U.S. Environmental Protection Agency, Washington, DC. 201460. March 23, 2000. https://archive.epa.gov/pesticides/trac/web/pdf/trac3b012.pdf.

Plummer M. JAGS: A program for analysis of Bayesian graphical models using Gibbs sampling. *Proceedings of the 3rd International Workshop on Distributed Statistical Computing (DSC 2003)*. ISSN 1609–395X. Vienna, Austria. 20–22 March, 2003.

Pouillot R, Delignette-Muller ML. Evaluating variability and uncertainty separately in microbial quantitative risk assessment using two R packages. *International Journal of Food Microbiology* 142(3) (2010), 330–340.

Pouillot R, Hoelzer K, Chen Y, Dennis S. Estimating probability distributions of bacterial concentrations in food based on data generated using the most probable number (MPN) method for use in risk assessment. *Food Control* 29(2) (2013), 350–357.

Wenzl T, Haedrich J, Schaechtele A, Robouch P, Stroka J. Guidance document on the estimation of LOD and LOQ for measurements in the field of contaminants in feed and food. EUR 28099. Publications Office of the European Union, Luxembourg, 2016. ISBN 978-92-79-61768-3. doi: 10.2787/8931.

163

7

Understanding Uncertainty and Variability in Risk Assessment

Régis Pouillot and Laurent Guillier

Contents

Variability is the law of life, and as no two faces are the same, so no two bodies are alike, and no two individuals react alike and behave alike under the abnormal conditions which we know as disease.

William Osler

We can name things only with uncertainty, and our words become certain only when they cease to refer to actual things.

Albert Camus

7.1 INTRODUCTION

Risk analysis has become an integral part of policy making. It is recommended that the two major components of risk analysis, i.e. risk assessment and risk management, are carried out separately (Codex Alimentarius Commission [CAC], 1999). This separation was designed to allow risk assessment to make an independent contribution following its own method, mostly relying on science. Science and policy making work in different epistemological cultures. Scientific knowledge is always provisional, open to criticism, while policy requires knowledge that is final and that permits making decisions. This difference between approaches of knowledge for risk assessment and risk management highlights the need for scientists to communicate clearly on what they believe to be the most likely estimate for a risk question but also to express the limits of their current knowledge. The concept of uncertainty is therefore of primary importance for risk assessors.

Most risk assessment models for biological and chemical hazards in food need to take into account, and do take into account, the phenomenon of variability. Indeed, most foods do not contain any harmful bacteria or chemical hazards. The consumption of a particular food is not "good" or "bad" as a whole because of the potential presence of hazards. The variety of the food chain enables consumers to be provided with foods of various microbiological and chemical qualities. Moreover, there is a natural heterogeneity among individuals in terms of food consumption, suscepti-bility toward the hazards, etc.

Uncertainty and variability are, then, two key concepts in the risk analysis framework. While it is recommended to consider and evaluate variability and uncertainty separately (CAC, 1999), these two concepts are nevertheless frequently misunderstood or misused.

This chapter gives a focus on the two concepts. It first reviews the usual definitions found in the field of risk analysis. Then, some practical cases are presented in order to help the reader to understand and distin-guish both concepts. Next, some practical recommendations are provided for characterizing uncertainty and variability.

7.2 SOME CLASSICAL DEFINITIONS

7.2.1 Variability

Variability refers to the inherent heterogeneity or diversity of data in an assessment. It is "a quantitative description of the range or spread of a set of values" (US Environmental Protection Agency [EPA], 2011) and is often expressed through statistical metrics, such as variance, standard deviation, quantiles, or interquartile ranges, that reflect the variability of the data.

A useful way to think about sources of variability is to consider these four broad categories:

1. Spatial variability, i.e. variability across locations
2. Temporal variability, i.e. variability over time
3. Intra-individual variability, i.e. variability within an individual
4. Inter-individual variability, i.e. variability across individuals (U.S. EPA, 2011)

Variability is observed in each component of risk assessments for biological and chemical hazards in food. In the hazard characterization component (briefly speaking, the "dose–response"), the individual response to a given dose of a given strain is frequently highly variable. For *Listeria monocyto-genes* infection, for example, the underlying condition of the consumer has been described as playing a major role in the probability of him or her being infected (Goulet et al., 2012). Similarly, the virulence of *L. monocytogenes* strains has been described as highly variable, and more and more data are available on this subject, while pathogenicity is being further explored, and next generation methods allow better characterization of strains. Variability is even more obvious in the exposure assessment component of a risk assessment: for most contaminants, the great majority of the food servings are not contaminated, while few, or very few, are. Among the contaminated units, variability is also high, and contamination is usually described with heavy-tailed distributions (lognormal distribution, notably). For bacterial contamination, exponential growth may be the origin of the heterogeneity of the contamination levels and the presence of extremely high levels of contamination. The variability in serving size or frequency of consumption is also an element that explains the exposure assessment variability.

Consequently, the risk characterization component of a risk assessment, a function of the hazard characterization and the exposure assessment, will lead to a highly variable output. Because of variabilities in individual susceptibility, in strain virulence, in the capacity of strains to grow in the food, in the contamination process, in food preparation and storage, and in serving size, the probability of observing an adverse effect following the consumption of a product is extremely variable from serving to serving, from individual to individual, from day to day, from region to region, etc. The consequences of this adverse effect can also be variable, notably according to the individual age or underlying conditions, leading to an additional variability in the risk.

Variability cannot be reduced, but it can be better characterized. As an example, we illustrate in Figure 7.1a the (theoretical) distribution of the probability of a disease associated with the consumption of a given food for a given, say, chemical contaminant. The probability can span from 10^{-16} to 10^{-4} in this population. After further studies, it appears that the population can be divided into two subpopulations, with one subpopulation showing on average and for most of the individuals (but not always) a higher risk (Figure 7.1b) than the other subpopulation (Figure 7.1c). This knowledge does not reduce the variability of the risk in the overall population, but now, the risk is better characterized. An example could be a subpopulation of individuals living close to a source of contamination or a subpopulation of heavy consumers of a given product. Characterizing the population with the higher risk and mitigating the risk specifically in this subpopulation could help reduce the overall risk more efficiently than mitigating the risk over the whole population.

7.2.2 Uncertainty

Uncertainty refers to a lack of data or an incomplete understanding of the context of the risk assessment decision. It can be either qualitative or

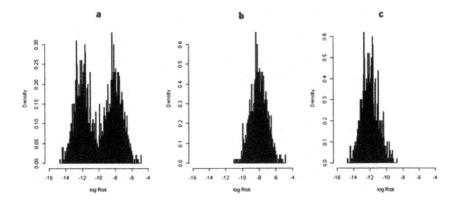

Figure 7.1 Example of a better characterization of variability. The distribution of the risk in the whole population (a) results from the mixing of two populations with clearly distinct risks (b and c). Knowing this doesn't reduce the risk in the population, but it can help to better characterize the risk.

quantitative (U.S. EPA, 2011). Uncertainty has been further characterized as scenario uncertainty, model uncertainty, or parameter uncertainty (U.S. EPA, 2011). Just as variability is everywhere, uncertainty is everywhere.

Scenario uncertainty affects how an exposure scenario is defined, the selection of exposure parameters, exposure routes and pathways, populations of concern, chemicals of concern, and the selection of appropriate models. It arises from descriptive errors, aggregation errors, errors in professional judgment, and incomplete analysis (U.S. EPA, 2011). By definition, because those uncertainties are linked to an error or a bad knowledge of the process, they are usually not identified and are therefore difficult to quantify. In our domain of risk assessment for biological and chemical hazards in food, a scenario uncertainty frequently arises from considering and modeling only the "baseline" situation, while the incidents are not linked to some extreme of this baseline situation but to a completely different situation that is not, and cannot be, predicted by the model. An example could be the assessment of the risk based on a model considering the use of a veterinary drug in beef cattle (modeling extreme situations of overdosage and lack of withdrawal period), while this drug is actually being used on another, non-authorized species, and most of the risk is linked to this unknown route of exposure. Whatever the quality of the model in beef cattle, there is a scenario uncertainty that should be considered for a reliable risk assessment.

Model uncertainties arise because of gaps in the scientific theory that is required to make predictions on the basis of causal inferences. Common types of model uncertainties in various risk assessment–related activities include oversimplified representations of reality, failure to account for correlations, and use of surrogate variables (U.S. EPA, 2011). These uncertainties are difficult to quantify and sometimes difficult to identify.

Risk assessments depict reality interpreted through mathematical representations that describe major processes and relationships. Uncertainty in parameter estimates stems from a variety of sources, including measurement errors, non-representativeness of the sample, random errors, etc. (U.S. EPA, 2011). Some of these uncertainties may be quantified. Risk assessors mostly focus on this source of uncertainty.

Uncertainty can be reduced or, theoretically, eliminated with more or better knowledge and data. A simple example, for uncertainty in parameters, is the mean contamination of a given product. The knowledge of this parameter will be refined with more samples.* Sampling more samples will not reduce the variability of the contamination from sample to sample, but it will refine the estimate of this variability distribution and its associated statistics (mean, standard deviation, and quantiles). Another example is the serving size. Whatever the size of the studied population, some individuals will eat large servings, and others will eat smaller ones. This is variability, and it is irreducible. If a motivated researcher measures, on a given day, the serving size of a given food for a whole population, he or she will have a perfect knowledge of the serving size distribution, hence eliminating the uncertainty for that parameter, that day, without reducing the observed variability.

The different categories of uncertainty were described in an ontology of uncertainty (Rona-Tas, 2017) for both data and model uncertainty. Figure 7.2 illustrates the ontology related to the different forms of uncertainty associated with data.

Note that in most domains other than risk assessments for biological and chemical hazards in food (e.g. finance, engineering, etc.), uncertainty is the major factor considered in the risk assessment. As a consequence, most of the references and software manuals consider that the distributions that are modeled reflect only uncertainty. This probably contributes to the confusion regarding these two concepts in our domain of interest. In the Society for Risk Assessment glossary (SRA, 2018), there is no entry for "variability."

---

* This is true only if the sample is representative of the population of interest. A biased sample leads to biased estimates, even if the sample is very large, and sampling more samples will not help.

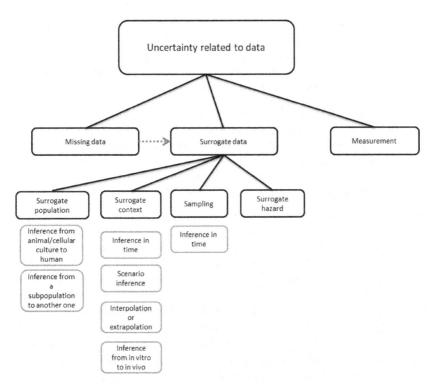

Figure 7.2 Ontology related to the different forms of uncertainty associated with data (Rona Tas, 2017).

7.3 UNDERSTANDING UNCERTAINTY AND VARIABILITY

7.3.1 A Proposal to Better Understand and Consider Variability and Uncertainty

The previous definitions are well known. Variability and uncertainty seem to be two very separate concepts on paper. However, in practice, there is still a lack of understanding of these two concepts and some difficulties in considering them.

A proposal to understand, consider, and separate variability and uncertainty in the food safety risk assessment framework, and to avoid mixing all sorts of uncertainty and variability distributions, is to use the following reasoning: most risk assessment scopes are to evaluate one or a few estimates linked to the risk and the *uncertainty* surrounding them.

171

These estimates are not variable, but in order to evaluate them, we have to consider and integrate the *variability* of the risk in the population.

The risk assessor will first have to define its statistic of interest (maybe with the help of the risk manager) as precisely as possible. It is best to choose an estimate that is not variable. Examples are:

- What is the expected cost from cancers linked to the consumption of inorganic arsenic from rice in San Theodoros each year?
- What is the expected proportion of individuals exposed to a dose of lead higher than the reference dose in San Theodoros?
- What is the expected number of cases of listeriosis next year from the consumption of smoked salmon in San Theodoros?

These numbers may be considered unique. They are not variable. The actual number of cancers linked to inorganic arsenic from rice may vary from area to area in San Theodoros or from year to year, but the expected number of cases in all San Theodoros in a random year does not, because it integrates all this variability. The expected number of cases of listeriosis from the consumption of smoked salmon in San Theodoros integrates a large number of sources of variability: from individual to individual, from subpopulation to subpopulation, the production type of cold salmon (hot smoked and cold smoked salmon), the producers of smoked salmon in San Theodoros, etc., and we definitely need to consider all this variability to estimate the expected number of cases of listeriosis next year from the consumption of smoked salmon in San Theodoros. But once all those dimensions of variability are integrated, this number is not variable.

However, this number is uncertain: the estimate of this expected number of cases will not be known perfectly, and the risk assessor will have to provide an uncertainty range (or some measure of uncertainty).

The question of interest could be "What is the annual variability of the risk of listeriosis in this country?" It is possible to answer this question, but it will actually be much more tricky. Indeed, in this case, the answer will not be a single value but at least two (mean and standard deviation of the distribution of the expected number of cases expected each year) or even a whole empirical distribution (the year-to-year distribution of the expected number of cases). In order to answer this question, the analysis will have to separate phenomena that are variable from year to year (prevalence level, for example) and phenomena that are variable within each year (the serving-to-serving contamination, that is, the serving contamination given the distribution of the contamination of each year and the serving size) and will have to integrate, separately, those two sources of variability (maybe using a second-order Monte-Carlo simulation with both dimensions reflecting variability: one reflecting the within-year and the other one the year-to-year variability). Mixing all the variabilities (inter- and intra-year) in a Monte-Carlo simulation and assuming that the resultant distribution is a year-to-year variability is wrong (see Box 7.1).

BOX 7.1   A FREQUENT ERROR: MIXING DIFFERENT LEVELS OF VARIABILITY

Assume a model leading to a year-to-year variability distribution of the contamination of the product. Once this distribution of contamination is specified, one could simply, through a Monte-Carlo simulation, multiply the values obtained with the individual variability distribution of the serving size, inject the results into a dose–response model, and multiply the results by the number of servings per year.

The final distribution will not be a year-to-year variability distribution of the number of cases. Indeed, mixing an individual variability (serving size) with an annual variability (contamination) distribution is wrong. Considering the final distribution as a year-to-year variability distribution of the number of cases would assume that in a given year, all individuals consume a fixed serving size.

Iteration	Mean annual contamination (units per gram)	Serving contamination (units/gram)	Serving size (gram)	Ingested dose (a × b) (units)	Probability of illness	Number of servings	Number of cases
Level of variability	Year-to-year	Individual-to-individual	Individual-to-individual	???	???	Year-to-year	???
Iteration #1	3	2	10	20	.01	100	1
#2	0	0	20	0	0	120	0
#3	8	8	15	120	.05	100	5
...	...	...	...	...	...	...	...

Combining distributions in the variability dimensions that reflect various levels of variability (year-to-year vs. individual-to-individual or lot-to-lot) should be done with great caution, and the final results may be uninterpretable.

173

7.3.2 Why Is There a Lack of Understanding?

There is clearly a lack of understanding and confusion around the notion of variability and uncertainty in risk assessment models for biological and chemical hazards in food.

Both notions, (parameter) uncertainty and variability, are usually expressed as distributions and when facing a set of data, variability and uncertainty are actually very difficult to disentangle. Consider the estimation of the prevalence of contaminated smoked salmon in San Theodoros. The prevalence varies from year to year. Moreover, the knowledge about this prevalence and its variability is uncertain. As an example, consider some products tested over a 5-year period with 6/100, 3/100, 9/100, 11/100, and 7/100 positive samples. Does the observed variation in the estimated prevalence correspond to a year-to-year variability? Or does it correspond to sampling variability? Or to uncertainty? Various models can be built from the same data. Under some assumptions, a complete model that derives a year-to-year distribution of prevalence and its uncertainty is possible (see Box 7.2).

BOX 7.2   MULTIPLE WAYS TO CONSIDER A SET OF DATA

Prevalence data were collected during 5 years, with 100 samples per year, and led to 6, 3, 9, 11, and 7 positive samples.

Hypothesis 1: the prevalence is constant from year to year: these estimates of prevalence are five measures of the same values. Data can then be gathered. The best estimate is $p = (6 + 3 + 9 + 11 + 7)/(100 + 100 + 100 + 100 + 100) = 36/500 = 7.2\%$.

Simulating variability only: use $p = 7.2\%$. To model the number of positive units in a lot of size N, use a binomial(N, 7.2%) (under the assumption of independence of contamination). Note that there is no underlying reason to use a beta distribution to model variability.

Simulating variability and uncertainty in a second-order Monte-Carlo simulation. Using a Bayesian reasoning (Vose, 2008), a prior distribution as a beta(0.5, 0.5) (Miconnet et al, 2005), one can simulate the *uncertainty* distribution of the prevalence of contaminated product using a $p \sim$ beta(36 + 0.5, 500 – 36 + 0.5). There is no variability distribution for p.

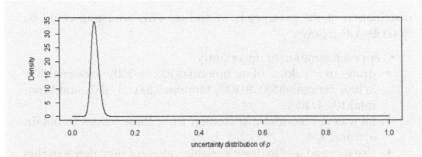

Hypothesis 2: the prevalence varies from year to year : there is one measure of prevalence per year. If we assume that the prevalence follows a *variability* distribution from year to year that is a beta distribution, the observations $(6, 3, 9, 11, 7)$ are the results of a beta binomial distribution.

Simulating variability only: the best estimate for the parameters of the underlying beta *variability* distribution for p_y are $\alpha_1 = 61.7$, $\alpha_2 = 796$ (using maximum likelihood estimation).* To model the number of positive units in a lot of size N in a year at random, draw p_y from a beta$(61.7, 796)$; then draw the number of positive units out of an N sample for this year from a binomial(N, p_y) distribution (see figure below).

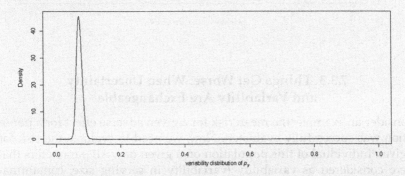

Simulating variability and uncertainty in a second-order Monte-Carlo. Using, e.g. parametric bootstrap, we could derive an uncertainty

---

* Maybe using exp(coef(vglm(cbind(y <- c(6,3,9,11,7), 100-y) ~ 1, betabino-mialff))) from the vglm function of the VGAM package in R.

distribution of the parameters α_1 and α_2. This bootstrap could be described as follows:

- For each iteration of uncertainty
- draw five values of a binomial(100, 6/100), binomial(100, 3/100), binomial(100, 9/100), binomial(100, 11/100), and binomial(100, 7/100).
- Fit a beta binomial distribution from these values to obtain $\alpha_1^{(u)}$ and $\alpha_2^{(u)}$.
- Use $\alpha_1^{(u)}$ and $\alpha_2^{(u)}$ to draw variable values of prevalence in this uncertainty dimension.

The figure below shows the results of this two-dimension Monte-Carlo simulation.

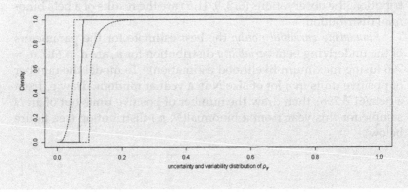

7.3.3 Things Get Worse: When Uncertainty and Variability Are Exchangeable

Consider an example: the mean risk for a given adverse effect for a population was successfully evaluated. Now, we need to evaluate the risk for a given individual of this population on a given day. All parameters that were considered as variability (variability in serving size, contamination, and susceptibility of the individual) were integrated in the estimate of the population risk. However, those parameters should no longer be considered as variability for the "individual" evaluation. The serving of this individual is of a given, unique, not variable (but unknown) size; the contamination level is of a given, unique, not variable (but unknown) level; and the individual is of a given, unique, not variable (but unknown) susceptibility. If nothing is known about this individual, then the distribution that was considered as variability should now be considered as uncertainty. The uncertainty will then be extremely large.

Now, assume that we happen to know that the individual is an 85-year-old man. With this additional information, we may use a refined dose–response (i.e. the dose–response for elderly people), a more refined serving size estimate, etc. The uncertainty will be reduced, but some will remain (actual serving size, actual contamination, etc.).

In the example of Figure 7.1, if the average risk of the population has to be defined, data from Figure 7.1a should be used and averaged. Figure 7.1a reflects the variability of the risk in the population. Now, assume that we want to evaluate the risk for one individual of the population. The actual risk is one of the values in the figure, from 10^{-14} to 10^{-4}, but we don't know which one. Figure 7.1a then reflects the uncertainty of the estimate of the individual risk.

A more relevant example was developed in the quantitative assessment of relative risk to public health from foodborne *Listeria monocytogenes* among selected categories of ready-to-eat foods (Food and Drug Administration/Food Safety and Inspection Service [FDA/FSIS] 92003). When the risk ranking was estimated, the dose–response considered the variability in the strain infectivity, because cases in the population are linked to variable strains. Now, when facing an outbreak, the estimate of the risk should no longer consider this variability distribution. Indeed, the outbreak is linked to a single strain, which has a fixed (but unknown) virulence value. If no information about the strain virulence is available, the virulence variability distribution should now be considered as uncertainty.

7.4 VARIABILITY AND UNCERTAINTY IN PRACTICE

7.4.1 Conduct an Uncertainty Analysis

Various uncertainty analysis approaches are used by different actors in the field of risk assessment (Bundesinstitut für Risikobewertung [BFR], 2015, European Food Safety Agency [EFSA], 2018, Hart et al., 2010). These approaches include useful elements to consider when constructing an appropriate generic uncertainty analysis framework. Recently, the French Agence Nationale de Sécurité Sanitaire de l'Alimentation, de l'Environnement et du Travail (ANSES) (2016, 2017) proposed an iterative approach in five stages adapted from the one proposed by EFSA (2018). The five steps are presented in Figure 7.3.

The first step aims at identifying and describing all uncertainties encountered throughout the evaluation process. It aims to provide a general view of the sources of uncertainty and should allow the selection of sources of uncertainty to be further addressed in subsequent steps of the uncertainty analysis.

At this stage, all sources of uncertainty are identified and described, using different descriptors, more commonly referred to as dimensions (Walker et al., 2003):

- Its location, i.e. where the source of uncertainty is in the appraisal process, if the uncertainty arises from (not taking into account) the variability, from the limit of knowledge, etc.

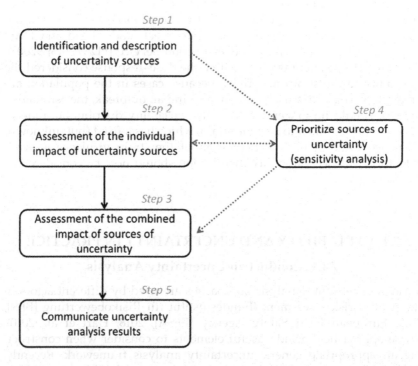

Figure 7.3 Uncertainty analysis approach adapted from ANSES (2017) and EFSA (2018).

- Its level, i.e. where the source of uncertainty is placed on the spectrum of the kind of uncertainty it induces, ranging from irreducible ignorance to inaccuracy
- Its direction, i.e. the (estimated) meaning of its influence on the result, usually expressed using the symbols "+" and "−" or "="*

7.4.2 Considering Uncertainty and Variability Separately

7.4.2.1 Characterize the Variability/Uncertainty from the Data

The variability distribution should describe how a parameter varies from one individual to the other or from one region to the other or from one period of time to the other. Characterizing this distribution requires theoretical knowledge or observed data collected from this distribution. Contrary to most other scientific domains, where the statistic of interest is mostly the mean, a risk assessor is interested in the whole distribution (and notably the extreme observations). Specific data, looking for and characterizing those extreme situations, should usually be collected.

Three situations can occur now. First, the parametric distribution to use for this parameter (normal, lognormal, gamma, etc.) is known because of theory or precedent. In that case, the parameters of this distribution have to be inferred from the observed data. Second, if there is no specific knowledge of the underlying distribution, various parametric distributions can be fitted to the observed data, and the one that best fits the data and that is reasonable with regard to the data can be chosen. Lastly, it is possible to sample directly from your observed data. There are pros and cons for each of these methods, but this is outside the scope of this section.[†]

Usually, not all individuals will have been sampled. There will be some (parameter) uncertainty around the distribution specified. A part of this parameter uncertainty can be estimated using statistical procedures such as Bayesian inferences and (parametric and non-parametric) bootstrap.

At the end of the process, a distribution of hyperparameters reflecting the uncertainty of variability parameters is obtained from raw data. The

---

[*] A good example of such an uncertainty analysis was developed by Boué et al. (2017) during the development and application of a probabilistic risk-benefit assessment for infant feeding.

[†] See Vose (2008) for example.

variability of the parameter of interest is then derived conditionally to a given set of uncertain parameters (see Box 7.3).

BOX 7.3 PRINCIPLE OF DERIVATION OF UNCERTAINTY AND VARIABILITY DISTRIBUTION

From a dataset $X \to$ derive an uncertain parameters distribution $Y = f(X) \to$ simulate variability distribution $Z = g(Y)$

Example:

Dataset $X \to I = 1, ..., I$ bootstrap samples from X, $X^{(i)} \to$ estimates of $m^{(i)}$, the mean of $X^{(i)}$, and $s^{(i)}$, the standard error from $X^{(i)} \to$ conditionally to a couple use $Z \sim$ normal distribution $(m^{(i)}, s^{(i)})$

7.4.2.2 Use Monte-Carlo Simulations to Integrate Variability

It can be shown that under light assumptions, the expected number of cases of illness in a population is given by the number of servings (in the population of interest, in the period of interest) multiplied by the arithmetic mean* (over servings, in the population of interest, during the period of interest) risk per serving. This mean risk per serving can sometimes be obtained by calculations, but most of the time, the calculations are inextricable, as they consist of a (too) complex integration of distributions. One solution is then to use a Monte-Carlo simulation to evaluate this mean risk: the principle is simply to simulate the distribution (from serving to serving) of the risk, and theory says that when the number of simulated values tends to infinity, the mean of the resulting values tends to the actual mean that would be obtained mathematically.

The Monte-Carlo method is appealing because one can think of each iteration as the simulation of "one possible serving." A simple way to ensure that the reasoning and the parameterization of the model are correct is then to ask two questions:

i. Could each of the simulated servings occur in real life? As an example, if a distribution used in the simulation can lead to an impossible serving size, temperature, or concentration of bacteria/chemical in the product, this distribution should be reevaluated.[†]

---

* Note: not the geometric mean, not the mean of the log, not the median ....
[†] Maybe bounded.

This is of particular interest once it is understood that we are evaluating an arithmetic mean and that this mean can be especially sensitive to an extreme value.

ii. Is each iteration in the model representative of my serving-to-serving distribution? As an example, if the model considers various production types, or various manufacturers, the weight of each production type (manufacturer) has to be represented in the simulations. If a producer provides 50% of the cold smoked salmon in San Theodoros, this producer should account for 50% of the simulations.*

One will then ask when to stop the level of simulation: if the objective is to simulate the expected number of cases, should one simulate each serving with the actual number of bacteria (as an integer: 0, 1, 2, …) and then simulate if the consumer is sick or not sick following the ingestion of this dose? Actually, this would generally need an extremely large number of simulations, especially for products that are not frequently contaminated. In fact, we don't actually want to *simulate* illnesses, but just to estimate an arithmetic mean of the risk. Basic knowledge of probability theory and statistics is useful here: as an example, the mean risk can be evaluated from the proportion of contaminated servings and the mean risk resulting from contaminated servings; the mean is then Probability($n > 0$) * (mean risk$|n > 0$). Moreover, most of the dose–response models (exponential, beta Poisson, probit, ...) already integrate a part of the variability: the exponential dose–response integrates the serving-to-serving distribution of contaminant given the mean contamination under the assumption of a Poisson distribution; the beta Poisson distribution integrates this distribution and also an individual-to-individual variability of sensibility (or a serving-to-serving variability of virulence) under the assumption that this distribution follows a beta distribution, etc. In conclusion, any mathematical integration that is available can be discarded from the Monte-Carlo simulation.

It is important to notice that this Monte-Carlo simulation integrating variability leads to one single value (the expected number of cases of illness in the population). This number is the number estimated using your current assumptions regarding your underlying parameters. The result of a risk assessment that does not consider uncertainty is a single value integrating variability.

---

* A more complex simulation framework can help.

7.4.2.3 Characterize the Uncertainty

Characterization of the uncertainty of each input of a risk model and assessment of their impacts on output is part of the formal process of uncertainty analysis (see Figure 7.3). Uncertainty characterization of inputs is usually done together (but separately) with variability characterization. The integration of uncertainty for model output depends on the number of uncertain inputs.

7.4.2.3.1 *Limited Number of Uncertain Values/Scenario*

If there is a limited number of uncertain parameters, with a limited number of possible values (e.g. the proportion of susceptible individuals in the susceptible population is 0.25 or 0.33, according to two sources), or if there is a limited number of alternative scenarios to be tested, a simple way to consider uncertainty is to run the Monte-Carlo integration (of the variability dimension) once for each uncertain value. In the preceding example, the output would then be an estimation of the expected number of cases conditionally to a proportion of the susceptible population of 0.25 and another estimate conditionally to a proportion of the susceptible population of 0.33.

7.4.2.3.2 *Second-Order Monte-Carlo Simulation*

The second-order Monte-Carlo simulation is just a generalization of the process described previously when there is a large number of uncertain values/scenario.

The Monte-Carlo simulation in the variability dimension provides, from a set of input variable parameters conditionally to uncertain parameters, one value of the estimate. It can be written as $Y = f(X)$, where Y is the estimate and X is a set of parameters considered in this simulation as being known exactly (no uncertainty within a variability Monte-Carlo simulation). In order to estimate how the estimation of Y changes according to other sets of X, a Monte-Carlo simulation can be derived. The "variability" Monte-Carlo simulation is then embedded within an "uncertainty" Monte-Carlo simulation (Figure 7.4) (Frey 1992; Vose, 2008, Pouillot and Delignette-Muller, 2010).

The final result is then an uncertainty distribution of expected number of cases. Using the median (or the mean*) of this distribution provides a central estimate of the expected number of cases; the 2.5th and the 97.5th

---

* The use of the median or the mean can be discussed.

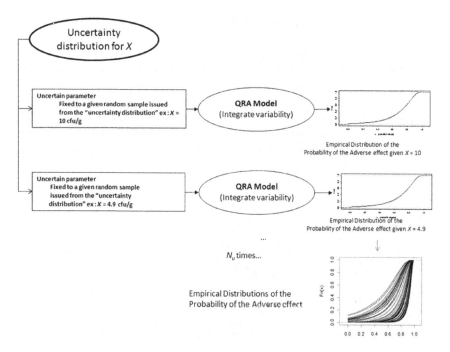

Figure 7.4 Illustration of the second-order Monte-Carlo simulation.

percentile of this distribution provide a "credible interval" around this central estimate.

7.4.2.3.3 Classical Errors

The two main errors when handling variability and uncertainty are getting rid of (a part of) uncertainty and mixing the two dimensions in a single-dimension Monte-Carlo simulation.

To illustrate the possible impact of this second error, let's consider a simple situation where one wants to assess the probability of exposure being above a certain limit (e.g. 4 µg) and the situation is considered as "safe" if fewer than 10% of the exposures are below this threshold. The ingested dose is simply calculated by multiplying the serving size (*cons*) by the concentration of hazard (*conc*). Serving sizes are variable among the population of interest: a dataset of 240 serving size measurements is available to characterize this variability. The concentration of the hazard in the foods of interest is also variable. Six values are available to characterize the variability of the hazard concentration. A gamma(*s*, *r*) and a

183

normal(m, sd) distribution are fitted to the consumption and the concentration data, respectively.

In Situation 1 (Figure 7.5), variability was taken into account but not uncertainty; that is, m, sd, s, and r parameters were considered as fixed values. A one dimensional Monte-Carlo simulation was used to assess the variability of ingested doses. In this situation, the probability of being exposed to more than 4 µg is 0.11. The situation is considered as "unsafe" (more than 10% of servings >4 µg). This information is correct, but there is no information on the confidence associated with this probability.

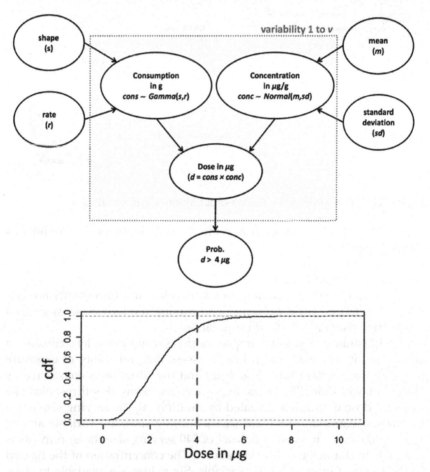

Figure 7.5 Variability only is considered. This is not wrong, but uncertainty is not assessed.

Because only six values for hazard concentration are available, the confidence interval might be large.

In Situation 2 (Figure 7.6), the uncertainty distribution of parameters of the two distributions was characterized by bootstrap. However, a one-dimensional Monte-Carlo simulation was (erroneously) used. For each simulation, random values of r, s, m, and sd were sampled, a single value of $cons$ and $conc$ was drawn from gamma and normal distributions, and a

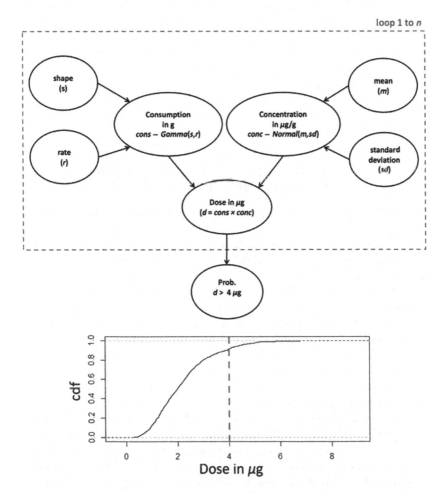

Figure 7.6 Variability and uncertainty are considered but are mixed in a one-dimensional Monte-Carlo simulation. This is wrong, and the results are misleading.

single dose d was calculated. From the resulting n doses, the probability of being exposed to more than 4 µg was evaluated. In this (incorrect) way of estimation, the probability of exceeding a dose of 4 µg was estimated to be 0.085. The situation is, here, considered as "safe" (fewer than 10% of servings >4 µg). Moreover, there is no uncertainty associated with that probability, although the uncertainty of inputs has been characterized.

In Situation 3 (Figure 7.7), uncertainty is separated from variability. A second-order Monte-Carlo simulation framework is used. The probability of being exposed to more than 4 µg is uncertain. The median probability is equal to 0.109, and the 95% credibility interval is (0.057–0.162). Note here that the credible interval includes the value of 0.10. The message for the risk manager should then be (as expected): you would need more data to conclude in this situation.

In practice, some resources exist to conduct second-order Monte-Carlo simulations. Box 7.4 introduces those that are freely available.

BOX 7.4   RESOURCES FOR SECOND-ORDER MONTE-CARLO SIMULATIONS

We will limit here the description of the resources that are available freely.

AuvTool (http://foodrisk.org/resources/display/57) is a tool to develop two-dimensional Monte-Carlo simulations. It was developed by the U.S. EPA in 2002. It includes bootstrap simulation and helps deal with variability and uncertainty, mixture distributions, measurement error, and censored data.

The FDA iRisk tool is a web-based system (https://irisk.foodrisk. org/) designed to analyze data concerning microbial and chemical hazards in food and return an estimate of the resulting health burden on a population level. Version 4.0 and up allows the user to perform second-order Monte-Carlo simulations to consider variability and uncertainty separately.

Monte-Carlo simulations can actually be developed using any computing language. R (https://cran.r-project.org/) is a free software environment for statistical computing. "mc2d" (https://cran.r-project.org/web/packages/mc2d/index.html) was developed specifically as an R package to help develop second-order Monte-Carlo simulations (Pouillot and Delignette-Muller, 2010).

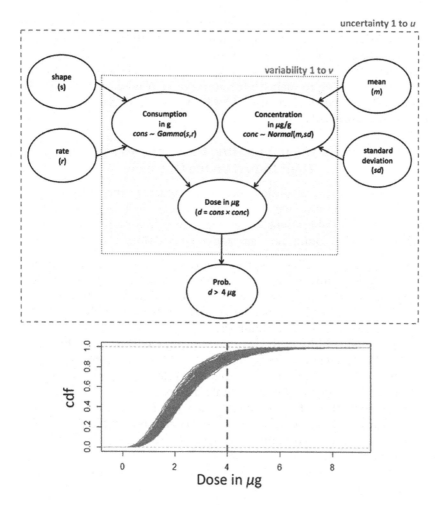

Figure 7.7 Variability and uncertainty are considered and assessed separately in a two-dimensional Monte-Carlo simulation.

7.4.3 Prioritize the Different Sources of Uncertainty

The next step aims to order the different sources of uncertainty according to their impact on the risk. In a context of quantitative evaluation, and particularly when it comes to dealing with the uncertainty of input variables of models, it is customary to rely on sensitivity analysis. This type of analysis quantitatively measures the contribution of the uncertain variables of a model to the variations of its outputs (Augustin, 2011, Ellouze et al., 2010, Saltelli, 2002). Thus, the sensitivity analysis allows the input variables that have a strong influence on the outputs of the model to be distinguished from those that have a lesser influence, and thus to classify the uncertain variables according to their contribution to the overall uncertainty. This prioritization is particularly useful for recommendations to reduce uncertainty (Augustin, 2011).

7.4.4 Communicating Uncertainties and Their Impact on the Outcome

Communicating uncertainties and their impact on the outcome of an assessment is the last step. This step is crucial as it will help decision making. Recent publications are giving it an increasingly important role (EFSA 2018). This communication step must describe:

- The transparency of the process: it is recommended to document the uncertainty by making a concise synthesis describing the main sources of uncertainty, the additional uncertainties whose impact could not be evaluated, and the subjective choices performed during the risk analysis.
- The intelligibility of the results by various stakeholders: this can be done through a communication based on visual tools allowing the representation of the results in the form of graphs and the communication of the uncertainty matrix.
- The adaptation of communication to a wider audience (managers, scientists, or the general public): this is mainly to avoid the use of too specialized vocabulary, which could lead to a misinterpretation of the results.

7.5 DISCUSSION

Considering variability and uncertainty separately is an international recommendation. It is important to evaluate the impact of these two elements. Nauta (2000) showed the impact of separating these two concepts. Mixing variability and uncertainty can lead to biased estimates. However, few risk assessments currently consider uncertainty and variability separately.

The most important step of the process is to identify and qualify the sources of uncertainty through an uncertainty analysis. Sometimes, a quantitative risk assessment can be developed, and a part of the data uncertainty can be estimated. Usually, the variability dimension can be integrated to estimate the actual statistic of interest (expected number of cases in a given population). The results of such quantitative risk assessment consist of a single number, representing the number of cases given a set of uncertain parameters. Uncertainty can then be considered by evaluating the model multiple times, for multiple sets of uncertain parameters, either one by one or through a Monte-Carlo simulation.

REFERENCES

ANSES (Agence nationale de sécurité sanitaire de l'alimentation, de l'environnement et du travail). (2016). Prise en compte de l'incertitude en évaluation des risques : Revue de la littérature et recommandations pour l'Anses. Available at https://www.anses.fr/en/system/files/AUTRE2015SA0090Ra.pdf.

ANSES (Agence nationale de securite sanitaire de l'alimentation, de l'environnement et du travail). (2017). Illustrations et actualisation des recommandations pour l'évaluation du poids des preuves et l'analyse d'incertitude à l'Anses. Available at https://www.anses.fr/fr/system/files/AUTRE2015SA0089Ra-2.pdf.

Augustin, J. C. (2011). Global sensitivity analysis applied to food safety risk assessment. *Electronic Journal of Applied Statistical Analysis* 4(2):255–264.

Boue, G., Cummins, E., Guillou, S., Antignac, J. P., Le Bizec, B., & Membre, J. M. (2017). Development and application of a probabilistic risk-benefit assessment model for infant feeding integrating microbiological, nutritional, and chemical components. *Risk Analysis* 37(12):2360–2388.

Bundesinstitut für Risikobewertung. (2015). Guidelines on uncertainty analysis in exposure assessments. Available at https://www.bfr.bund.de/cm/350/guidelines-on-uncertainty-analysis-in-exposure-assessments.pdf.

Codex alimentarius Commission. (1999). *Principles and Guidelines for the Conduct of Microbiological Risk Assessment*. Rome: FAO Edition: 6.

EFSA (European Food Safety Authority) Scientific Committee, Benford, D., Halldorsson, T., Jeger, M. J., Knutsen, H. K., More, S., Naegeli, H., Noteborn, H., Ockleford, C., Ricci, A., Rychen, G., Schlatter, J. R., Silano, V., Solecki, R., Turck, D., Younes, M., Craig, P., Hart, A., Von Goetz, N., Koutsoumanis, K., Mortensen, A., Ossendorp, B., Martino, L., Merten, C., Mosbach-Schulz, O., & Hardy, A. (2018). Guidance on uncertainty analysis in scientific assessments. European Food Safety Agency (European Food Safety Authority). *EFSA Journal* 16(1):5123, 39 pp. doi: 10.2903/j.efsa.2018.5123.

Ellouze, M., Gauchi, J. P., & Augustin, J. C. (2010). Global sensitivity analysis applied to a contamination assessment model of *Listeria monocytogenes* in cold smoked salmon at consumption. *Risk Analysis* 30(5):841–852.

189

Food and Drug Administration/Food Safety and Inspection Service. (2003). *Quantitative Assessment of Relative Risk to Public Health from Foodborne Listeria monocytogenes Among Selected Categories of Ready-to-Eat Foods.* Food and Drug Administration, United States Department of Agriculture, Centers for Disease Control and Prevention: 541.

Frey, H. C. (1992). *Quantitative Analysis of Uncertainty and Variability in Environmental Policy Making.* American Association for the Advancement of Science/U.S. Environmental Protection Agency.

Goulet, V., Hebert, M., Hedberg, C., Laurent, E., Vaillant, V., De Valk, H., & Desenclos, J. C. (2012). Incidence of listeriosis and related mortality among groups at risk of acquiring listeriosis. *Clinical Infectious Diseases* 54(5):652–660.

Hart, A., Gosling, J. P., Boobis, A., Coggon, D., Craig, P., & Jones, D. (2010). Development of a framework for evaluation and expression of uncertainties in hazard and risk assessment. Final Report of Food Standards Agency Project: T01056.

Miconnet, N., Cornu, M., Beaufort, A., Rosso, L., & Denis, J. B. (2005). Uncertainty distribution associated with estimating a proportion in microbial risk assessment. *Risk Analysis* 25(1):39–48.

Nauta, M. J. (2000). Separation of uncertainty and variability in quantitative microbial risk assessment models. *International Journal of Food Microbiology* 57(1–2):9–18.

Pouillot, R., & Delignette-Muller, M. L. (2010). Evaluating variability and uncertainty separately in microbial quantitative risk assessment using two R packages. *International Journal of Food Microbiology* 142(3):330–340.

Rona-Tas, A., Cornuéjols, A., Blanchemanche, S., Duroy, A., & Martin, C. (2017). Enlisting supervised machine learning in mapping scientific uncertainty expressed in food risk analysis. *Sociological Methods & Research* 0049124117729701.

Saltelli, A. (2002). Making best use of model evaluations to compute sensitivity indices. *Computer Physics Communications* 145(2):280–297.

SRA. (2018). Society for Risk Analysis Glossary. https://sra.org/sites/default/files /pdf/SRA%20Glossary%20-%20FINAL.pdf

U.S. Environmental Protection Agency. (2011). *Exposure Factors Handbook 2011 Edition (Final).* Washington, DC: U.S. Environmental Protection Agency.

Vose, D. (2008). *Risk Analysis: A Quantitative Guide.* Chichester: Wiley and Sons.

Walker, W. E., Harremoës, P., Rotmans, J., van der Sluijs, J. P., van Asselt, M. B., Janssen, P., & Krayer von Krauss, M. P. (2003). Defining uncertainty: A conceptual basis for uncertainty management in model-based decision support. *Integrated Assessment* 4(1):5–17.

8

Application of Sensitivity Analysis Methods in Quantitative Risk Assessment

Araceli Bolívar and Fernando Pérez-Rodríguez

Contents

8.1 INTRODUCTION

With the term sensitivity analysis (SA), we refer to a group of statistical techniques that are used in the field of risk assessment and related scientific areas (e.g. economics, mathematics, civil engineering, etc.) to know how a model operates and which factors in it are the most significant. For this purpose, systematic changes are performed in the model, and

191

the effect on results is analysed (Ragsdale, 2008; Saltelli, 2002b). There are several SA techniques, which can be based on mathematical or numerical indexes, statistical tests or graphical representations. In all cases, SA can generate useful information to assess the relative importance of each model variable and even to rank them according their impact on a specific outcome (financial risk, public health outcome, economical profit, etc.).

Building a model is a process that unavoidably leads to several assumptions and interpretations of the data handled. Therefore, model structure, methods and variables cannot be considered as the optimal solution, still less the unique solution (European Commission, 2003). This fact brings with it an important source of uncertainty in the model. Hence, its assessment and subsequent reduction should be understood as a necessary task to develop and build any type of model.

The analysis of sensitivity is an asset to understand the influence of the variability in our model on the final output. In addition, it is also an effective tool to identify data gaps in the modelling process. By knowing this, risk managers, for example, can allocate resources in a more efficient manner to reduce uncertainty.

The application of SA to two-dimensional probabilistic models, in which variability and uncertainty are described separately, can entail certain complexities; however, at the same time, it also allows sources of uncertainty to be identified and the effect of specific factors on the final output (e.g. microbial risk) to be better isolated. This aspect is key to assessing whether a specific management action will result in the desired outcome; otherwise, it will be mostly dominated by uncertainty. In those cases in which two-dimensional models are studied, the SA techniques should be adequately selected to avoid equivocal conclusions produced because of using the wrong method. The work by Mokhtari and Frey (2005) presented an analysis of the performance of specific SA techniques (i.e. analysis of variance [ANOVA]) to cope with the complexities derived from two-dimensional models, providing some advice on the drawbacks and advantages of the assessed techniques.

In the context of microbial food safety, if we are interested in inferring or carrying out food safety interventions on the basis of results, a simple prediction or estimate, e.g. number of illness cases or hospitalizations, would not be enough. Conversely, we should know which variables are determinant in our output and how we should proceed to respond to risk management questions. By applying SA, a deeper knowledge on the factors influencing risk can be achieved, providing an information basis to prioritize control measures or derive specific process control activities by setting critical limits and process parameters within a Hazard Analysis Critical Control Point (HACCP) scheme. From a more public health perspective, SA applied to quantitative microbial risk assessment (QMRA) models could aid in providing a quantitative insight into how changes in food regulation or food safety strategies, based on variables and uncertainty included in the risk assessment, could turn into significant improvements in microbial food safety. Thus, SA should be considered a critical step in microbial risk assessment, especially when model output is intended to be deployed for risk management activities at both governmental and operational level. The present chapter aims to provide a general review of the main SA methods applied to the context of QMRA, describing typologies and classifications and showing the main characteristics, drawbacks and strengths of the selected methods. In addition, specific examples are presented illustrating the use of the different SA methods addressed herein.

8.2 DESCRIPTION OF METHODS AND APPROACHES FOR SENSITIVITY ANALYSIS

In this section, we present an overview of the different classifications of sensitivity analysis methods described in the general literature with further focus on those most widely used in food risk assessment.

Traditionally, SA methods have been structured into two main groups: local SA methods and global SA methods. This classification is based on whether output variability is obtained by varying the inputs around a reference (nominal) value (local SA) or across their entire feasible space (global SA) (Pianosi et al., 2016). Additionally, screening tests can be considered as another type of SA techniques, although they can also be included in either local or global methods (Saltelli et al., 2000). These techniques are helpful in screening the most influential model inputs with relatively low computational effort, especially in models that are computationally expensive (Frey et al., 2004).

Another distinction often made in literature refers to the sampling strategy used to estimate the sensitivity indexes. In this context, SA can be divided into "one-factor-at-a-time" (OAT) methods and "all-factors-at-a-time" (AAT) methods as suggested by Pianosi et al. (2016). The first class consists of analysing the effect of varying one model input factor at a time while keeping the others constant. In contrast, the second type, AAT methods, are based on varying all the input factors simultaneously. While local SA typically uses OAT sampling, global SA can use either OAT or AAT strategies (Pianosi et al., 2016).

The most popular SA methods applied in literature are either local or OAT analyses (Saltelli and Annoni, 2010). The main advantage of these approaches is that they are easier to implement compared with global methods (Frey and Patil, 2002). However, they do not take into account the simultaneous interaction between input variables, representing one of the major shortcomings in comparison with the global or AAT methods (Saltelli et al., 2008). Some studies have illustrated the implementation of global SA to the assessment of the bacterial contamination of a food, showing the relevance of this approach to study models used in the frame of QMRA (Augustin, 2011; Ellouze et al., 2010).

In the context of risk assessment in foods, most studies follow the classification proposed by Frey and Patil (2002). These authors classified SA methods according to the type of QMRA model to be analysed, that is, (1) mathematical methods for deterministic frameworks, (2) statistical methods for probabilistic frameworks and (3) graphical methods for visual representation of a method's results. In this chapter, the three different categories are presented in Sections 8.2.1 through 8.2.3. A summary of SA methods and their main characteristics is provided in Table 8.1.

It should be noted that most of the methods described in the present chapter are used after model simulation. If a model is linear, Cullen and Frey (1999) referred to simple approaches that can be used prior to simulation to obtain a rough indication about which inputs contribute most to output variability. These methods are based on summary statistics of model inputs, such as the sum of variances of the inputs (also known as Gaussian or first order approximation) or the coefficient of variation.

There are some aspects that risk managers should consider when applying a certain method over another. Those are the characteristics of the model, the objective of the model and the specifications of the software tool used to compute the model. No single method perfectly addresses all these criteria, and different sensitivity indexes can provide different rankings for the studied risk factors (Pérez-Rodríguez, 2006). Frey and Patil (2002) provided a review of 10 SA methods that deserve consideration in the context of QMRA. Some studies have evaluated the capabilities of these methods based on their application to particular QMRA models (Frey et al., 2003; Patil and Frey, 2004). According to these works, combination and comparison between various SA methods could help to provide more robust findings. Frey et al. (2004) have developed a detailed document providing guidance to food safety practitioners concerning the

Table 8.1 Summary of Methods, Characteristics and Software/Packages to Perform Sensitivity Analysis

Classification	Method	Acronyms	Characteristics	Computational Issues	Software/Package
Mathematical	Nominal range Sensitivity analysis	NRSA	Easy to understand; No interaction among model inputs, applied to linear models	Potentially time-consuming	
	Break-even analysis	BEA	Complex for models with many inputs / decision options	Potentially time-consuming	
Probabilistic	Analysis of variance	ANOVA	Model independent approach	Time-consuming for a large number of inputs with interactions	SPSS, SAS, STATISTICA (and many others)
	Classification and regression tree	CART	Model independent approach; Not applicable to two-dimensional probabilistic models	Presentation of results depends on the specific software used	R (tree package), S-PLUS, SAS
	Regression analysis (sample and rank)	RC, SRC, SRRC	Strong assumptions (normality of residuals, linear relationship)	SRC, SRRC: computationally intensive	SimLab, MATLAB®, R (sensitivity package)
	Correlation analysis (sample and rank)	CC, PCC, RCC, PRCC	PCC, PRCC measure unique contribution of an input	CC, RCC: easy to compute; PCC, PRCC: computationally intensive	@Risk, Crystal Ball, MATLAB, R (sensitivity package) SAS
Graphical	Tornado plot				@Risk
	Scatter plot				@Risk, MATLAB

application of SA to QMRA models. Despite the great variety of methods, only two are widely used in QMRA studies: correlation analyses (statistical method) and scatter plots (graphical method) (Pérez-Rodríguez et al., 2007).

8.2.1 Mathematical Methods

Mathematical methods are suitable tools to perform SA for **deterministic models**; therefore, they are also called deterministic methods. These techniques involve evaluating the variability of the output with respect to a range of variation of a single input.

8.2.1.1 Nominal Range Sensitivity Analysis (NRSA)

NRSA (considered as local SA) investigates the effect of an input on the model output, in which the input changes, in a deterministic manner, across its entire range of plausible values, while all other inputs are held constant at their baseline (nominal) values (Cullen and Frey, 1999). This method implicitly assumes that the model is linear.

NRSA provides a numerical measure of sensitivity or in other words, a quantitative ranking of inputs. For that, a nominal point-estimate and a range of plausible values (minimum and maximum) must be previously identified for each selected input. Therefore, the sensitivity index (S) for a certain input (x) is calculated as follows:

$$S = \frac{Y_{x_{maximum}} - Y_{x_{minimum}}}{Y_{x_{nominal}}} \tag{8.1}$$

where Y_x is the value of the output variable obtained at the maximum, minimal or nominal values of the selected input x. A nominal point could be intended to represent the mean, median, minimum, maximum, or a particular percentile of a distribution for a given input.

Two specific cases of NRSA are differential sensitivity analysis (DSA) and the difference in the log-odds ratio (ΔLOR). Conceptually, the previous one and these two methods are equivalent, in the sense that a perturbation of an input is propagated forward through the model, and the corresponding change in the model output is observed.

The range of variation of the inputs for NRSA is typically larger than that for DSA. For NRSA, the choice of ranges of variation for each input might typically be based upon a 95% probability range (Frey et al., 2004). In DSA, the input variable is varied by a very small perturbation (e.g. ±1%

or 5%) of the nominal point-estimate. Thus, the sensitivity index may be calculated as

$$S = \frac{Y_{x_{nominal}+\Delta x} - Y_{x_{nominal}-\Delta x}}{Y_{x_{nominal}}} \qquad (8.2)$$

where Δx is the small deviation in the nominal point-estimate of the selected input x and $Y_{x_{nominal}}$ the same as in Equation 8.1. For large and complex models that involve numerical differentiation calculations (e.g. partial derivatives), automatic differentiation techniques, such as automatic differentiation of Fortran (ADIFOR), can greatly facilitate the implementation of derivative-based analyses (Helton and Davis, 2003).

The ΔLOR method is a useful measure of sensitivity when the model output is a probability (Menard, 2001). In this case, the change in the output variable is evaluated based upon minimum and maximum values of an input (holding all other inputs at their nominal values). For example, if the model output is the probability of illness (P), the ΔLOR measures the sensitivity to an input by calculating the log-ratio of the P occurring to the P not occurring $(1 - P)$:

$$S = log\left[\frac{P_{x_{maximum}}}{1 - P_{x_{maximum}}}\right] - log\left[\frac{P_{x_{minimum}}}{1 - P_{x_{minimum}}}\right] \qquad (8.3)$$

where P_x is the output variable at the maximum or minimum values of the selected input x. The value of ΔLOR can be positive or negative. In the exemplified case, if ΔLOR is positive, the changes in the input under study enhance the probability of illness, and if negative, the changes cause a reduction. The greater the magnitude of ΔLOR, the greater is the influence of the input.

The three methodologies described can be also viewed as screening techniques, since they are typically used for a first screening of the most influential model inputs. The major drawback of these methods is that they are based upon a linearity assumption and cannot account for non-linear interactions among inputs, making them prone to underestimating true model sensitivities (Frey and Patil, 2002).

8.2.1.2 Break-Even Analysis (BEA)
This method is aimed at identifying a set of values of the inputs (break-even points) that provide a model output for which decision-makers would be indifferent among the various risk management options (EFSA, 2018).

BEA is commonly used when the output is expressed as a dichotomous variable. For example, in risk assessment studies, BEA requires the output to be characterized as acceptable or unacceptable risk. The results of BEA can be graphically represented if only two input factors are varied (Patil and Frey, 2004).

BEA represents a useful conceptual tool for evaluating the impact of uncertainty on the preference of policy makers or decision-makers and for assessing the robustness of a decision to change in inputs. As an illustrative example (Figure 8.1), we have hypothesized a risk assessment model estimating the exposure to a biological hazard from consumption of a food product. In our example, the two most influential factors on the output (i.e. number of illness cases per serving in a certain population) are mean refrigerator temperature (°C) and initial contamination level (log colony forming units[CFU]/g), and the decision variable was whether a food safety objective (FSO) of 100 CFU/g at the time of consumption was exceeded or not. In practice, the BEA is based on uncertainty intervals (previously established) for the worst-case scenario of the two selected factors, while the other inputs are kept at their nominal values.

This example shows the potential of the BEA method to inform the decisions to be made by risk managers. As shown in Figure 8.1, it would

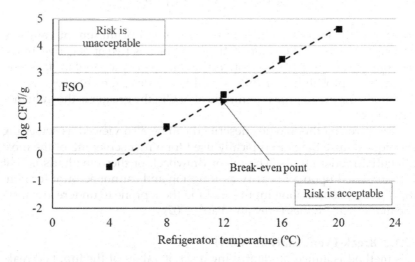

Figure 8.1 Graphical representation of BEA results from a hypothetical QMRA model. Square symbols represent the worst-case values of the input "initial contamination level" at a X nominal value (log CFU/g).

be possible to indicate to policy makers that a certain initial contamination level should be fixed to avoid exceeding the FSO for a worst-case refrigerator temperature of 12 °C. The biggest drawback of this method is that it cannot distinguish the relative importance of the sensitive inputs (Frey and Patil, 2002).

Patil and Frey (2004) demonstrated the application of the different mathematical methods described in this section with a probabilistic *Vibrio parahaemolyticus* risk assessment model as a case study. Since NRSA, ΔLOR and DSA require point-estimates of the output, distributions were replaced with point values. NRSA and ΔLOR identified the same two most sensitive inputs, which were then used to perform the BEA method. In this case, BEA was applied to plot an indifference or break-even line based on a combination of values of the two selected inputs to decide whether or not the risk was acceptable.

8.2.2 Statistical Methods

In the food safety domain, QMRA models are usually generated within a probabilistic framework, in which inputs are defined by probability distributions. For these cases, statistical methods are proposed to carry a sensitivity analysis. These techniques, which are also referred to as variance-based methods, entail running simulations using Monte Carlo or similar sampling techniques (e.g. Latin Hypercube sampling). The impact of the variance of inputs on the output distribution is then evaluated, while the effect of simultaneous interactions among multiple inputs is considered (Frey et al., 2004).

There is a myriad of statistical SA methods that could be applied in food risk assessment. The purpose of this section is not to carry out a comprehensive review but to illustrate those practices appraised to be relevant and applicable to QMRA studies. Typically, a food safety process risk model is characterized by nonlinearities, interactions between inputs, continuous and categorical inputs, and threshold and saturation points (Frey, 2002). Considering the structure of the model (linear, nonlinear), we have subcategorized some of the most relevant methods into two groups: (1) model independent methods and (2) model dependent methods. For the sake of clarity, the following section will use this terminology to refer the methods being independent (or dependent) from assumptions regarding the functional form of the relationships between inputs and outputs. Examples of methods belonging to the first group are ANOVA and classification and regression tree (CART). Such methods are likely to be more promising than others since an ideal SA method should be model independent. In addition, ANOVA and CART have been identified as powerful methods able to deal with the typical features of QMRA models (Mokhtari and Frey, 2005; Mokhtari et al., 2006a). Hence, attention is given to these techniques in Section 8.2.2.1. The second group of methods includes regression and correlation analyses. The applicability of these techniques to QMRA studies is briefly presented in Section 8.2.2.2.

8.2.2.1 Model Independent Methods
8.2.2.1.1 ANOVA
ANOVA is a parametric statistical method that can be used to investigate whether the output variation is significantly associated with the variation of one or more inputs. In ANOVA, inputs are referred to as "factors" and the output is referred to as "response variable". Factors are grouped into ranges of values of equal length, the so-called factor levels. In brief, ANOVA uses the F-test to determine whether a statistically significant difference exists among mean responses for main effects or interactions between factors (Archer et al., 1997). The higher the F value for a factor, the more sensitive the response variable is to that factor (Mokhtari and Frey, 2005).

One of the main advantages of ANOVA is that both categorical (e.g. food category, population group) and continuous inputs (e.g. distribution of storage temperature of the food product) can be addressed. A continuous factor can be treated as a categorical input by splitting the range into a number of mutually exclusive intervals and considering these intervals as factor levels (Busschaert et al., 2010). For example, a dynamic storage temperature profile ranging from 0 to 12 °C can be divided into three discrete levels: low (0–4 °C), mild (4–8 °C) and abuse (8–12 °C).

Previous QMRA studies have used ANOVA to assess the effect of one or multiple factors (and their interactions) on the response variable (i.e. probability of illness or infection) (Busschaert et al., 2010, 2011; Membré et al., 2008; Mokhtari and Frey, 2005). These works have demonstrated the robustness of the ANOVA method in models with a two-dimensional probabilistic framework, that is, models in which uncertainty about the risk estimate is characterized separately from variability. The pseudorandom samples from the Monte Carlo simulation are used as the data set for ANOVA (Busschaert et al., 2010). In the case of two-dimensional risk models, the second-order Monte Carlo decomposition is performed considering, separately, the variability associated with different uncertainty realizations (Mokhtari and Frey, 2005). Model simulations are usually carried out by applying sampling techniques implemented in MS Excel (Microsoft © Redmond, USA) and its add-ins or general statistical packages (see Chapters 7, 9 and 12). Examples of add-in software packages are Analytica (Lumina Decision Systems ©, Campbell, USA) and @Risk (Palisade ©, New York, USA).

ANOVA is often preferred over other variance-based techniques (e.g. Sobol or Fourier amplitude sensitivity test [FAST]) since it is conceptually easier to understand within a reasonable time for a non-expert (Membré et al., 2008). In addition, it is a statistical method, which scientists are very familiar with due to its widespread application for data analysis. The method is robust against the violation of certain assumptions, such as the normality test. In turn, it may be sensitive to measurement errors in the factors or the existence of correlations between them. In addition, the method can become computationally intensive if there is a large number of factors (Pérez-Rodríguez, 2006). A number of commercial software tools (e.g. SPSS, Statistica, Statgraphics and SAS) and programming environments (e.g. R project and MATLAB) allow the application of ANOVA.

8.2.2.1.2 *Classification and Regression Tree (CART)*

Tree-based models are statistical models that explain the variation in a single output by one or more input variables (De'Ath and Fabricius, 2000). If the output is categorical, the model is defined as a "classification tree", and if it is continuous, the model is referred to as a "regression tree". The combination of both types of model is the so-called CART. It can be defined as a non-parametric method, in which values of an output are divided into homogeneous groups based on classification rules that depend on cut-off values of selected inputs (Frey et al., 2004). The principles of this method are given by Breiman et al. (1984).

Trees are represented graphically and are composed of different split points or cut-off values, named nodes: the root node (located at the top of the tree and representing undivided data), intermediate nodes (successively splitting data) and terminal nodes or leaves (representing predicted mean values of the output). A tree is grown until a termination criterion at a node is reached (Pappenberger et al., 2006).

In the field of QMRA, CART is used to identify ranges of key inputs that are associated with high exposure, risk or both. The work developed by Mokhtari et al. (2006a) illustrates the application of this method to a QMRA model describing the contamination risk from farm to table by *Escherichia coli* O157:H7 in ground beef. They applied the CART method to specific modules in the exposure assessment, in particular to portions in the slaughter module and preparation modules. For instance, when CART was applied to the slaughter module, data were first divided into two datasets based upon chilling effect. In the root node, a cut-off value of 2.2 log-units was identified for splitting data, so the tree describes contamination scenarios in which the chilling effect is lower and greater than (or equal to) 2.2 logs.

CART provides faster identification of important inputs and critical limits from a risk management point of view. It also allows the identification of interactions between sensitive inputs without any assumptions with respect to the model structure, and it responds appropriately to the existence of thresholds in the model. However, CART is only applicable to one-dimensional risk models, and since it is not a standard analysis technique, the representation of results may depend on the specific software package used (Frey et al., 2004).

8.2.2.1.3 Other Potential Methods
There are many other statistical methods that are promising but not yet widely used in studies of QMRA. Sobol's method, FAST and the mutual information index (MII) are examples of methods that do not assume a specific relationship between model variables. They can also identify the main model input effects as well as interaction effects between them. An example of the application of FAST and MII to a QMRA model is given by Patil and Frey (2004). The first application of Sobol's method in the field of QMRA was performed by Mokhtari et al. (2006b) to a one-dimensional probabilistic risk model. More recent studies have estimated the Sobol sensitivity indexes in two-dimensional SA (Busschaert et al., 2011; Rigaux et al., 2014; Vásquez et al., 2014) using Saltelli's algorithm (Saltelli, 2002a; Saltelli et al., 2010) and computed with R sensitivity analysis packages. However, these methods often involve a certain computational complexity associated with the calculation of sensitivity indexes (high-dimensional integration).

8.2.2.2 Model Dependent Methods
The application of the statistical methods mentioned in this section is supported by the existence of a mathematical relation between output and inputs, which can be measured by different **regression indexes**, in many cases assuming a linear relationship. The regression coefficients (RCs) in the construction of a regression model can be used to assess the combined effects of multiple variables on the output. The magnitude of the RCs is commonly normalized by the ratio of the standard deviations of model

variables, known as standardized regression coefficients (SRCs). The coefficient of multiple determination (R^2) is an indicator of goodness of fit of a regression model. In the phase of setting up a model, it can be used to determine factors most influential on the dependent variables (EFSA, 2018). The main drawback of R^2 is that it is not a reliable index when there is a nonlinear relationship among variables (Frey and Patil, 2002).

An alternative to the regression analysis is to calculate **correlation coefficients** (CCs) or partial correlation coefficients (PCCs) (Helton and Davis, 2003). Calculation of CCs, such as the sample (or Pearson) correlation, is a simple but formal method to measure the strength of the linear relationship between an individual input and the output (Helton and Davis, 2002). When the regression model is constructed in a stepwise manner (i.e. stepwise regression analysis), PCCs can account for the unique linear relationship between inputs and outputs that cannot be explained by other variables already included in the regression model (Cullen and Frey, 1999). Although correlation analyses are not able to provide insight regarding possible interaction effects between inputs, they are broadly considered conventional SA methods and thus can be used as a benchmark for comparison with other SA methods (Mokhtari and Frey, 2005).

For nonlinear but monotonic models, **the rank regression approach** can be used to linearize the underlying relationship between model variables. With the rank transformation, both the input and output values are replaced by their ranks, and then, the above-mentioned regression and correlation indexes are applied to these ranks (Helton and Davis, 2002). In this way, rank correlation coefficients (RCCs), standardized rank regression coefficients (SRRCs) and partial rank correlation coefficients (PRCCs) are used instead of CCs, SRCs and PCCs, respectively.

Rank-based approaches are more compatible with the mathematical structure of QMRA models (i.e. highly nonlinear but monotonic) (Patil and Frey, 2004). Among these techniques, the rank (Spearman) correlation is commonly applied in QMRA studies (Pérez-Rodríguez et al., 2006; Vásquez et al., 2014; Zwietering and van Gerwen, 2000). It is typically used to rank model inputs based on their influence on output. The Spearman's correlation coefficient (ρ) determines the strength of the monotonic relationship between an input (X_i) and output (Y) by using rank-transformed data. It is defined as

$$\rho = \frac{\sum_{k=1}^{n}\left(x_i^k - \overline{x_i}\right)\left(y^k - \overline{y}\right)}{\sqrt{\sum_{k=1}^{n}\left(x_i^k - \overline{x_i}\right)^2 \cdot \sum_{k=1}^{n}\left(y^k - \overline{y}\right)^2}} \tag{8.4}$$

where

    n is the sample size (i.e. number of iterations in the simulation).

    x_i is the rank of the ith input Xi.

    y is the rank of the output Y.

    k stands for the kth iteration.

The value of the coefficient varies between -1 and 1. The higher the value of the coefficient, the more sensitive is the input. The correlation input/output may be influenced by strong correlations between input variables. As mentioned earlier, PCCs can be used as an alternative for accounting for relationships between other input variables (Hamby, 1994).

These methods are commonly used by practitioners because they are often included in commercial software packages, such as @Risk, of widespread use in risk assessment. Their main drawback is that the robustness of results largely depends on the functional form (underlying model) and the relationship among inputs and output (Pérez-Rodríguez et al., 2007).

8.2.3 Graphical Methods

Graphical methods are based on the visual analysis of the relationship between input and output variables in a QMRA model by using charts, graphs or plots (Frey and Patil, 2002; Patil and Frey, 2004). These methods are independent of the type of QMRA model and can be applied to both deterministic and probabilistic frameworks. The main application is to serve as an exploratory analysis of the simulated and modelled variables, identifying dependences between inputs and output in the model.

Scatter plot is a graphical method describing the relationship between input and output (Figure 8.2). This is carried out by plotting all the values simulated from one input versus the respective output values (Cullen and Frey, 1999). The use of scatter plots, in both deterministic and probabilistic models, can shed light on the dependences between variables that are hardly detected by other SA methods. Moreover, through the visual analysis of scatter plots, specific regions can be identified in inputs where the dependence is more intense or changes in one input result in a significant increase or decrease of the output (Vose, 2000). This SA technique is recommended as a preliminary approach

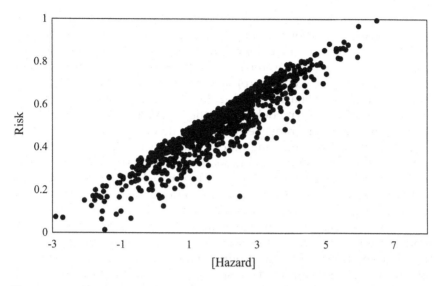

Figure 8.2 An example of a scatter plot representing the simulated pathogen concentration (input) versus the corresponding relative risk (output) and exhibiting a linear relationship (i.e. straight line) between variables.

upon which the most suitable SA technique can be selected for further analysis. As a drawback, scatter plot analysis requires that data, in probabilistic models, are retrieved and organized for graphical representation, which can be a laborious task if the simulation software does not provide this feature. Note that simulations can be performed with thousands of iterations, producing a huge amount of data to be handled, which is, in many cases, difficult to accomplish manually without software support.

Scenario analysis, sometimes called "what-if analysis" or conditional SA, is not based on calculating specific numerical indexes because of the type of outcome obtained by the method; instead, it is aimed at graphically representing the effect of changes in an input on the model output. This is performed by fixing the rest of inputs to specific values (e.g. mean) and modifying the value of the targeted input in the variable range. This technique, unlike NRSA, enables consideration and comparison of the effect of distinct point-estimate values for the fixed inputs; for example, considering, in addition to the mean, the maximum value or specific percentiles. The selected values from the input are represented versus output, generating a response curve, which can be visually analysed to derive conclusions on the relationship and dependences between variables. The most common approach in scenario analysis is to combine key input variables making reference to three possible cases: (a) worst-case or conservative scenario; (b) most likely or base scenario; (c) best-case or optimistic scenario (EFSA, 2018). Scenario analysis can be also applied in probabilistic models, fixing the targeted input variable at different values within its range (e.g. selecting them randomly or based on specific statistics) and allowing other inputs to vary during simulation. In that way, the variability and random effect of inputs are considered in the impact of one specific input on the model output. As stated for scatter plot, conditional SA can provide information about nonlinearity trends between variables, saturation points (i.e. asymptotic behaviour) and thresholds (i.e. shift points). As a drawback, this cannot always rank inputs based exclusively on the nature of the response curve. In contrast to NSRA, conditional SA can account for correlations between inputs or nonlinear interactions in model response, but it has limitations because of the combinatorial explosion of possible cases (Frey and Patil, 2002). Most published QMRA studies include what-if scenarios for the purpose of exploring risk mitigation strategies by identifying factors contributing to high exposure or risk.

A well-known method for graphically representing results from the correlation methods is the Tornado graph. In this graphical method, CCs are represented in bars having a length proportional to the coefficient value. Often, in the same graph, several inputs are represented and ranked from the largest to the smallest value, from the top down. The form produced by the bar distribution looks like a tornado; hence the name of the graph. An example of Tornado analysis is represented in Figure 8.3.

More specifically to QMRA, it is common to reflect graphically the effect of each step along the food chain on a specific exposure outcome (i.e. prevalence and concentration) by drawing lines or bars, which usually go up and down or do not change depending on whether the specific step increases or reduces the level of the microbiological hazard. This type of graphical representation can aid risk managers to evidence those steps that most contribute to a high exposure to the studied hazard or identify those that can have a relevant role in the reduction of the hazard (Zwietering and van Gerwen, 2000).

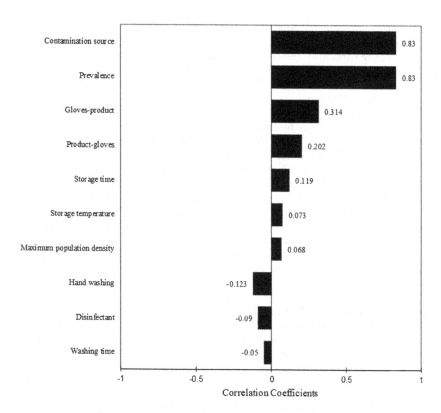

Figure 8.3 An example of a Tornado graph representing the correlation coefficients calculated for the inputs of a probabilistic exposure model estimating the final concentration of the product at the moment of consumption (output).

REFERENCES

Archer, G., Saltelli, A., Sobol', I.M. 1997. Sensitivity measures, ANOVA like techniques and the use of bootstrap. *Journal of Statistical Computation and Simulation*, 58(2), 99–120.

Augustin, J.-C. 2011. Global sensitivity analysis applied to food safety risk assessment. *Electronic Journal of Applied Statistical Analysis*, 4, 255–264.

Breiman, L., Friedman, J.H., Olshen, R.A., Stone, C.J. 1984. *Classification and Regression Trees*. Belmont, CA: Wadsworth.

Busschaert, P., Geeraerd, A.H., Uyttendaele, M., Van Impe, J.F. 2010. Sensitivity analysis of a two-dimensional risk assessment. *IFAC Proceedings Volumes*, 43(6), 323–328.

207

Busschaert, P., Geeraerd, A.H., Uyttendaele, M., Van Impe, J.F. 2011. Sensitivity analysis of a two-dimensional quantitative microbiological risk assessment: Keeping variability and uncertainty separated. *Risk Analysis* 31(8), 1295–1307.

Cullen, A.C., Frey, H.C. 1999. *Probabilistic Techniques in Exposure Assessment. A Handbook for Dealing with Variability and Uncertainty in Models and Inputs.* New York: Plenum.

De'Ath, G., Fabricius, K.E. 2000. Classification and regression trees: A powerful yet simple technique for ecological data analysis. *Ecology*, 81(11), 3178–3192.

EFSA (European Food Safety Authority). 2018. The principles and methods behind EFSA's guidance on uncertainty analysis in scientific assessment. *EFSA Journal*, 16(1), 5122. doi: 10.2903/j.efsa.2018.5122.

Ellouze, M., Gauchi, J.-P., Augustin, J.-C. 2010. Global sensitivity analysis applied to a contamination assessment model of *Listeria monocytogenes* in cold smoked salmon at consumption. *Risk Analysis*, 30(5), 841–852.

European Commission. 2003. Risk assessment of food borne bacterial pathogens: Quantitative methodology relevant for human exposure assessment (final report). Available at: http://ec.europa.eu/food/fs/sc/ssc/out308_en.pdf.

Frey, H.C. 2002. Introduction to special section on sensitivity analysis and summary of NCSU/USDA workshop on sensitivity analysis. *Risk Analysis*, 22(3), 539–545.

Frey, H.C., Mokhtari, A., Tanwir, D. 2003. Evaluation of selected sensitivity analysis methods based upon applications to two food safety process risk models. Prepared by North Carolina State University for office of risk assessment and cost-benefit analysis. U.S. Department of Agriculture, Washington, DC. Available at: www.ce.ncsu.edu/risk.

Frey, H.C., Mokhtari, A., Zheng, J. 2004. Recommended practice regarding selection, application, and interpretation of sensitivity analysis methods applied to food safety process risk models. Prepared by North Carolina State University for Office of Risk Assessment and Cost-Benefit Analysis. U.S. Department of Agriculture, Washington, DC. Available at: www.ce.ncsu.edu/risk.

Frey, H.C., Patil, R. 2002. Identification and review of sensitivity analysis methods. *Risk Analysis*, 22(3), 553–577.

Hamby, D.M. 1994. A review of techniques for parameter sensitivity analysis of environmental models. *Environmental Monitoring and Assessment*, 32(2), 135–154.

Helton, J.C., Davis, F.J. 2002. Illustration of sampling-based methods for uncertainty and sensitivity analysis. *Risk Analysis*, 22(3), 591–622.

Helton, J.C., Davis, F.J. 2003. Latin hypercube sampling and the propagation of uncertainty in analyses of complex systems. *Reliability Engineering & System Safety*, 81(1), 23–69.

Membré, J.-M., Kan-King-Yu, D., Blackburn, C.W. 2008. Use of sensitivity analysis to aid interpretation of a probabilistic *Bacillus cereus* spore lag time model applied to heat-treated chilled foods (REPFEDs). *International Journal of Food Microbiology*, 128, 23–33.

Menard, S. 2001. *Applied Logistic Regression Analysis. Sage University Papers Series on Quantitative Applications in the Social Sciences*, 7–106. Thousand Oaks, CA: Sage. ISBN: 0761922083.

Mokhtari, A., Frey, H.C. 2005. Sensitivity analysis of a two-dimensional probabilistic risk assessment model using analysis of variance. *Risk Analysis*, 25(6), 1511–1529.

Mokhtari, A., Frey, H.C., Jaykus, L.A. 2006a. Application of classification and regression trees for sensitivity analysis of the *Escherichia coli* O157:H7 food safety process risk model. *Journal of Food Protection*, 69(3), 609–618.

Mokhtari, A., Moore, C.M., Yang, H., Jaykus, L., Morales, R., Cates, S.C., Cowen, P. 2006b. Consumer-phase *Salmonella enterica* serovar Enteritidis risk assessment for egg-containing food products. *Risk Analysis*, 26(3), 753–768.

Pappenberger, F., Iorgulescu, I., Beven, K.J. 2006. Sensitivity analysis based on regional splits and regression trees (SARS-RT). *Environmental Modelling & Software*, 21(7), 976–990.

Patil, S.R., Frey, H.C. 2004. Comparison of sensitivity analysis methods based on applications to a food safety risk assessment model. *Risk Analysis*, 24(3), 573–585.

Pérez-Rodríguez, F. 2006. Evaluación cuantitativa del riesgo microbiano en productos cárnicos cocidos: Modelos de contaminación cruzada y su impacto sobre la gestión del riesgo. PhD Thesis, University of Cordoba.

Pérez-Rodríguez, F., Todd, E.C.D., Valero, A., Carrasco, E., García, R.M., Zurera, G. 2006. Linking quantitative exposure assessment and risk management using the food safety objective concept: An example with *Listeria monocytogenes* in different cross-contamination scenarios. *Journal of Food Protection*, 69(10), 2384–2394.

Pérez-Rodríguez, F., Van Asselt, E.D., García-Gimeno, R.M., Zurera, G., Zwietering, M.H. 2007. Extracting additional risk managers information from a risk assessment of *Listeria monocytogenes* in deli meats. *Journal of Food Protection*, 70(5), 1137–1152.

Pianosi, F., Beven, K., Freer, J., Hall, J.W., Rougier, J., Stephenson, D.B., Wagener, T. 2016. Sensitivity analysis of environmental models: A systematic review with practical workflow. *Environmental Modeling & Software*, 79, 214–232.

Ragsdale, C.T. 2008. *Spreadsheet Modeling and Decision Analysis: A Practical Introduction to Management Science*, 5th edition. Mason, OH: Thomson South-Western.

Rigaux, C., André, S., Albert, I., Carlin, F. 2014. Quantitative assessment of the risk of microbial spoilage in foods. Prediction of non-stability at 55°C caused by *Geobacillus stearothermophilus* in canned green beans. *International Journal of Food Microbiology*, 171, 119–128.

Saltelli, A. 2002a. Making best use of model evaluations to compute sensitivity indices. *Computer Physics Communications*, 145(2), 280–297.

Saltelli, A. 2002b. Sensitivity analysis for importance assessment. *Risk Analysis*, 22(3), 579–588.

Saltelli, A., Annoni, P. 2010. How to avoid a perfunctory sensitivity analysis. *Environmental Modelling & Software*, 25(12), 1508–1517.

Saltelli, A., Annoni, P., Azzini, I., Campolongo, F., Ratto, M., Tarantola, S. 2010. Variance based sensitivity analysis of model output. Design and estimator for the total sensitivity index. *Computer Physics Communications*, 181(2), 259–270.

Saltelli, A., Chan, K., Scott, E.M. 2000. *Sensitivity Analysis. Probability and Statistics Series*. New York: John Wiley & Sons. ISBN: 9780470743829.

Saltelli, A., Ratto, M., Andres, T., Campolongo, F., Cariboni, J., Gatelli, D., Saisana, M., Tarantola, S. 2008. *Global Sensitivity Analysis: The Primer*. Chichester: John Wiley and Sons Ltd.

Vásquez, G.A., Busschaert, P., Haberbeck, L.U., Uyttendaele, M., Geeraerd, A.H. 2014. An educationally inspired illustration of two-dimensional Quantitative Microbiological Risk Assessment (QMRA) and sensitivity analysis. *International Journal of Food Microbiology*, 190, 31–43.

Vose, D. 2000. *Risk Analysis: A Quantitative Guide*. New York: John Wiley & Sons, Inc., pp. 326.

Zwietering, M.H., Van Gerwen, S.J.C. 2000. Sensitivity analysis in quantitative microbial risk assessment. *International Journal of Food Microbiology*, 58(3), 213–221.

Section II

Microbial Risk Assessment

9

Quantitative Methods for Microbial Risk Assessment in Foods

Winy Messens, Marios Georgiadis, Caroline Merten, Kostas Koutsoumanis, Matthias Filter, Carolina Plaza-Rodriguez, and Fernando Pérez-Rodríguez

Contents

9.1 INTRODUCTION

According to the Codex Alimentarius Commission, risk assessment is a scientifically based process consisting of four steps: (i) hazard identification, (ii) hazard characterization, (iii) exposure assessment, and (iv) risk characterization (CAC 1999). As opposed to qualitative microbial risk assessment (MRA), which provides the risk outcome as descriptive categories, quantitative microbial risk assessment (QMRA) provides numerical expressions to quantitatively assess the adverse health effects resulting from exposure to pathogens. QMRA provides a quantitative basis to support decision-making that aims to reduce food safety risks (FAO/WHO 2006) and plays an important role within international food safety policies (Havelaar, Nauta, and Jansen 2004).

Unlike quantitative chemical risk assessment, the methodology used for QMRA must consider that the microbial counts can change (increase or decrease) during the consecutive phases of the food chain. In addition, the characteristics of the various microorganisms addressed and the differences between the (susceptible) population groups make the implementation of QMRA a challenging undertaking (Voysey and Brown 2000, FAO/WHO 2006).

The elements and approaches needed to perform a QMRA span from microbial counts obtained on foods by enumeration methods to predictive tools applying mathematical models to understand the relationship between the occurrence of pathogens in a food matrix (both prevalence and concentration) and their public health risk. The development of QMRA models is the most advanced in terms of resource requirements and complexity, whether using a deterministic approach (in which the variables are represented by single-point estimates) or a stochastic approach (in which probability distributions are used to describe variables). The latter is generally considered to be the most suitable, as it provides confidence intervals (CIs) of the risk assessment output and therefore, in general terms, gives more realistic estimates. However, it is often complex and difficult to generate (FAO/WHO 2006). This complexity is reflected in the number and diversity of risk assessment software tools currently available. It should be clear that the outcome of such an exercise is dependent on the inputs and functional forms of the models used, which often originate from assumptions that need to be carefully considered and justified.

The generation process of QMRA models is founded on background knowledge (Haberbeck et al. 2018), which consists of data and models that can be retrieved from different sources. Sources include scientific studies (published and/or unpublished), monitoring and surveillance data, laboratory diagnostic data and data from disease outbreak investigations, data from food consumption surveys, national and international risk assessments, and expert opinion (FAO/WHO 2006, Haberbeck et al. 2018). New types of information can be incorporated into QMRA modelling. For example, the omics-based mechanistic inputs may fill key knowledge gaps by providing full genome coverage as well as providing new perspectives on strain diversity and physiological uncertainty (Brul et al. 2012). The emerging field of metagenomics offers the potential to fully characterize the genome content of microorganisms in a sample and therefore is likely to have a great influence on how QMRA will be carried out in the future (Brul et al. 2012).

Importantly, the characterization of uncertainty and variability can increase the understanding of the outputs of the risk assessment, informing decision-makers about the reliability of the obtained results and guiding them in the process of decision-making (Nauta 2000). The degree of credibility of the output of a QMRA model depends largely on the quality and quantity of the data used as well as the appropriateness of the model structure and assumptions made. Careful consideration of the entire modelling process, including the validity of the assumptions and related uncertainties, is, therefore, very important.

Transparency and consistency should play a major role throughout the QMRA process, which should be fully documented and systematically described, including a representation of the strengths and limitations of the model (data quality, assumptions, model structure, variability and uncertainty, and other important attributes of the assessment). This will contribute to the application and re-use of the available knowledge (data and models). Several additional resources have been proposed to enhance transparency and to facilitate knowledge exchange (and therefore the entire QMRA process), such as the establishment of harmonized data formats, the development of consistent rules for knowledge annotation, and the creation of open access food safety knowledge repositories (Plaza-Rodríguez et al. 2018, Haberbeck et al. 2018).

9.2 QUANTITATIVE RESOURCES FOR RISK ASSESSMENT

As mentioned before, QMRA consists of four steps. The hazard identification step, the food matrix, and the consumers considered (population groups) determine the risk assessment question to be addressed by the QMRA. The hazard characterization step considers the pathogenicity of the microorganism and usually includes a dose/response (DR) relationship model. The exposure assessment step estimates the dose ingested and usually considers the prevalence and concentration of the pathogens in the food at the moment of consumption, the consumption frequency, and the serving size. Predictive microbiology can play an important role in the exposure assessment to describe the growth or death of pathogens based on the properties of the food matrix and the environment. Finally, the risk characterization is a function of the hazard characterization and the exposure assessment. The outcome of the QMRA may be expressed in different ways, ranging from the number of human cases to the economic impact or the burden of the disease in a specific population and during a specific period. The execution of each of these four steps is a complex process and is resource demanding. The use of diverse background knowledge, data, and mathematical models from various sources is essential. For example, QMRA models require a lot of, ideally good-quality, data in order to determine or model the values of their input variables. In the following subsections, the main data sources for QMRA will be discussed, using as an example the EU data framework to illustrate how data can be integrated into a quantitative approach. Data sources relevant to the hazard identification step are presented in Section 9.2.1 (data on human infections or diseases). Data for the hazard characterization principally concern DR modelling and are discussed in Section 9.2.4. Data and information for the different components of exposure assessment are addressed in Sections 9.2.2 (information on pathogen occurrence in animals, food and feed, and growth of microorganisms) and 9.2.3 (food consumption data).

9.2.1 Human Data

9.2.1.1 Data on Human Cases

Within the framework of the Zoonoses Directive 2003/99/EC,[1] the EU Member States (MS) are required to collect data on the occurrence of zoonoses, zoonotic agents, antimicrobial resistance (AMR) and foodborne outbreaks. The European Food Safety Authority (EFSA), in collaboration with the European Centre for Disease Prevention and Control (ECDC), responsible for the surveillance of communicable diseases in humans, coordinates the collation and analysis of these data to publish findings in annual EU Summary Reports (EUSR) (example: EFSA and ECDC, 2018). Since 2019, these have been referred to as EU One Health Zoonoses Reports (see EFSA and ECDC, 2019).

The data collection on human diseases is conducted in accordance with Decision 1082/2013/EU[2] on serious cross-border threats to health. The case definitions are described in Decision 2012/506/EU.[3] Since 2008, data on human cases have been received via the European Surveillance System (TESSy), maintained by ECDC. The human data are published annually in the EU One Health Zoonoses Reports and are available in the interactive Surveillance Atlas.[4] In addition, annual epidemiological reports are available on the ECDC website.[5] A few European Economic Area (EEA) countries also submit their data.

Boelaert et al. (2016) distinguished three main categories of monitoring data. The first category includes data that are comparable across MS because the monitoring schemes are harmonized and the data originate from a census sampling. These data may be used for trend analysis at the EU and MS level and to assess the indirect impact of control and eradication programmes. The second category consists of less harmonized data that are not comparable across MS. A third category comprises data that are not comparable between MS as these are generated by non-harmonized monitoring schemes across MS and for

217

which no mandatory reporting requirements exist. Therefore, according to Boelaert et al. (2016), the reported findings must be interpreted with extreme caution.

Data on human cases can be used in the hazard identification and characterization step of an MRA. For example, in the Scientific Opinion of the BIOHAZ Panel on *Listeria monocytogenes* contamination of ready-to-eat (RTE) foods and the risk for human health in the EU (EFSA BIOHAZ Panel, 2018a), human data were used to describe the epidemiology of human listeriosis in the EU/EEA and pregnancy-associated human listeriosis cases (hazard identification) as well as the clinical picture of reported human listeriosis cases (hazard characterization). The surveillance of human listeriosis is focusing on severe invasive forms of *L. monocytogenes* infection, mostly manifested as septicaemia, meningitis or spontaneous abortion. In this Scientific Opinion, human data were also used to analyze the trend of human invasive listeriosis notification rates in the EU/EEA using time series analysis (TSA) considering confirmed cases in the total population as well as in defined age–gender groups.

9.2.1.2 Data on Foodborne Outbreaks

The annual EU One Health Zoonoses Reports present the sources of foodborne disease outbreaks caused in humans by consuming contaminated food. These data represent the most comprehensive dataset available for assessing the burden of foodborne outbreaks in the EU/EEA and the contributing risk factors. According to Boelaert et al. (2016), these outbreak data belong to the abovementioned second category of data. Although harmonized specifications on the reporting of foodborne outbreaks at the EU level have been applied since 2007, due to the lack of mandatory harmonization of the national foodborne outbreak investigation systems, differences in the number and type of reported outbreaks are to be interpreted with caution as these may not necessarily reflect differences in the level of food safety among MS; rather, they may indicate varying sensitivity of the national surveillance systems. Foodborne outbreaks are distinguished as 'weak-evidence' or 'strong-evidence' outbreaks based on the strength of evidence implicating a particular food vehicle. In the case of 'strong-evidence' outbreaks, more detailed information is collected, including the food vehicle and its origin, the nature of evidence linking the outbreak cases to the food vehicle, the extent of the outbreak, the place of exposure, the place of origin of the problem and contributory factors. The technical specifications for reporting foodborne outbreaks in the EU can be found in EFSA (2014a).

Foodborne outbreak data can be used in the hazard identification step of an MRA to identify the food categories that caused, for example, most outbreaks, human cases or hospitalized cases. In the abovementioned Scientific Opinion, data on 'strong-evidence' foodborne outbreaks caused by *Listeria* from 2008 to 2015 were used to conclude that 37 strong-evidence foodborne outbreaks caused by *L. monocytogenes* were reported in the EU/EEA with 525 human cases, 182 hospitalizations, and 37 deaths during that time period. The outbreaks were usually small. The 'meat and meat products' food category was responsible for 11 outbreaks causing 126 cases. 'Fish and seafood' and 'dairy' food categories were responsible for seven and four outbreaks, causing 40 and 44 cases, respectively. These three food categories caused 59% of the strong-evidence foodborne outbreaks and 40% of the human cases, pointing to these categories as the highest-risk products in relation to human listeriosis in the EU population.

9.2.2 Animal/Food Data

9.2.2.1 EFSA Monitoring Data

According to Annex I of the Directive 2003/99/EC,[1] monitoring shall cover the following eight zoonoses and zoonotic agents: brucellosis and agents thereof, campylobacteriosis and agents thereof, echinococcosis and agents thereof, listeriosis and agents thereof, salmonellosis and agents thereof, trichinellosis and agents thereof, tuberculosis due to *Mycobacterium bovis*, and Shiga toxin-producing *Escherichia coli* (STEC). Article 4 of Chapter 2 of the same directive prescribes that monitoring shall take place at the stage(s) of the food chain most appropriate to the zoonosis or zoonotic agent concerned, that is, at the level of primary production and/or at other stages of the food chain, including in food and feed. For food, monitoring schemes for *Salmonella*, *L. monocytogenes* and STEC are carried out, underpinned by Commission Regulation (EC) 2073/2005[6] on microbiological criteria (EC, 2005). Specific rules for the coordinated monitoring programmes and for the food business operators (FBO) are laid down in Articles 5 and 6, respectively, of Chapter 2 of the Zoonoses Directive, while Article 8 of Chapter IV details rules for epidemiological investigation of foodborne outbreaks. The reporting requirements are described in Annex IV of the same directive.

According to the Directive 2003/99/EC,[1] 'MS must ensure that the monitoring system provides relevant information at least with regard to a representative number of isolates of *Salmonella* spp., *Campylobacter jejuni* and *Campylobacter coli* from cattle, pigs and poultry and food of animal origin derived from those species'. Commission Implementing Decision 2013/652/EU[7] sets up priorities for the AMR monitoring from a public health perspective, lists the collection of representative isolates for monitoring AMR from specific animal populations and food categories, and lays down detailed requirements on the harmonized monitoring and reporting of AMR (see for example EFSA and ECDC, 2019).

The *L. monocytogenes* MS food monitoring data originate from their reporting obligations under Commission Regulation (EC) No 2073/2005.[6] According to Boelaert et al. (2016), these data belong to the abovementioned second data category. In the EU One Health Zoonoses Reports, the results of *L. monocytogenes* testing in RTE food samples are evaluated with reference to the *L. monocytogenes* Food Safety Criteria for the different RTE food categories across the food chain, applying certain assumptions where appropriate. More information can be found in EFSA and ECDC (2016) and EFSA BIOHAZ Panel (2018a).

These monitoring data can be used in the exposure assessment step of an MRA to inform about the occurrence of a pathogen in food animals or food. In the abovementioned Scientific Opinion, it was concluded that during the time period 2008–2015, non-compliance at processing ranged from 3.5% to 9.6% for 'RTE fishery products', from 0.9% to 6.8% for 'RTE products of meat origin other than fermented sausage', and from 0% to 0.6% for 'RTE products of meat origin, fermented sausage'. Non-compliance ranged from 0.2% to 1.8% for 'soft and semi-soft cheese' and from 0 to 0.3% for 'hard cheese'. At retail, non-compliance was generally lower and for most of the years was below 1%. The lower level of non-compliance at retail was, at least partly, explained by the application of different limits for retail and processing.

9.2.2.2 EU-Wide Baseline Survey Data

Estimates of the occurrence (prevalence and concentration) of a pathogen in specific food items should refer to a defined population of food items of interest and ideally, be obtained by representative sampling of items from this population. Surveys aiming to obtain such representative samples from well-defined populations of interest need to be carefully designed and executed. EU-wide baseline surveys (BLS) have been carried out on the occurrence of zoonotic agents in food and animal populations, at primary production and at retail level. These surveys provide representative and comparable data across MS and a few EEA countries.

An example of such a survey is the EU-wide BLS on *L. monocyto-genes*, which was conducted in 2010 and 2011, in RTE food items from three categories: packaged (not frozen) smoked or gravad fish, packaged heat-treated meat products, and soft or semi-soft cheeses, excluding fresh cheeses. More information about the survey (including specification of food items included in each of these categories) can be found in Commission Decision 2010/678/EU.[8] Results were obtained from 3053 fish samples tested at the time of sampling and end of shelf-life (using paired samples from the same batch) as well as 3530 meat product and 3452 cheese samples tested at the end of shelf-life.

EFSA published the results of this survey in two Scientific Reports (EFSA 2013 and 2014b). The Part A Report (EFSA, 2013) describes the EU-wide *L. monocytogenes* prevalence and concentration (colony forming units [CFU]/g) for the three RTE food categories. Beyond the presentation of sample descriptions and summary statistics, statistical analysis focused on the prevalence of contaminated samples and the proportion of samples with *L. monocytogenes* counts greater than 100 CFU/gram.

These EU-wide BLS data can be used in the exposure assessment step of an MRA to inform about the occurrence of *L. monocytogenes* in food at retail. They have been used for the exposure assessment in Pérez-Rodríguez et al. (2017) and EFSA BIOHAZ Panel (2018a). The CIs estimated for the prevalence and concentration were employed to represent the initial contamination levels (including the related uncertainty) used to simulate, in a two-dimensional model, the *L. monocytogenes* growth from retail to consumption.

9.2.2.3 EU Rapid Alert System for Food and Feed Data

Commission Regulation (EU) No 16/2011[9] lays down the implementing measures for the requirements of Regulation (EC) No 178/2002[10] around the Rapid Alert System for Food and Feed (RASFF). This is established as a system facilitating the notification of food/feed safety alerts among the competent authorities of MS. RASFF might typically be used for notification of food batches where sampling and analysis as a result of companies' own checks, border control, official control on the market, etc. has detected non-conformance with regard to a criterion or where food batches have been implicated in illnesses. The RASFF system is primarily a communication facility enabling many food safety risks to be averted before they can cause further harm to European consumers. It is not an epidemiological surveillance system but provides some understanding of the types of hazards typically detected in particular foods.

Considering the abovementioned limitations, RASFF data may be employed, in the hazard identification step of an MRA, to identify the foods where a pathogen has been recovered during the period or to extract information for the exposure assessment on the concentrations of a pathogen in a particular food. In the abovementioned Scientific Opinion (EFSA BIOHAZ Panel, 2018a), a search on the RASFF database[11] was conducted, on 13 December 2016, from 2008 onwards, using as product category 'food' and hazard 'L. monocytogenes'. In total, 91% of the 760 notifications with *L. monocytogenes* were related to food considered to be RTE. The following RASFF food product categories were most commonly notified: 'fish and fish products' (N = 282), 'milk and milk products' (N = 186), and 'meat and meat products other than poultry' (N = 112). Together, these categories accounted for 87% of 690 notifications related to RTE foods. It was stressed that RASFF notifications are not based on fully harmonized notification criteria and are not representative of the *L. monocytogenes* occurrence in specific food products or the distribution of foodborne outbreaks associated with *L. monocytogenes* in specific foods. The average *L. monocytogenes* concentration found among RASFF notifications related to RTE foods was 2.61, 2.46 and 2.34 $\log_{10}$ CFU/g for the categories 'milk and milk products', 'fish and fish products' and 'meat and meat products other than poultry', respectively. The respective highest maximum concentrations were 6.25, 5.32, and 4.75 $\log_{10}$ CFU/g. The concentration was >2 $\log_{10}$ CFU/g in approximately 80% ('fish and fish products') and 65% ('milk and milk products' 'and 'meat and meat products other than poultry') of notifications among RTE foods.

9.2.2.4 Other Resources, e.g. Data/Models from Scientific Literature
Other data sources (such as data from scientific literature) than the abovementioned databases may need to be consulted for additional animal-food data. For example, in the abovementioned Scientific Opinion, information derived from the extensive literature search (ELS) by Jofré et al. (2016) on the occurrence and levels of contamination of *L. monocytogenes* in RTE foods and risk factors for *L. monocytogenes* contamination of various RTE foods was used.

In addition, predictive microbial models can be incorporated into a QMRA model within the exposure assessment step, helping to determine microbial dynamics and elucidate prevalence and concentration at each step along the food chain. In this sense, predictive microbial models can be used to describe the microbial responses to environmental conditions (e.g. temperature) and product characteristics (e.g. pH, water activity (a_w))

and organic acids concentration) during the different stages of the food production chain, such as storage and processing.

9.2.3 Consumption Data

Food consumption data are used at the exposure assessment step. The quality of such data is important in order to obtain realistic and accurate estimates of the amount (frequency and serving size) of food consumed so as to estimate the human exposure to pathogens. These data are often unavailable for specific subpopulations (which may be defined by age, gender, underlying conditions, etc.) in specific geographic locations. Therefore, data from different subpopulations may need to be used, considering the assumptions behind the applicability of such data. Uncertainty also needs to be explicitly considered in these cases.

Additionally, data may not exist for specific kinds of foods (but may exist for more aggregated categories) or may not be available at the desired granularity and/or time periods. In such cases, judgements must be made about the applicability of the available data. This is true also when available data may be used as proxy data for other variables on which information is not available. For example, when no food consumption data are available for a specific food or food category, import/export and production data associated with the food chain under study can be used to approximate consumption patterns (i.e. apparent consumption; see Sections 9.2.3.2 and 9.3.5). Sometimes, such data cannot be considered quantitatively but can be used to make qualitative assessments or deduce possible trends.

9.2.3.1 EFSA Consumption Data

The EFSA Comprehensive European Food Consumption Database[12] contains data on food consumption patterns across the EU. It provides detailed information on food consumption for a number of European countries for defined food categories and specific population groups. EFSA currently uses the food classification system FoodEx2 'Exposure hierarchy' with seven levels to categorize foods and beverages included in the database. Summary statistics from the database enable screening for chronic and acute exposure to substances and organisms that may be found in the food chain. In the database, dietary surveys and food consumption data for each country are divided by category. Categories include age, from infants to adults aged 75 years or older, and food group (>2500) and type of consumption, covering both regular and high consumption, thus allowing calculations to be tailored to each category of consumer. Statistics on chronic and acute food consumption are available for the total population ('all subjects' and 'all days') or consumers only, and in grams per day (g/day) or grams per day per kilogram of body-weight (g/kg bodyweight per day).

Both in the abovementioned Scientific Opinion and in Pérez-Rodríguez et al. (2017), the exposure assessment made use of consumption data for three RTE food categories sampled in the EU-wide BLS, extracted from the EFSA database. In this case, FoodEx1 categories ('exposure hierarchy' with four levels) were used to identify eating occasions for soft and semi-soft cheese, cooked meat products, sausage and pâté. As FoodEx1 was not detailed enough, the original national food descriptors were used to identify eating occasions for smoked and gravad fish. Information related to the sur-veys included country; survey; survey starting and end date; total number of subjects; and total number of days for which consumption events were reported. Summary statistics were reported for the following population strata and food groups: age class (i.e. 1–4, 5–14, 15–24, 25–44, 45–64, 65–74, and ≥75 years old), gender and food group (i.e. smoked fish, gravad fish, cooked meat products, heat-treated sausages, pâté, and soft and semi-soft cheese). Food consumption summary statistics were extracted to derive the consumption frequency and serving size: total number of eating occasions; total amount (grams) consumed in all eating occasions; mean number of eat-ing occasions per day in all days; mean, medium, 25th percentile and 75th percentile for the number of eating occasions per day in consuming days only; and mean, medium, 25th percentile and 75th percentile for the amount (grams) per eating occasion in consuming days only. These consumption data were then used to assess the exposure of these 14 population strata to *L. monocytogenes* by consumption of the foods belonging to the six food groups.

9.2.3.2 Other Consumption Data

Other international organizations may also publish data that can be used directly or as a proxy in order to deduce food consumption data. The Food and Agriculture Organization/World Health Organization (FAO/WHO) Global Individual Food consumption data Tool (FAO/WHO GIFT)[13] makes publicly available food-based indicators, derived from gender and age disaggregated data on individual food consumption, in order to strengthen nutrition information systems. Food consumption data disag-gregated by gender and age can help in assessing the dietary exposure to food safety hazards, identifying the main food sources of these hazards and carrying out risk assessment.

The FAO also publishes food and agriculture data on production, trade, food supply, etc. In EFSA BIOHAZ Panel (2018a) and Pérez-Rodríguez et al. (2017), a 'rough' calculation of smoked salmon consumption in the EU was done using information from the workspace FAO Fishery and Aquaculture Statistics via the application FishstatJ (FAO, 2016). Several assumptions were made, and it was stressed that the results could only be considered a proxy for smoked salmon consumption; attention was drawn to the uncertainty that would characterize any conclusions (EFSA BIOHAZ Panel, 2018a). As indicated before, a judgement must be made each time on whether and how the available data can be used in QMRA. In any case, the uncertainty associated with the specific assessment needs to be addressed and may sometimes be quite large.

9.2.3.3 Consumer Behaviour

Consumer behaviour, including handling and storage of foods at the domestic level, can have an important impact on risk. For example, exposure may increase due to cross-contamination among unpackaged foods in the refrigerator or as a result of actions that may allow (increased) growth, i.e. improper storage temperatures and times. In addition, since risk may need to be considered for different subgroups of the population, the consumer differences in food handling among population groups can be potential drivers for changes in the occurrence of a pathogen.

9.2.4 Dose–Response Data

A hazard characterization or DR model quantifies the likelihood of a response (for example, infection, illness or death) given a certain dose of a pathogen in a food. This likelihood is dependent on the interaction of host, pathogen and food matrix issues, often referred to as the infectious disease triangle (Buchanan et al., 2000). An accurate DR model should reflect these interactions. However, this is a challenge for researchers and modellers because of the high complexity of these interactions and the impossibility of undertaking voluntary human feeding of pathogens. In general, the response is infection or illness, but occasionally it is death or life quality metrics (see Section 9.3.5 and Chapter 3). Since there are important limitations to developing experiments in animal models and specifically human volunteer studies, epidemiological data (i.e. outbreak and sporadic cases, e.g. taken from annual EU One Health Zoonoses Reports) become crucial for developing DR models or refining and adjusting models to represent specific populations or susceptibilities. The DR model described in Pouillot et al. (2015) is based on a log-normal exponential model that derives the probability of illness given an expected dose (number of *L. monocytogenes* CFU per serving) and the *r* parameter, which represents the probability that one single CFU will cause illness (by surviving the different barriers and multiplying in a favourable site of infection). This depends on the characteristics of the host and the strain of *L. monocytogenes*. The model was built on an existing exponential DR model and refined on the basis of French epidemiological data, survey data and outbreak investigation data associated with an outbreak of listeriosis linked to consumption of ice-cream. This enabled the model to estimate risk for specific susceptible populations and account for the outbreak outcomes. For further detail about how data can be used to derive DR models, readers are referred to Chapter 14.

9.3 RISK MODELLING PROCESS AND MODEL INTEGRATION

9.3.1 From Data to Risk: Data Treatment

As mentioned before, QMRA requires a large amount of diverse data. Modelling also requires the use of specific mathematical equations and assumptions, which need to be substantiated in each case. Data may originate from diverse sources and often come from the scientific literature. Employing a systematic review methodology can increase the transparency and credibility of the literature review (EFSA, 2010). According to EFSA (2010), 'A systematic review is an overview of existing evidence pertinent to a clearly formulated question, which uses pre-specified and standardised methods to identify and critically appraise relevant research, and to collect, report and analyse data from the studies that are included in the review.' However, it needs to be noted that not all questions are suitable to be answered using a systematic review, while sometimes, a systematic review may not be worthwhile or feasible (EFSA, 2010). Higgins and Green (2011) is an important reference for the methodology and practice of systematic reviews. Some specific guidance for the use of systematic reviews to obtain information when conducting risk modelling is given in EFSA (2010).

A systematic review is often conducted following a protocol (a plan that details the methods and approaches), preferably agreed upon before the review begins. It details the methods and approaches in such a way that these can be agreed upon beforehand and optimized. This is usually an iterative process, in which methodologists and subject matter experts should be involved. It allows standardization of approaches and decisions by all involved, and it introduces transparency in the various judgements, allowing the users of the review to assess it and, should they wish, to reproduce or update it (EFSA, 2010). EFSA (2010) lists eight steps that need to be followed and documented when conducting a systematic review: 1. preparing the review, 2. searching for research studies, 3. selecting studies for inclusion or exclusion in the review, 4. collecting data from the included studies and creating evidence tables, 5. assessing methodological quality of included studies, 6. synthesizing data from included studies – meta-analysis, 7. presenting data and results, and 8. interpreting results and drawing conclusions.

It should be noted that the studies that are included in the systematic review (i.e. those that have remained in the process after step 3) need to be critically appraised with respect to their validity, both internal and external (Higgins and Green 2011). Assessment of 'risk of bias' can be facilitated using critical appraisal tools (CATs). For more details on assessment of risk of bias, the reader is referred to Higgins and Green (2011).

The synthesis of the evidence from the individual studies in a systematic review can be done either in a descriptive (narrative) way or by using statistical techniques (meta-analysis) (EFSA, 2010; OHAT/NTP, 2015). When a meta-analysis is feasible and appropriate (for some criteria, the reader is referred to Higgins and Green 2011 and to OHAT/NTP, 2015), it can provide several advantages compared with using the individual-studies estimates: it yields an analysis with increased power and precision, supports the assessment of uncertainty and allows the investigation of differences in findings between studies (Higgins and Green 2011; EFSA, 2010). It may also enable addressing issues not addressed in individual studies (Higgins and Green 2011). On the other hand, increased attention is required concerning the use and correct application of appropriate statistical methods and the potential of existing bias in the studies to be considered for the meta-analysis (EFSA, 2010).

Systematic reviews often are resource-intensive, sometimes requiring an important time commitment. Various tasks of systematic reviews can be automated using informatic systems (Tsafnat et al., 2014). Jaspers et al. (2018) investigated the use of machine learning techniques (MLTs) for the automation of several steps of the systematic reviews conducted at EFSA using various methodologies. They developed an application that can be used to automate step 3 of the review, specifically the screening

227

for relevance of abstract and full-text of the publication using several methodologies and evaluating their performance. The tool is available at https://shiny-efsa.openanalytics.eu/.

Information for risk assessment could also originate from other sources. The use of epidemiological information can play an important role in risk assessment (FAO/WHO, 2006). Descriptive epidemiological studies can provide information on the distribution of the outcomes of interest (which could include pathogens, diseases or other characteristics) in specifically defined populations of interest. The *L. monocytogenes* EU-wide BLS mentioned in Section 9.2.2.2 is an example of such a study. When using the results of such surveys, it is important to consider the target populations of interest and whether a representative sample has been obtained from those populations. Observational epidemiological studies can provide information on potential risk factors for defined outcomes of interest and therefore, indicate variables that should also be considered in the modelling.

Another source of information is the risk profiles for specific combinations of hazards and food items. These are intended to describe a microbial food safety problem and its context by gathering existing information on hazards, exposure levels, and other relevant data in order to provide a general picture of the problem and identify potential risk management options, if any. As mentioned by Codex Alimentarius, the risk profile could give enough information to support specific management actions (WHO/FAO, 2008). Several examples of risk profiles can be found in the EFSA context: the Scientific Opinions on the public health hazards to be covered by inspection of meat (bovine animals) (EFSA BIOHAZ Panel, 2013), the application of control options and performance objectives and/or targets for *Campylobacter* at different stages of the food chain of broiler meat production (EFSA BIOHAZ Panel, 2011) and public health risks associated with food-borne parasites (EFSA BIOHAZ Panel, 2018b). Risk profiles are also used as a 'screening' tool, for example to identify hazards, describe main exposure pathways or define risk outcomes (EFSA, 2015). Risk profiles can be a rich source of information for a quantitative risk assessment, which usually requires much more data and more exhaustive data processing and analysis. The type of information available will depend on the specific focus of the risk profile, which can be more oriented to exposure levels, predictive models, hazard descriptions, data gap identifications, etc.

9.3.2 Deterministic Models vs. Stochastic Models

As described in the Introduction, risk assessment models can be constructed as deterministic or stochastic. In the deterministic approach, all input variables are assumed to take one specific value, while in the stochastic approach, a range of values with associated plausibility (in the form of a statistical probability distribution) are used as input.

The deterministic approach is much simpler than the stochastic one, since it does not require the definition of a probability distribution that summarizes the uncertainty and/or variability associated with the specific variable. Moreover, it is easier computationally, since it does not require the use of sampling techniques (like Monte Carlo sampling) and often can provide results as simple solutions to closed form equations. The single-point value used can be a 'best guess' based on previous experience or expert opinion or can be a measure of central tendency (or some pre-defined percentile) of a relative frequency distribution of the variable of interest derived from available data. Sometimes, even if probability distributions for the modelled variables are available, these are not used as distributions, but 'point-estimates' are derived from them for simplicity in the modelling process.

The most commonly used measures of central tendency are the mean and the median; however, on some occasions, the mode may be used. When pre-existing data are used, the mean may not be a good choice when there are extreme values in the dataset, in which case the median might be a better value to use. The mode, being the value with the highest plausibility, may also be a reasonable choice on some occasions. It needs to be noted, though, that in cases where extreme values are plausible or the distributions of potential values are skewed or bimodal, it would be even more important to avoid using a deterministic approach but instead to use the entire distributions in a stochastic approach.

A more conservative approach would be to use a more extreme value from the 'relative frequency' distribution, for example the maximum (or minimum) possible value that the variable can take, or to use the 95th (or 5th) percentile of the relative frequency distribution, which would estimate the value for which the probability of the actual variable value being higher (or lower) would be 5%. This can be considered a more conservative approach to risk assessment, since it can account for more extreme possible values of the variable, which may result in higher risk estimates from the model. This may or may not be desirable depending on the nature of the problem and the objectives of the risk assessment. However, it still disregards the variability and uncertainty of the process and therefore fails to give a complete picture for the modelled variable.

In all these cases, the advantage of simplicity in defining the variable (only one value) and executing the model is seriously outweighed by the omission of modelling the related variability and uncertainty and, therefore, the increased chance of excluding representative and realistic risk scenarios from the model outcome. An 'intermediate' solution would be to use several values for a variable covering a range of possible values and observe the effect on the outcomes. However, this would still miss the specification of the probability with which the used (or other) values would occur and therefore, would not offer an adequate improvement in the process. Such an approach would also become complicated, in practice, when more than one or two variables need to be modelled.

Stochastic modelling, therefore, should be preferred when feasible, because it allows the explicit modelling of the input values, including the associated variability and uncertainty. In these cases, a probability distribution can be constructed for each variable to describe the possible values it can take with the associated plausibility. Such distributions are usually continuous probability distributions, but discrete distributions may also be used. These distributions would need to be constructed based on previous knowledge of the characteristics of the variable and the biology of the process that it describes. Sometimes, information may exist in the literature, or previous measurements or assessments may be available. In those cases, it may be possible to construct empirically a frequency distribution of those results. Subsequently, a theoretical probability distribution can be selected as a model (Vose, 2000) by a fitting process based on a parametric distribution for that empirical distribution, or the latter can be sampled directly at the next step (non-parametric bootstrapping). Expert knowledge elicitation (EKE) can also be used when the information from the literature or other sources is scarce or non-existent. For details and guidance on EKE for risk assessment in food, the reader is referred to EFSA (2014c).

When the associated uncertainty is high, this needs to be reflected in the constructed distributions (see Chapter 7). These distributions need to be examined carefully to understand what they mean for the modelled variables and to assess whether they reflect the current knowledge and the experts' assessments. Ultimately, the use of any such distribution should be considered an assumption of the analysis, which will need to be evaluated before the analysis is carried out. It is recommended to investigate the effect that these input distributions have on the final outcomes of the model by evaluating how these are impacted by changing the input distributions one at a time. Scenarios can also be constructed for different values/combinations of the input variable values, and distributions and separate outcomes can be presented for each of these scenarios. The techniques used to assess the impact of uncertainty and variability of variables on model outcome are referred to as sensitivity analysis. A detailed description of the most relevant sensitivity analysis methods applied in QMRA is provided in Chapter 8.

9.3.3 Prevalence and Concentration

The exposure to the pathogen is often described within a QMRA by prevalence and concentration, which are also key for food safety management. Concentration is usually defined as the number of cells, measured as colony forming units, per unit size of food product (grams, millilitres, kilograms, litres, etc.), while prevalence is the proportion of product units containing the pathogen in a well-defined population of units.

The measurement and expression of prevalence can be specifically defined in different ways throughout the modelling process according to the unit of the product considered in each step. This is because along the food chain, unit size can change; for example, the initial product size can be different from the finished product or the portion consumed. In the case of liquid foods such as milk, in early steps, the unit is normally based on the container unit, which will decrease to the dimensions of a carton or bottle of milk when the product is finished. The effect of the reduction in product unit size (known as partitioning) on prevalence and concentration should be considered in the risk modelling process through mathematical approaches (Nauta, 2005). Microorganism distributions in the food unit (clustering, low numbers, etc.) and food unit sizes should be accounted for when modelling the partitioning effect.

Furthermore, mixing ingredients can result in a larger unit size. This can impact the prevalence and concentration. A statistical approach is needed to analyze whether the concentration distributions for the different small units (i.e. ingredients) are or are not similar based on either the Monte Carlo analysis or the Central Limit Theorem (CLT) (Pérez-Rodríguez and Zwietering, 2012). When partitioning and mixing are included in the model, prevalence and concentration are defined in a more 'mechanistic' way by explaining the influence of the different processes or events on them.

Concentration is usually defined based on microbiological testing, in which microbial counts are extrapolated to the whole sample and product under the assumption that microorganisms follow a homogeneous distribution. Concentration data can be left and right censored depending on the limit of quantification and the analytical method used. An example of left-censored data is when no colonies are detected upon plating 0.1 mL homogenized sample on agar. This does not necessarily mean that no microorganisms are present but that probably, their level is below a certain value, e.g. <100 CFU/g when using a 1:10 dilution from the original sample. Right-censored data are less common in microbiology. When decimal dilutions are properly performed, exact or approximate counts can be obtained. However, the presentation of concentration data is sometimes right censored, for example when concentrations are given as intervals, and the maximum values are expressed as e.g. >10,000 CFU/g. In QMRA, estimated counts are used to account for the pathogen concentrations by applying either point-estimates (i.e. mean, mode, 95th percentile, etc.) or probability distributions. The latter are preferred, since they enable the variability observed in the system studied, between product units, batches, industries, etc., to be incorporated into the model. If a regression analysis is performed using censored data, the use of specific methods should be considered in which specific distributions (e.g. uniform distribution) or representative values (e.g. midpoint) are assumed for the censored region (see Chapter 6).

The most commonly used distribution to describe between-batch variability is the log-normal distribution, with a mean and standard deviation derived from the respective statistics of the distribution of the concentration values transformed into logarithmic scale. As the normal distribution has an infinite domain, models should limit maximum and minimum values in the distribution to avoid unrealistic values or scenarios that are not biologically plausible. A maximum value is defined according to the observed counts or estimates, or based on the maximum population reached by microorganisms, usually for bacteria around 9–10 log CFU/g. The distribution should also be constrained on the left side to a minimum value if it is intended to exclusively represent concentration values. This minimum value can be set based on the detection limit of the microbiological method used or the theoretical minimum concentration present in a specific food product, considering its weight or size. For example, in a 200 g product, the minimum theoretical concentration would be 1/200 g (0.005 CFU/g or −2.30 log CFU/g). Other distributions have been suggested to represent concentration, such as the Poisson distribution in its different forms (generalized and zero-inflated Poisson), the gamma distribution, the Poisson-log-normal distribution and the negative binomial distribution. Some are discrete distributions, thus being able to account for the discrete nature of microorganisms, while most of them allow the description of both concentration and prevalence, since they include zero in the distribution domain. A review of these distributions and their adequacy for representing microbial concentration in QMRA, based on specific criteria, is provided by Jongenburger et al. (2012a, 2012b).

Probability distributions to describe prevalence can refer to different reference populations, affecting the model interpretation and its use. If prevalence is referred to the whole population or country, the distribution should be better interpreted as uncertainty, since it is assumed that the actual prevalence (p) in the population is one value, and its variation is the reflection of the lack of knowledge on p. The Bayesian approach can be applied instead of the classical frequentist statistics, making use of the subjective probabilities to represent uncertainty on p, developing a deductive approach from particulars into the general (i.e. the population). An important advantage of the Bayesian methods, unlike frequentist methods, is that these can still be precise enough when no data are available. Bayesian inference is, therefore, applied to prevalence by assuming a prior belief on p, which can be based on prior knowledge obtained from surveys looking at the absence/presence of the pathogen or expert opinion. National and EU monitoring programmes are recommended, since they can often provide a broader and representative picture of the variables, provided they have been designed with this purpose. Results collected from RASFF or derived from recalls can lead to a more biased estimate, leading to higher prevalence values than the actual one. Nonetheless, caution should be applied when considering external studies, which may include non-homogeneous conditions and thus produce a biased estimation. If one decides to use a Bayesian approach, special attention should be paid to the choice of prior distributions that are used to express one's prior beliefs about the parameter. Common choices include a uniform (0, 1) distribution, which expresses a large uncertainty about the parameter, and beta (α, β) distributions, which can be induced into a variety of shapes by choosing appropriate values for their parameters α and β. Beta distributions can be constructed by defining two parameters analytically ($\alpha = s + 1$, $\beta = n - s + 1$), with n being the total number of samples analyzed and s the number of positive samples obtained. This distribution is widely used to represent the uncertainty or random variation of prevalence (Vose, 2000). Note that this an empirical method to construct the distribution, and therefore, it should be assessed afterwards whether the distribution represents our prior belief about uncertainty. Other approaches can be considered to characterize uncertainty on prevalence based on classical statistics such as the Clopper–Pearson intervals, approximation of the normal distribution to the binomial distribution as a specific case of the CLT (i.e. de Moivre–Laplace theorem) or simply fitting a suitable probability distribution to the observed prevalence values.

The Clopper–Pearson method, often referred to as an 'exact' method, is based on the cumulative probabilities of the binomial distribution. Due to the discrete nature of this distribution, the estimated intervals can fail to cover, with precision, all population proportions (Clopper and Pearson, 1934).

The normal approximation to the binomial distribution can also be implemented to elucidate uncertainty intervals on p. The most commonly used formula for a binomial confidence interval relies on approximating the distribution of error about a binomially distributed observation, $\hat{p}$, with a normal distribution. However, although the approximated distribution is frequently incorrectly considered as a binomial distribution, the error distribution itself is not binomial, and hence other methods (see later) are preferred.

The approximation is usually justified by the CLT, indicating the conditions under which the statistic $\hat{p}$ would have an approximately normal distribution. In this case, the $(1 - \alpha)$ CI for the $\hat{p}$ can be obtained by the formula

$$\hat{p} \pm z_{\left(1-\frac{\alpha}{2}\right)}\sqrt{\frac{1}{n}\hat{p}\left(1-\hat{p}\right)} \tag{9.1}$$

where $\hat{p}$ is the proportion of successes in a Bernoulli trial process estimated from the sample, $z_{\left(1-\frac{\alpha}{2}\right)}$ is the $(1 - \frac{1}{2}\alpha)$ percentile of a standard normal distribution, α is the probability of type I error and n is the sample size.

The CLT applies poorly to this distribution with a sample size below 30 or where the proportion is close to 0 or 1. The normal approximation fails when the sample proportion is exactly zero or exactly one. However, in such cases, other methods exist to provide an upper or a lower confidence limit, respectively.

An example was developed to illustrate the differences in outcomes from these methods using prevalence data for *L. monocytogenes* in different fish species of smoked and gravad fish products in the EU. Data were extracted from the EU-wide BLS as referred to in Section 9.2.2.2 (EFSA, 2013). The abovementioned methods were applied as explained, and 95%CIs were derived. As shown in Table 9.1, 95%CIs were similar for the Clopper–Pearson approach ('exact' method) and Bayesian methods. Indeed, 'exact' intervals are well approximated by the Bayesian approach based on a uniform prior distribution (Tuyl, 2011). The 'exact' method has already been deployed in risk assessment studies to derive prevalence CIs (Banach et al., 2016). On the contrary, CIs derived from the Normal approximation approach were slightly different, though values tended to converge as the number of datapoints increased, as can be observed with the salmon data, with the larger number of samples. This illustrates the dependence of the Normal approximation approach on the number of samples used, making the use of the Bayesian methodology preferable. It can be easily defined in a stochastic environment, based on different prior knowledge, and then updated if new data sources are provided. The probability distributions for *L. monocytogenes* prevalence in different fish species derived from the Bayesian approach are depicted in Figure 9.1, including an overall (i.e. total) prevalence distribution built using all samples.

9.3.4 Models and Modelling Approaches

QMRA can be built on experimental data. However, when such data are not available or are uncertain (i.e. with high experimental error), predictive microbiology models can be applied to represent the effect of specific steps of the (farm-to-fork) process on the fate of the microorganism. The use of predictive modelling is particularly relevant for those steps involving the inactivation or growth of microorganisms. In inactivation studies, high lethality levels usually result in microbial concentrations below the limit of detection, which hampers determination of the effect of such a process because of the lack of counts. Also, predictive modelling is very valuable for considering the effect of household practices and factors (e.g. domestic storage), for which available data are often limited due to sampling constraints.

Exposure models usually combine data, probability distributions for model variables and predictive microbiology models (see Chapter 11) accounting for the fate of the microorganisms along the various food stages. Predictive models can be used to predict the impact of a specific food technology process by defining model variables through probability distributions or point-estimate values (see Section 9.3.2). If no predictive

235

Table 9.1 Estimation of the Uncertainty of the *Listeria monocytogenes* Prevalence in Different Fish Species of Smoked and Gravad Fish Based on Data from the EU-Wide Baseline Survey (BLS) on the *L. monocytogenes* Prevalence of Certain Ready-to-Eat Foods (EFSA, 2013) Using Different Mathematical Approaches

Fish Species	*Listeria* Prevalence Results[a]			Informed Prior Beta Distribution[b]				Normal Approximation to Binomial Distribution[c]			Exact Confidence Interval[d]	
	s	n	p	α	β	Upper Limit	Lower Limit	Mean	Upper Limit	Lower Limit	Upper Limit	Lower Limit
Salmon	226	1859	0.1216	227	1634	0.1372	0.1075	0.1216	0.1364	0.1067	0.1373	0.1071
Mackerel	24	410	0.0585	25	387	0.0856	0.0397	0.0585	0.0813	0.0358	0.0858	0.0379
Herring	17	183	0.0929	18	167	0.1438	0.0590	0.0929	0.1350	0.0508	0.1446	0.0551
Other fish	23	275	0.0836	24	253	0.1224	0.0565	0.0836	0.1164	0.0509	0.1229	0.0538
Mixed fish	21	326	0.0644	22	306	0.0965	0.0426	0.0644	0.0911	0.0378	0.0968	0.0403
TOTAL	311	3053	0.1019	312	2743	0.1131	0.0916	0.1019	0.1126	0.0911	0.1131	0.0914

[a] Data as extracted from the EU-wide BLS. n: number of tested units; p: prevalence (s/n); s: number of positive units.

[b] Uncertainty described by a beta distribution (α, β) using the Bayesian approach $(\alpha = s + 1; \beta = n - s + 1)$. The distribution simulation output was used to estimate 95% two-sided confidence intervals of prevalence, denoted as Upper limit and Lower limit.

[c] Normal approach based on the Central Limit Theorem to estimate 95% two-sided confidence intervals of prevalence.

[d] Based on the calculation of exact (95% two-sided) confidence intervals (Clopper and Pearson, 1934).

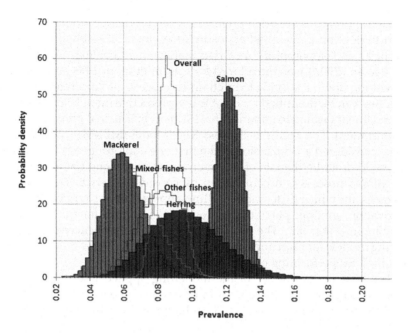

Figure 9.1 Simulated beta distributions obtained for *Listeria monocytogenes* prevalence in different fish species of smoked and gravad fish products based on data from the EU-wide baseline survey (BLS) on the *L. monocytogenes* prevalence of certain ready-to-eat foods (EFSA, 2013). A global distribution was created with the total number of prevalence data.

models are available, and the values of the variables and their influence are uncertain, a stochastic approach, in which probability distributions are used, can be more suitable. These distributions can be obtained from either a mathematical fitting method, based on available data, or an EKE process (EFSA, 2014c). The use of probability distributions enables the incorporation of the uncertainty associated with each factor, variable or function parameter, providing a more representative picture of the microbial event and attendant risk (Lammerding and Paoli, 1997). When probability distributions are used to represent variability and uncertainty without distinction, the model is called a first-order model (one-dimensional model), while if different distributions are used to represent the variability and the uncertainty of one specific variable, the model is considered to be a second-order model (two-dimensional model) (Vose, 2000). In the latter case, model simulation should be performed in two dimensions, which usually requires more computational resources and time.

Two approaches are often preferred when predictive models are combined to develop a microbial exposure assessment, possibly due to their simplicity and applicability in a risk management context. The Process Risk Model (PRM) was introduced by Cassin et al. in 1998 and consists of breaking down a specific food chain or process into discrete events or steps that can be described by combining predictive models in a sequential and linear fashion to obtain a final estimation (i.e. final prevalence and concentration). The Modular Process Risk Model (MPRM) (Nauta, 2008) can be considered an expansion of the first one and proposes a structured process to describe the fate of the microorganisms along the food chain. The MPRM process is accomplished by considering six basic microbial processes, for which different modelling approaches are applied: growth, inactivation, mixing, partitioning, removal, and transfer (or cross-contamination) (Table 9.2). The author presents a comprehensive review of existing models for each microbial process, describing how models can be combined for performing exposure assessment.

The level of complexity of the modelling applied is usually related to the existence (or lack thereof) of information on specific steps (process variables, hazard prevalence, effect on hazard of specific steps, etc.) and

Table 9.2 Qualitative Effect of the Six Basic Processes Proposed by Nauta (2001) and Nauta et al. (2001) on Prevalence (Effect on the Fraction of Contaminated Units), Concentration (Effect on the Total Number of Cells over All Units) and Unit (Physically Separated Quantity of Product) Size

Process	Effect on the Fraction of Contaminated Units	Effect on the Total Number of Cells over All Units	Effect on Unit Size
Growth	=	+	=
Inactivation	−	−	=
Mixing	+	=	+
Partitioning	−	=	−
Removal	−	−	=
Cross-contamination	+	=/+	=

(Adapted from Nauta, M.J., in *Microbial Risk Analysis of Foods*, Washington: ASM Press, 99–136, 2008)
=: no effect; +: increase; −: decrease.

the complexity of the process to be modelled in the various steps. The latter may be due to the existence of multiple food processing pathways and variables (i.e. different sequences of events for food processing and handling), different contamination sources (e.g. contamination by air or contact surface, contamination from different ingredients) and bidirectional and interrelated processes (e.g. transfer between food and contacting surface) (Pérez-Rodríguez et al., 2008). A higher level of detail or the consideration of more specific phenomena (influence of pre-culture conditions on lag time, phenotypic and genotypic diversity, stress resistance and adaptation, etc.) in the modelling process implies that customized mathematical algorithms should be used to incorporate loops or interdependence between variables and logical functions, making the model more complex both in the modelling stage and in the interpretation of the model outcomes (e.g. by risk managers). The increase in the number and interdependencies of variables and model parameters can introduce new sources of uncertainty (e.g. incorporating variability in the physiological state of cells), increase the number of modelling assumptions and make the model more prone to errors during both model programming and outcome interpretation. As suggested by Zwietering (2009), a complex approach is not always better than a simple model. Although simple models may be more imprecise, they are often very useful to identify the main factors with fewer resources and to detect errors in more complex analyses by comparing approaches. Besides, complex models can be employed as benchmarks to validate the simpler approach when real-world data are not available or accessible, provided the complex model is a mechanistically validated model, or to get insight into more specific phenomena (e.g. bacterial transfer from blade to meat slice during slicing). Therefore, both levels of complexity are needed and should be combined in a balanced fashion, adapted to the type of the food safety questions to be answered, accounting for those pathways and phenomena that are deemed to be relevant for the model purpose.

To address model complexities, a stepwise approach for QMRA was proposed by van Gerwen et al. (2000), in which the risk assessment process is structured in two levels of detail. In the first level, the microbial responses along the food chain are represented in a very broad way by determining the logarithmic effect of the targeted food processes or steps on the pathogen population (exposure assessment). The combination of all the potential effects is then introduced into a linear DR model, which can be defined e.g. by considering the minimum infective dose. The outcome is expressed in a semi-quantitative way by considering different risk categories based on established categorization criteria (ranging from very high to very low). The second level focuses on those food phenomena identified as relevant in the previous level, considering quantitative variables to define the process and applying predictive models to reflect the effect of each food stage along the food production–distribution–consumption chain. The model inputs can be defined as point-estimates, or ranges, allowing a potential variation range of the input variable, e.g. temperature differences between domestic refrigerators, to be considered. The output from the exposure assessment is expressed in terms of prevalence and concentration of microorganisms. When this is inputted into a DR model, which can comprise variability and uncertainty of host and pathogen virulence, the probability of infection or illness is estimated. The authors did not refer to the use of probability distributions, though stochastic variables could be incorporated into the second level to better describe variability and uncertainty in important risk factors.

The use of Bayesian techniques has been proposed as an alternative to classical methods to define risk model variables. By combining prior probability distributions of the model variables, based on prior knowledge, with the likelihood of the data, conditional on the variables, one can derive posterior probability distributions for these variables. The posterior distributions might then be used to make inferences about the variables. The advantage of using the Bayesian approach is that probability distributions can be updated as new data are added (reports, new experiments, etc.) allowing, in addition, a better characterization of the uncertainty of the variable. In addition, the Bayesian approach can run more efficient probabilistic sensitivity analyses than the standard Monte Carlo analysis using a lower number of model realizations (Oakley and O'Hagan, 2004). However, despite its potential, few attempts to accomplish a Bayesian QMRA comprising all steps have been made to date (Albert et al., 2008), while the Bayesian studies are more focused on specific elements or risk assessment components, such as deriving kinetic parameters for predictive growth models (Delignette-Muller et al., 2006) or identifying contamination sources (i.e. exposure assessment) (Smid et al., 2011).

Bayesian network (BN) analysis is put forward as a suitable approach to overcome the lack of data, one of the most critical aspects in QMRA. As pointed out by Greiner et al. (2013), a QMRA model can be entirely developed by using BN based on the same mathematical model structures as those usually applied in typically stochastic QMRA. BN analysis consists of graphically designing interdependences between variables, determining how certain variables can influence others by creating a cause–effect structure (Beaudequin et al., 2015). A schematic example of BN applied to QMRA is presented in Figure 9.2. This BN modelling scheme allows backward analysis to be developed, whereby the input variable values required to obtain a specified outcome can be elucidated.

Foods can undergo variations in temperature, physico-chemical conditions and composition due to specific technological processes or in general, because of the variations along the food chain. The time dependence of these changes should not be ignored when a QMRA is built, since it can have an impact on the microbial response and on prevalence and concentration. Drying or fermentation processes are examples of this, in which variables such as a_w or moisture content and pH change over time. When risk models are intended to represent these phenomena in detail, dynamic conditions should be considered in the modelling process. In these cases, environmental variables (temperature, pH, a_w, etc.) should be modelled as an array of values reflecting the variable's change over time (with the values of the variable at each time interval) and its effect on microbial behaviour (at each time interval). Differential equations are to be implemented to properly reflect this time-dependent phenomenon, considering appropriate methods for solving differential equations, such as fourth-order Runge–Kutta (Hairer et al., 1989).

There are some cases in which the risk process is composed of a sequence of events in time in which each event implies a change in

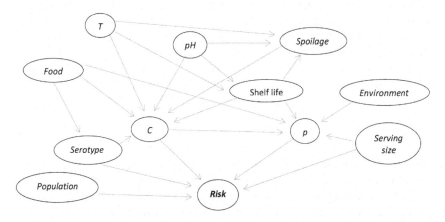

Figure 9.2 Example of a hypothetical Bayesian network reflecting the causal relationship between the main variables and the consumer risk associated with a specific microbiological pathogen and food category. C: concentration; *p*: prevalence; *T*: storage temperature.

241

the system state that affects the next event. This is the case with food cross-contamination, in which the order of events (i.e. when it happens) influences the number of microorganisms transferred in each cross-contamination step. For example, cross-contamination outcomes would be different depending on whether a contaminated working surface is used after or before cleaning or whether the contact with a food product happens immediately after the surface was contaminated or later on, and the length of time over which a microbial population can decay due to environmental stress or even increase because of the contamination of the surface with another contaminated product. For modelling these processes, the discrete-event simulation (DES) method can be applied, in which the operations of a system are simulated as a discrete sequence of events in time (Choi and Kang, 2013).

9.3.5 Model Integration: Population Risk versus Individual Risk

As specified in Section 9.2.4, the DR output can be expressed in various ways, such as the probability of illness, infection, death or hospitalization. In most cases, risk estimates are obtained directly from the application of the DR model (e.g. probability of illness) or from its transformation by applying an adjustment or conversion factor, e.g. to consider unreported cases or to derive the probability of death based on illness or infection estimates (Pérez-Rodríguez et al., 2007). More holistic risk metrics measure the impact on population health including the social and economic costs associated with foodborne diseases, though they require further calculations. Concepts such as disability-adjusted life years (DALY), accounting for the years lived with disability plus the years of life lost, are widely applied in epidemiology and can also be used for microbial risk assessment (Byrne et al., 2008; Sales-Ortells et al., 2015). A drawback of applying such risk metrics is that additional information is needed for their calculation (i.e. duration of illness and pathogen-specific mortality rate). This could hamper their use in those contexts where this information is missing or not well described. In such cases, the improvement of surveillance systems could help to overcome these limitations, providing a suitable knowledge basis for DALY estimation. For further detail about the use of risk metrics in the context of microbial risk assessment, readers are referred to Chapter 3.

Individual risk accounts for the risk associated with the ingestion of pathogenic cells by one individual, which can be expressed per serving, per a specific number of servings or per time period (e.g. per week, month or year). In the latter case, the consumption frequency should be used to estimate the number of servings during a time period. These data can be derived or extrapolated from food consumption surveys. The probability resulting from the DR model for each serving can be multiplied, if independence between eating events is assumed, to estimate the overall probability per individual associated with a specific consumption pattern. If no immunization memory among exposures is assumed, DR model could be applied to the total dose, which is defined as the sum of individual doses. In the specific case of the exponential DR model and other DR models that approximate linearity at low dose values, results are the same or similar regardless of whether doses are summed up or applied individually in the DR equation (Haas et al., 1999). For further information on the mathematical fundamentals of DR modelling for microbial hazards, readers can refer to Chapter 14. If the individual probability of getting ill considers the frequency of occurrence of the pathogen (as not all consumed servings are usually contaminated), the prevalence of the pathogen should be incorporated in the calculations. Therefore, in individual risk ($risk_i$), the conceptual risk function can be expressed as follows (Equation 9.2):

$$risk_i = f\left(p, C, s, [r_1, r_2 \ldots r_n]\right) \tag{9.2}$$

where
    p is prevalence
    C is concentration
    s stands for the serving size
    r_n is a set of n parameters describing the DR relationship

The product of C by s is equal to the dose.

Population risk refers to the risk measured or estimated for a collection of individuals defined by a set of common characteristics. This is usually expressed as the number of cases of illness or death (i.e. burden of illness) for a country, geographical region and/or population group (defined by age, immunological status, etc.). Also, other more informative risk measures could be used, such as those mentioned for individual risk, i.e. DALY. To determine the population risk, the probability distribution describing risk variability (Figure 9.3) is integrated for the entire population so that variability in the population risk is collapsed into one single value corresponding to the burden of illness of the targeted population. To do this, the total number of exposures (e_t) in the population should be incorporated, which is usually approximated by the total number of contaminated servings to which the population is exposed. The approach

243

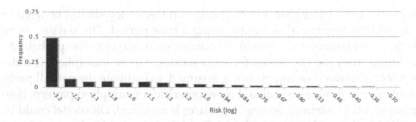

Figure 9.3 Example of a frequency distribution describing risk variability, where the x-axis represents the probability of getting ill (expressed in log scale) and the y-axis the frequency of risk values for a hypothetical microbial hazard in an unspecified food.

for estimating the population risk can be represented conceptually by Equation 9.3.

$$\text{population risk} = \sum_{i=1}^{i=e_t} \int_0^1 \text{risk}_i \qquad (9.3)$$

The estimated number of cases can be variable when a two-dimensional model is used, in which uncertainty and variability are simulated separately. The variation in the burden of illness represents, in this case, the model error derived from the uncertainty of input variables.

There are different methods to estimate the number of servings; the most common ones are based on extrapolating the number of servings from food consumption surveys or calculating the apparent consumption using as a basis the import and export statistics of the product in the country or population. An example of use of consumption data for estimating the population risk can be found in the risk assessment of *L. monocytogenes* in different RTE food categories performed by Pérez-Rodríguez et al. (2017). In this study, EU consumption data were collected and combined with apparent consumption estimates, as a proxy, to provide a good approximation of the number of servings associated with specific food categories. Nonetheless, the determination of the number of servings is often subject to a large uncertainty due to the lack of detail on specific food types in the food consumption data (food typology definition), short duration of the surveys and limited representativeness of the sample, among other things.

The impact of the variation in the number of servings (representing the uncertainty range on this variable) on the risk outcome is assessed and compared with the impact of the variation of other model variables in the following example. To this end, a straightforward risk assessment was performed using the exponential DR model and basic parameters and variables, whose values are presented in Table 9.3. In this example, the effect of three risk variables was investigated: serving size, number of servings and mean concentration at the moment of consumption. As shown in Figure 9.4, the change in the variable values, based on a multiplicative factor, affected the risk outcome (i.e. number of cases). Serving size and number of servings exhibited a similar behaviour, while increases in the mean concentration value resulted in a larger increase in risk, especially when the variable increased above one order of magnitude with respect to its baseline value (i.e. ×10). Below this multiplication value, changes in the number of servings appeared to produce a greater influence on risk than the concentration changes, as evidenced in the first points of Figure 9.4. Nonetheless, the results illustrate the importance of performing an accurate estimation of the number of servings to lead to a more reliable risk assessment.

9.3.6 Modelling and Simulation Tools

As explained earlier, QMRA models are built by combining predictive microbiology models, stochastic models (i.e. probability distributions) and DR models in a systematic manner in order to determine the risk

Table 9.3   Parameters and Variables Used to Calculate the Population Risk Associated with a Hypothetical Microbial Hazard and Food

Variable	Unit		Value/Distribution
r^a	Probability of one cell causing illness		1.0×10^{-12}
Concentration[b]	log CFU/g	C	Normal $(0.2; 2; 1/s; 9)^c$
Prevalence[d]	–	p	0.06
Serving size	g	s	50
Number of servings[e]	–	e_t	1.0×10^8

[a] Coefficient of an exponential dose response model $(1 - e^{(-d \cdot r)})$, where $d = C \cdot s$.
[b] Concentration distribution in the contaminated food units at the moment of consumption.
[c] Normal (mean; SD; min; max).
[d] Proportion of contaminated food units at the moment of consumption.
[e] Number of portions (eating occasions) of a hypothetical food consumed in 1 year.

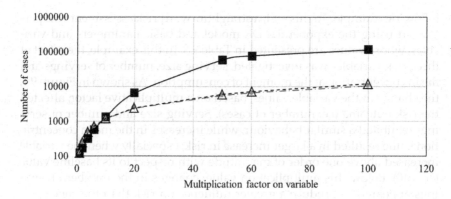

Figure 9.4 Number of cases simulated from a scenario analysis consisting of applying a multiplication factor (×1–100) to the following variables: concentration at the moment of consumption (■——), serving size (●-----) and number of servings (△-----).

quantitatively. The predictive models intended for QMRA can be developed by generic statistical software such as the R software (R Development Core Team, 2009) and MATLAB® (MathWorks, Inc., Natick, USA) or more specific statistical software tools. A detailed description of predictive microbiology software applications is provided in Chapter 13.

Probability distributions can be generated based on the method of moments, by defining the parametric distribution with the moments calculated from the dataset (e.g. mean and standard deviation in the normal distribution), with a regression method or the Maximum Likelihood Estimator (MLE), the latter being preferable in many cases. Generic software or other software that is more specific to the area of risk assessment, such as @Risk (Palisade, New York, USA) or ModelRisk (Vose software, Antwerpsesteenweg, Belgium), which are Excel add-ons (Microsoft©, Redmond, USA), including a fitting module for probability distributions, can be deployed to derive probability distributions based on existing datasets. The software R includes a specific package called 'fitdistrplus', which was specifically designed for QMRA applications, enabling the selection and fitting of a parametric univariate distribution to a given dataset (Pouillot and Delignette-Muller, 2010).

Developing and running a QMRA model involves incorporating predictive models into a mathematical environment and simulating them according to some rules and constraints (i.e. correlations, model domains and biological conditions), considering the stochastic character-istics of specific variables or models. Several software tools are available to risk assessors to develop QMRA models. Most of them are generic tools; for example, statistical toolboxes such as the R or MATLAB soft-ware can be used to code entire QMRA models. These applications are especially powerful tools given their extensive development, the avail-ability of model libraries, their calculus capacity, and the flexibility pro-vided by the different programming functions and routines. Other very widely used tools are based on a spreadsheet approach, being designed to be used with software tools such as Microsoft Excel (Microsoft ©, Redmond, USA). The software @Risk has widespread application in the QMRA field and in other areas (e.g. insurance, cost risk analysis, the oil and gas industry, etc.). The application @Risk allows mathematical mod-els to be defined using an Excel spreadsheet, incorporating probability distributions, and connecting data and equations in a tabulated frame-work. Because of the restrictions arising from a tabulated format, Visual Basic for Applications (VBA) can be integrated to develop more complex models by calling specific macros (i.e. routines). The use of Bayesian net-works to develop QMRA models is a specific case that should be per-formed with more suitable software such as WINBUGS, which stands for Bayesian Inference Using Gibbs Sampling for Microsoft Windows (Lunn et al. 2000), enabling the characterization of parameter uncertainties in the modelling process.

To simulate QMRA models with stochastic variables, different meth-ods can be employed, such as symbolic analysis, the method of moments based on the moment properties, and numerical methods. The Markov Chain Monte Carlo (MCMC) method, a numerical method, is often used, since in many cases there is no exact solution for the model calculations, or they are too complex to be solved in a short time. In contrast to the two analytical approaches, the MCMC methods are an approximation technique, but increasing the number of iterations can lead to a closer convergence between simulated outcomes and theoretically defined distributions. The same generic software mentioned earlier can be used for MCMC, such as R and MATLAB. Alternatively, @Risk, ModelRisk or Analytica® (Lumina Decision Systems, Los Gatos, USA) can enable a more streamlined application with less training and programming knowledge; the two first tools are built in Excel, which is more familiar software to general users, whereas Analytica has a visual and graphical interface, enabling better interpretation and presentation of the model structure.

9.4 RISK ASSESSMENT OUTPUT INTERPRETATION: IMPORTANCE OF UNCERTAINTY ANALYSIS

The most important attribute of an MRA is that it is 'fit-for-purpose', i.e. that it answers the risk management question(s), improves the decision-making process and does so in a timely manner (Lammerding, 2007). Current thinking on the validity of risk assessments suggests that the dimensions of validity should be based on (a) time available, (b) resource requirements and (c) resolution of output. One of the most important components of the last dimension is the uncertainty analysis.

The simplest definition of uncertainty is the lack of or incomplete information (NRC, 2009). EFSA's Guidance document on uncertainty analysis in scientific assessment (EFSA Scientific Committee, 2018a) defines 'uncertainty' as a general term referring to all types of limitations in available knowledge that affect the range and probability of possible answers to an assessment question. Available knowledge refers to the knowledge (evidence, data, etc.) available to assessors at the time the assessment is conducted and within the time and resources agreed for the assessment. Sometimes 'uncertainty' is used to refer to a source of uncertainty and sometimes to its impact on the conclusion of an assessment. In science and statistics, concepts such as measurement uncertainty and sampling uncertainty should always be considered, while weaknesses in the methodological quality of studies used in assessments can be important sources of uncertainty. Uncertainties in how evidence is used and combined in an assessment – e.g. model uncertainty or uncertainty in weighing different and sometimes conflicting lines of evidence in a reasoned argument – must always be considered. The level of uncertainty depends on the quality, quantity and relevance of data as well as the reliability and relevance of models and assumptions. Uncertainty analysis is defined as the process of identifying and characterizing uncertainty about questions and/or quantities of interest in a scientific assessment (EFSA Scientific Committee, 2018b). A question or quantity of interest may be the subject of the assessment as a whole, or it may be the subject of a subsidiary part of the assessment that contributes to the final output (e.g. exposure and hazard assessment are subsidiary parts of risk assessment).

Uncertainty in risk assessment can be expressed in various ways, such as description of a source of uncertainty, qualitative description of the direction and/or magnitude of uncertainty, a precise or approximate probability, or a univariate or two-dimensional probability distribution.

Understanding the impact of uncertainty in the MRA outputs requires a close collaboration between risk assessors and risk managers. A standard set of guidelines for interpreting the results of a risk assessment, especially in relation to the uncertainty, does not currently exist. Due to the large variation in types of assessments and approaches, it is hard to be prescriptive. On the other hand, EFSA has developed guidance for risk assessors and communicators on how to communicate the various expressions of uncertainty, described in EFSA's Guidance document on uncertainty analysis in scientific assessments (EFSA Scientific Committee, 2019).

The following sections briefly present the main elements of uncertainty analysis as described in the recently published EFSA Guidance document on uncertainty analysis, a few examples on uncertainty analysis and their interpretation in risk assessment.

9.4.1 Elements in Uncertainty Analysis

The scientific opinion of EFSA on the principles and methods behind EFSA's Guidance on Uncertainty Analysis in Scientific Assessment (EFSA Scientific Committee, 2018b) presents the following elements of an uncertainty analysis:

Identifying uncertainties affecting the assessment. This is necessary in every assessment and should be done in a structured way to minimize the chance of overlooking relevant uncertainties.

Prioritizing uncertainties within the assessment. This step plays an important role in the planning of the uncertainty analysis, enabling the assessor to focus detailed analysis on the most important uncertainties and address others collectively when evaluating overall uncertainty. Often, prioritization will be done by expert judgement during the planning process, but in more complex assessments, it may be done using influence analysis or sensitivity analysis.

Dividing the uncertainty analysis into parts. In some assessments, it may be sufficient to characterize overall uncertainty for the whole assessment directly by expert judgement. In other cases, it may be preferable to evaluate uncertainty for some or all parts of the assessment separately and then combine them by either calculation or expert judgement.

Ensuring that questions or quantities of interest are well-defined. Each question or quantity of interest must be well-defined, such that the true answer or value could be determined, at least in principle. A well-defined (sub)-question refers to an outcome or quantity that could (in principle) be observed or measured without ambiguity in the real world or obtained from a defined scientific procedure. Each keyword requires an explicit definition, and the population, region, and time period of interest should be specified. For a variable quantity, the statistic(s) and/or quantities(s) are required. This is necessary to make the question or quantity a proper subject for scientific assessment and to make it possible to express uncertainty about the true answer or value clearly and unambiguously.

Characterizing uncertainty for parts of the uncertainty analysis. This is needed for assessments where assessors choose to divide the uncertainty analysis into parts, but it may only be done for some of the parts, with the other parts being considered when characterizing the overall uncertainty in the end of the scientific assessment.

Combining uncertainty from different parts of the uncertainty analysis. This is needed for assessments where the assessors quantify uncertainty separately for two or more parts of the uncertainty analysis.

Characterizing overall uncertainty. Expressing quantitatively the overall impact of as many as possible of the identified uncertainties and describing qualitatively any that remain unquantified.

Reporting uncertainty analysis. Required for all assessments.

Most of these elements are always required, while some depend on the needs of the assessment. Furthermore, the approach to each element varies, and sometimes the order in which they are addressed depends on the nature or type of each assessment.

Most of the elements of uncertainty analysis can be addressed at different levels of refinement and effort, analogous to the 'tiered approaches' used in some areas of risk assessment. Major choices include whether to assess all uncertainties collectively or divide the uncertainty analysis into parts; whether to quantify uncertainties fully using probability distributions, or partially using probability bounds; and whether to combine uncertainties by expert judgement or calculation. The uncertainty of an assessment is driven primarily by the quantity and quality of available evidence (including data and expert judgement) and how it is used in the assessment. Less refined options for uncertainty analysis are quicker and less complex, and characterize uncertainty less precisely. This may be sufficient for decision-making in many cases. More refined options characterize uncertainty in a more detailed way more rigorously and may result in more precise characterization of overall uncertainty, although this will always be subject to any limitations in the underlying evidence and assessment methods. However, more refined options are more complex and require more time and specialized expertise. In complex assessments, an iterative approach may be efficient, starting with simpler options and then using the results to target more refined options on the most important uncertainties, whereas in other assessments, there may be less scope for iteration.

The types of expression of uncertainty resulting from an uncertainty analysis are shown in Table 9.4 as presented in the EFSA guidance document on how to communicate on uncertainty. Different types of expression can be used in the same assessment (EFSA Scientific Committee, 2018b).

9.4.2 Interpretation of Uncertainty Analysis in MRA

MRA can provide a numerical description of the level of public health protection achieved by the current food safety management system (Koutsoumanis and Aspridou, 2016). In addition, MRA can be used to assess the risk reduction achieved by the implementation of specific control measures. The use of MRA for decision-making in this microbial risk management (MRM) process, however, requires careful interpretation of the uncertainty analysis in MRA by the risk managers.

Appropriate interpretation of uncertainty is very important in many areas of risk management. For example, when assessing compliance with food safety targets based on an appropriately designed MRA, the public health goal (i.e. Appropriate Level Of Protection –ALOP) can be translated into a microbiological limit at the time of consumption (i.e. Food Safety Objective – FSO). When the FSO is met, it is expected that the ALOP will be achieved. However, the actual degree of public health protection achieved also depends on the actual compliance with the FSO, which is characterized by high uncertainty at both country and regional level. For example, the observed number of listeriosis cases from the consumption of RTE foods might be substantially higher than the number of cases predicted by MRA assuming that the FSOs established for this pathogen are met (FAO/WHO, 2004). This difference could be explained by a significant degree of non-compliance with the FSOs. In this case, the interpretation of uncertainty can be supported by the development of scenario analysis

Table 9.4 Types of Expressions of Uncertainty Produced by Uncertainty Analysis When Following the Guidance on Uncertainty Analysis in Scientific Assessments (EFSA Scientific Committee, 2018a) and Included in the Conclusion, Summary or Abstract of a Scientific Assessment. The Same Assessment May Produce One or More of These Expressions (EFSA, 2019)

Type of Uncertainty Expression	Description
No expression of uncertainty	This occurs in two situations:
	When a standardized assessment procedure only takes into account standard uncertainties, its conclusion may be communicated without qualification of uncertainties (see EFSA Scientific Committee, 2018a,b for more explanation).
	When uncertainty is present in an assessment, but decision-makers or legislation requires an unqualified conclusion (e.g. safe, not safe or 'cannot conclude'), the basis for the conclusion does not need to be stated in the conclusion, summary or abstract, but it should be documented in the body of the assessment report or an annex. Expression/s of uncertainty may be included there.
Description of a source of uncertainty	Verbal description of a source or cause of uncertainty. In some areas of EFSA's work, there are standard terminologies for describing some types of uncertainties, but often descriptions are specific to the assessment in hand (EFSA Scientific Committee 2018b).
Qualitative description of the direction and/or magnitude of uncertainty	Words or an ordinal scale describing how much a source of uncertainty affects the assessment or its conclusion (e.g. low, medium or high uncertainty; conservative, very conservative or non-conservative; unlikely, likely or very likely; or symbols indicating the direction and magnitude of uncertainty: $---$, $--$, $-$, $+$, $++$, $+++$).
	Because the meaning of such expressions is ambiguous, EFSA's Guidance on Uncertainty recommends that they should not be used unless they are accompanied by a quantitative definition (EFSA Scientific Committee, 2018a).
Inconclusive assessment	This occurs in two situations:
	When decision-makers or legislation requires an unqualified conclusion but assessors judge there is too much uncertainty to give one and report that they cannot conclude. The basis for this uncertainty expression should be documented in the body of the assessment report or an annex, and may include one or more uncertainty expressions.
	When it is not required that conclusions must be unqualified, but the assessors are unable to give any quantitative expression of uncertainty, or where they judge that their probability for a conclusion could be anywhere between 0 and 100%. This should be accompanied by a qualitative description of the uncertainties (see description of 'A precise probability').

(Continued)

252

Table 9.4 (Continued)   Types of Expressions of Uncertainty Produced by Uncertainty Analysis When Following the Guidance on Uncertainty Analysis in Scientific Assessments (EFSA Scientific Committee, 2018a) and Included in the Conclusion, Summary or Abstract of a Scientific Assessment. The Same Assessment May Produce One or More of These Expressions (EFSA, 2019)

Type of Uncertainty Expression	Description
A precise probability	A single number (in EFSA outputs: a percentage between 0 and 100%) quantifying the likelihood of either:
	A specified answer to a question (e.g. a 'yes' answer to a 'yes/no' question).
	A specified quantity lying in a specified range of values, or above or below a specified value (e.g. 90% probability that between 10 and 100 infected organisms will enter the EU in 2019; 5% probability that more than 100 infected organisms will enter).
	Note that the term 'precise' is used here to refer to how the probability is expressed, as a single number, and does not imply that it is actually known with absolute precision, which is not possible.
An approximate probability	Any range of probabilities (e.g. 10–20% probability) providing an approximate quantification of likelihood for either:
	A specified answer to a question (e.g. a 'yes' answer to a 'yes/no' question).
	A specified quantity lying in a specified range of values, or above or below a specified value (e.g. 1–10% probability that more than 100 infected organisms will enter the EU in 2019).
	The probability ranges used in EFSA's approximate probability scale are examples of approximate probability expressions. Assessors are not restricted to the ranges in the approximate probability scale and should use whatever ranges best reflect their judgement of the uncertainty (EFSA Scientific Committee, 2018a).
A probability distribution	A graph showing probabilities for different values of an uncertain quantity that has a single true value (e.g. the average exposure for a population). The graph can be plotted in various formats, most commonly a probability density function (PDF), cumulative density function (CDF) or complementary cumulative density function (CCDF).
A two-dimensional probability distribution	In this guidance, the term 'two-dimensional (or 2-D) probability distribution' refers to a distribution that quantifies the uncertainty of a quantity that is variable, i.e. takes multiple true values (e.g. the exposure of different individuals in a population). This is most often plotted as a CDF or CCDF representing the median estimate of the variability, with confidence or probability intervals quantifying the uncertainty around the CDF or CCDF.

that explores different levels of compliance with the FSOs. In addition, interpretation should consider the relation between the stringency of an FSO and the level of compliance, since a more stringent FSO, for which verification might be more difficult, could lead to an increased level of non-compliance and an increase in the incidence of disease.

Another example for the interpretation of uncertainty analysis in MRA is provided by EFSA for risk ranking (EFSA BIOHAZ Panel, 2015). In most cases, risk ranking in food safety could be defined as the analysis and ranking of the combined probability of food contamination, consumer exposure and the public health impact of certain foodborne hazards. In a science- and risk-based system, resources for food safety should be deployed in a manner that maximizes the public health benefit achieved through risk reduction. Thus, risk ranking has been recognized as the proper starting point for risk-based priority setting and resource allocation, because it would permit policymakers to focus attention on the most significant public health problems and develop strategies for addressing them.

The analysis of uncertainty is a great challenge in risk ranking. Figure 9.5 presents an example of the uncertainty probability density distributions of model outputs for risks associated with three food–pathogen combinations. As shown, the rankings depend on what statistic is used to characterize a risk whose value is not known with certainty. If mean values are used (as a best guess), the three food–pathogen combinations

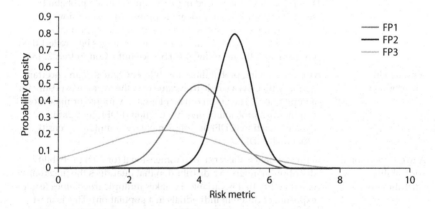

Figure 9.5 Uncertainty probability density distributions of the model outputs for risks associated with three food–pathogen (FP) combinations. Note that the x-axis uses a hypothetical risk metric; therefore, no units are used.

would be ranked 2–1–3. In contrast, if 99th percentiles are used (as a worst case), the order becomes 3–2–1.

Figure 9.6 presents another example, with two food–hazard combinations, A and B, in the presence of uncertainty on the parameters used to assess the associated risks for consumers, which propagate through the model, leading to uncertainties in risk estimates. In this case, risk calculations should reflect these uncertainties, and so should the ranking. For simplicity of illustration, log-normally distributed uncertainty is assumed to be directly affecting the risks for A and B.

Examining the distributions of the DALYs associated with A and B in Figure 9.7 for distribution A (DA) and distribution B (DB), respectively, one may observe that DA is much more uncertain than DB, but the expected value of DB is greater than DA. On the other hand, there is a range in which the DB percentiles are larger than those of DA. For example, if one were to perform the ranking based on the DALYs' 95th percentile values, the conclusion would be that combination A is more risky than B, contrary to what would happen if the rankings were based on the expected values. The drawback of comparing the expected values or specific percentiles lies in the loss of information about the distribution. In order to give a full account of the difference between the distributions of DA and DB. one has to consider the random variable DA–DB. whose probability distribution function (PDF) and cumulative distribution function (CDF) are shown in Figure 9.7.

In order to establish whether A is more risky than B, one can consider the probability that DA is greater than DB, i.e. rAB = 1 – P(DA–DB <0) (0) that DA is greater than DB. In this case, rAB = 1 – 0.678 = 0.322, which means that with a probability of 0.322, A is more risky than B. To decide on the relative importance of the two combinations A and B, one may

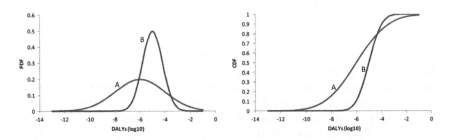

Figure 9.6 Probability density functions (PDF) and cumulative distribution functions (CDF) of the disability-adjusted life years (DALYs) for the random situations A and B. (From EFSA BIOHAZ Panel, 2015.)

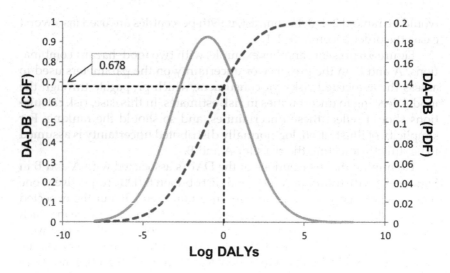

Figure 9.7 PDF and CDF of the random variable (DA–DB). The probability of DA–DB < 0 is 0.678. (From EFSA BIOHAZ Panel, 2015.)

choose a threshold (T) ranging from 0.5 to 1 on the rAB value, such that if rAB is larger than T, then A is more risky than B, and otherwise no conclusion can be drawn. Obviously, the lower the threshold, the higher the risk associated with the decision. However, the choice of a simple-valued threshold has some limitations when considering multiple combinations. These limitations can partially be overcome by referring the comparison to a threshold range [Tl, Tu] in such a way that for the two components A and B, the risk ranking will be based on the following:

If rAB > Tu, then A is more risky than B.
If rAB < Tl, then B is more risky than A.
If Tl < rAB < Tu, then A is equally risky as B.

These examples show that in order to interpret the results of a risk assessment, it is important that the influences of the variables, inputs and assumptions on the results are demonstrated. Decision-makers need as complete a picture as possible of the assessors' overall uncertainty to inform decision-making. Thus, risk assessment outputs should be combined with the characterization of the overall uncertainty that takes into account all identified sources of uncertainty, in all parts of the assessment, and also any dependencies between different sources of uncertainty (EFSA Scientific Committee, 2018b).

9.5 KNOWLEDGE EXCHANGE TO IMPROVE MICROBIAL RISK ASSESSMENT

In the last decades, the microbial food safety community has produced a great quantity of data that have been used to generate relevant, usable and meaningful answers to important questions, and even new 'processed' data. Based on these data and information, mathematical models aiming to describe the relationship between the occurrence of pathogens in food matrices and their public health impact have been created. These data and models were defined by Haberbeck et al. (2018) as 'data-driven knowledge' and 'model-based knowledge', respectively. Such knowledge is also the base for conducting the different steps of a QMRA. However, the broad application and re-usability of these data and methodologies are still hampered, as access to them and exchange of resources are usually difficult and time consuming.

A specific challenge in the area of food safety modelling is the integration of data, information and mathematical models from different sources, e.g. scientific publications, tools or software solutions. A strategy to tackle this challenge can be based on the establishment of harmonized information exchange formats and the development of open source software tools supporting such formats. This would also facilitate the development of consistent rules for knowledge annotation and the creation of open access knowledge repositories. In addition, already existing tools could source new data or models from such resources, promoting the application of newly generated knowledge in the QMRA process (Plaza-Rodríguez et al. 2018, Haberbeck et al. 2018).

9.5.1 Current Limitations of Knowledge Exchange

Current limitations in relation to the abovementioned knowledge exchange include (i) missing metadata on data-driven or model-based knowledge, preventing its proper application and re-use, (ii) missing transparency on model code/scripts, preventing quality control of existing knowledge, (iii) lack of harmonized information exchange formats, preventing efficient knowledge sharing, application and extension, and (iv) lack of open source software tools that improve a true knowledge exchange.

As described by Plaza-Rodríguez et al. (2018), transparency and consistency are two key elements for knowledge exchange, as they enable other researchers to check previous conclusions and build upon them. To support the exploitation of knowledge provided in scientific publications or software tools, it would therefore be necessary to share essential information, such as raw data, details on data collection procedures, criteria for data inclusion/exclusion, model code or script, and a detailed description of the matrix, microorganisms and environmental conditions (Plaza-Rodríguez et al. 2018). In addition, the detailed description of the steps followed within the model generation process, the references for all data used, the assumptions made during the model design and the uncertainties associated with the obtained predictions are of high importance for efficient knowledge exchange and re-usability.

One of the most important issues that hamper the application of already existing knowledge is the lack of essential metadata provided together with the data and models. For example, many predictive microbiology and QMRA models created in the past have been published as articles in scientific journals in non-reusable formats. To re-use them, it is necessary to extract information manually and re-implement the data/models into a software tool. However, these are not simple tasks, as they sometimes require a complete set of data or model specifications that are not always provided in the article in full detail (Plaza-Rodriguez et al. 2015). Moreover, not all the raw data used to generate the model are made available when a study is published, which prevents the re-use of such knowledge by others. In some cases, these data are just provided as a graphical representation (as in Black et al. 2008). Thus, to build new knowledge from such data-driven knowledge, researchers would need to digitize figures, which is labour intensive and prone to errors. In addition, many relevant metadata are frequently reported in the text, complicating the effort to get an overview, e.g. on the true range of applicability of a model. Model equations are normally provided as specifically formatted text embedded in scientific publications (as in Kusumaningrum et al. 2004), and the parameters obtained after fitting the formula to the raw data are included in the formula itself or in a table (as in van Asselt and Zwietering 2006). A model script, allowing the model to be used directly in a software tool for obtaining predictions, is not provided in many cases, and the steps followed for the model creation are just textually described (as in Haberbeck et al. 2015). Therefore, the re-implementation of models requires excessive time and effort and can be prone to failures and errors, making it sometimes difficult to reach the same outputs as those presented in the original paper (Plaza-Rodriguez et al. 2015).

The re-usability and application of available data and models for QMRA purposes have been greatly enhanced by the development of software tools that provide access to data and models and functionalities for generating new models and applying them within different scenarios (Tenenhaus-Aziza and Ellouze 2015). Some of these tools are available for free and as open access software. However, only a small fraction are available as open source software solutions providing full transparency up to the software implementation level (e.g. Predictive Microbial Modeling Lab (PMM-Lab); https://foodrisklabs.bfr.bund.de/pmm-lab/). A limitation of all non-open source software tools is that the software source code is not accessible to the end-users and cannot be checked, shared or re-used by them. This also hinders the communication between these tools, as the output of one tool cannot be imported directly as input into one of the others; e.g. a prediction made with FSSP (Food Spoilage and Safety Predictor; http://fssp.food.dtu.dk/) needs to be converted by the user into a format that can be imported by FDA-iRISK (FDA-iRISK® version 4.0; https://irisk.foodrisk.org/). In addition, available tools are implemented in different programming languages. The lack of harmonized information exchange formats and harmonized application programming interfaces (APIs) that could facilitate the communication between such independent software tools makes it currently impossible to programmatically combine models originating from different software tools and therefore prevents efficient knowledge sharing, application and model extension. Currently, the only way to re-use or combine existing models from different software tools consists in manually re-implementing the models of interest in a software tool that fits the needs of the end-user. As described before, this process is time consuming and error-prone.

9.5.2 Current Status of Knowledge Exchange

Over the last years, there has been increased awareness within the QMRA domain that efficient data, information and model exchange are important research objectives. For knowledge provided only as scientific publications, some authors offer to provide the model script upon request (Mylius, Nauta, and Havelaar 2007), or it is included directly as supplementary material (Schielke et al. 2011; Bartsch et al. 2019). In the latter cases, the Predictive Modelling in Food Markup Language (PMF-ML) has been used to provide the parameterized predictive microbial models in a software independent format. Specifications on PMF-ML were provided by Plaza-Rodriguez et al. (2015). In addition, publishing data, information or models into repositories like ComBase, FigShare or Zenodo (including EFSA's Knowledge Junction) is becoming popular.

A considerable number of software tools are nowadays available as commercial software tools (e.g. Dairy Products Safety Predictor, Listeria Meat Model, NIZO Premia and Sym'Previus), or as open access software accessible (i) via the internet (e.g. Baseline, ComBase, FDA-iRISK, Microbial Response Viewer, MicroHibro, Prediction of Microbial safety in Meat Product, etc.) or (ii) in a downloadable desktop version (e.g. FILTREX, FISHMAP, FSSP, GinaFit, GroPin and PMM-Lab). Some of them contain features facilitating knowledge sharing (e.g. ComBase, ICRA, FDA-iRISK, MicroHibro, PMM-Lab, FSK-Lab and openFSMR), supporting model generation (e.g. PMM-Lab, iPMP and R packages) and for model application/prediction (e.g. FSSP, MicroHibro, ComBase Predictor, PMP and many others). To improve transparency, some of these tools also provide access to the model source code with all the model parameter values (e.g. GroPIN, R packages, nlsmicrobio and PMM-Lab). However, there are currently only a limited number of tools that support the import and export of data and models into a common information exchange format.

The PMM-Lab software allows a direct information exchange between different software tools from the domain of predictive microbiology, e.g. from and to ComBase, from GroPIN into PMM-Lab and from PMM-Lab into R. The recently released FSK-Lab (Food Safety Knowledge Lab; https://foodrisklabs.bfr.bund.de/fsk-lab/) supports the import and export of models in the information exchange format Food Safety Knowledge Markup Language (FSK-ML), which enables models developed in different scripting languages to be shared.

9.5.3 Novel Initiatives to Improve Knowledge Exchange

The way of making use of scientific information within QMRA is changing. This is true due to the increasing understanding of the importance of transparency and reproducibility of risk assessment work and the need to re-use and build on top of existing knowledge.

Nowadays, the option to provide the raw data and the model script as supplementary material is encouraged by the editors of many scientific journals, as it provides transparency to the modelling process and facilitates the validation and increases the re-usability of published data and models. In this sense, some journals encourage authors to deposit raw experimental data at relevant data repositories and share the link within the article, or to deposit any research data (including raw and processed data, software code, algorithms, etc.) associated with the submitted manuscript in a free-to-use, open access repository like Mendeley Data (https://

data.mendeley.com/). The datasets are then listed and directly accessible to readers next to the published article online.

New solutions to improve knowledge exchange and re-use in QMRA are currently emerging. For example, the ComBase initiative has developed a new online data import functionality (ComBase Data Wizard; https://submit.combase.cc/login) that allows users to submit experimental data easily through its website.

A generic strategy to facilitate efficient information exchange was proposed in 2015 by Plaza-Rodriguez et al. (2015). This strategy included the establishment and adoption of harmonized information exchange formats. It also highlighted the need for software tools supporting such formats, the development of consistent rules for knowledge annotation and the creation of open access food safety knowledge repositories. On the basis of this strategy, the FSK-ML has been proposed to support the exchange of models available within the microbial risk assessment domain. FSK-ML defines a framework for sharing data-driven or model-based knowledge together with meaningful metadata in a machine-readable format. A specific feature of FSK-ML is that it supports model scripts in different scripting languages (e.g. R, Perl, Python or MATLAB). This is important to avoid the need for re-implementation of already existing legacy models (de Alba Aparicio et al. 2018). This is accomplished by the creation of a harmonized annotation 'around' the model (script) based on the annotation metadata schema and controlled vocabularies proposed by Haberbeck et al. (2018).

As described by Plaza-Rodríguez et al. (2018), one of the biggest benefits gained from harmonized information exchange formats is the opportunity to develop user-friendly model repositories with meaningful search and filter functions. These repositories allow users to download models and information about models, facilitating their re-use. Users can also contribute their own models that are compliant with the format and quality criteria implemented in the specific repository. An effort aimed at the establishment of a community-driven model repository is the RAKIP (Risk Assessment Knowledge Integration Platform) Initiative. The RAKIP Initiative emerged from a collaboration project carried out by three European food safety risk assessment institutions: ANSES, BfR and DTU Food. Within this project, new community resources that support knowledge integration and exchange within the predictive microbiology and QMRA areas were developed. For example, the FSK-ML format specification and a harmonized metadata schema for knowledge annotation were developed. With support from the AGINFRA+ research project, the

RAKIP Initiative is developing a web portal (a so-called Virtual Research Environment – VRE) that allows the food safety community to (i) contribute to the definition of harmonized terms and concepts, (ii) share files, software libraries, converter tools and software-specific import and export functions, promoting the adoption of FSK-ML, and (iii) illustrate how FSK-ML compliant models might be shared via a model repository with search and filtering functionalities. Currently, the RAKIP Initiative's focus is to improve existing resources like the FSK-ML format, the metadata schema for model annotations and the community web portal (https://foodrisklabs.bfr.bund.de/rakip-web-portal/).

In a related effort, EFSA and BfR established new open source, web-based model execution/simulation resources (https://foodrisklabs.bfr.bund.de/safetytools-efsa/). The services developed in this effort will also provide the opportunity to link with models published in EFSA's Knowledge Junction (https://zenodo.org/communities/efsa-kj). New functionalities in the open source software FSK-Lab now allow end-users to provide their models easily in the FSK-ML format. From experience with similar efforts in other scientific disciplines, e.g. Systems Biology, one can state that the broad adoption of these newly emerging resources will need time. However, it is important that the food safety community joins these efforts to maintain and update the developed resources in order to achieve a significant benefit for end-users from food safety authorities, food industry, consultancy companies or food research institutes.

The resources detailed in this section could facilitate transparent and consistent knowledge exchange, which in turn would help to improve microbiological food safety. This would be beneficial for the whole food safety community in many aspects, such as the easier application and re-use of existing knowledge by the food industry and authorities in the case of food safety emergencies and easier decision-making during risk management processes.

DISCLAIMER

The authors Winy Messens, Marios Georgiadis and Caroline Merten are employed with the European Food Safety Authority (EFSA). Winy Messens is employed in the Biological Hazards and Contaminants Unit, which provides scientific and administrative support to the Panel on

Biological Hazards. Marios Georgiadis is employed in the Assessment and Methodological Support Unit, which provides scientific and administrative support to EFSA's scientific activities, while Caroline Merten is employed in the Scientific Committee and Emerging Risks Unit, which provides scientific and administrative support to the Scientific Committee. However, the present chapter is published under the sole responsibility of the authors and may not be considered as an EFSA scientific output. The positions and opinions presented in this chapter are those of the authors alone and do not necessarily represent the views/any official position or scientific works of EFSA. To know about the views or scientific outputs of EFSA, please consult its website www.efsa.europa.eu.

NOTES

1. Directive 2003/99/EC of the European Parliament and of the Council of 17 November 2003 on the monitoring of zoonoses and zoonotic agents, amending Council Decision 90/424/EEC and repealing Council Directive 92/117/EEC. OJ L 325, 12.12.2003, pp. 31–40.
2. Decision No 1082/2013/EU of the European Parliament and of the Council of 22 October 2013 on serious cross-border threats to health and repealing Decision No 2119/98/EC. OJ L 293, 5.11.2013, pp. 1–15.
3. Commission Decision 2012/506/EU amending Decision 2002/253/EC laying down case definitions for reporting communicable diseases to the European Union network under Decision No 2119/98/EC of the European Parliament and of the Council. OJ L 262, 27.9.2012, pp. 1–57.
4. www.ecdc.europa.eu/en/surveillance-atlas-infectious-diseases
5. http://ecdc.europa.eu/en/publications/surveillance_reports/annual_epidemiological_report/Pages/epi_index.aspx
6. Commission Regulation (EC) No 2073/2005 of 15 November 2005 on microbiological criteria for foodstuffs. OJ L 338, 22.12.2005, pp. 1–26 as amended by Commission Regulation (EC) No 2019/229 of 7 February 2019.
7. Commission Implementing Decision 2013/652/EU of 12 November 2013 on the monitoring and reporting of antimicrobial resistance in zoonotic and commensal bacteria. OJ L 303, 14.11.2013, pp. 1–26.
8. Commission Decision No 2010/678/EU of 5 November 2010 concerning a financial contribution from the Union towards a coordinated monitoring programme on the prevalence of *Listeria monocytogenes* in certain ready-to-eat foods to be carried out in the Member States (notified under document C(2010) 7516).
9. Commission Regulation No 16/2011/EU of 10 January 2011 laying down implementing measures for the Rapid alert system for food and feed. OJ L 6, 11.1.2011, pp. 7–10.

10. Regulation No 178/2002 of the European Parliament and of the Council of 28 January 2002 laying down the general principles and requirements of food law, establishing the European Food Safety Authority and laying down procedures in matters of food safety. OJ L 31, 1.2.2002, pp. 1–24.
11. https://ec.europa.eu/food/safety/rasff_en
12. www.efsa.europa.eu/en/food-consumption/comprehensive-database
13. www.fao.org/gift-individual-food-consumption/en/

REFERENCES

Albert, I., Grenier, E., Denis, J.-B., Rousseau, J. 2008. Quantitative risk assessment from farm to fork and beyond: A global Bayesian approach concerning food-borne diseases. *Risk Analysis* 28(2):557–571; doi:10.1111/j.1539-6924.2008.01000.x.

Banach, J.L., Stratakou, I., van der Fels-Klerx, H.J., Besten, H.M.W. de, Zwietering, M.H. 2016. European alerting and monitoring data as inputs for the risk assessment of microbiological and chemical hazards in spices and herbs. *Food Control* 69:237–249; doi:10.1016/j.foodcont.2016.04.010.

Bartsch, C., Plaza-Rodriguez, C., Trojnar, E., Filter, M., Johne, R. 2019. Predictive models for thermal inactivation of human Norovirus and surrogates in strawberry puree. *Food Control* 96:87–97.

Beaudequin, D., Harden, F., Roiko, A., Stratton, H., Lemckert, C., Mengersen, K. 2015. Beyond QMRA: Modelling microbial health risk as a complex system using Bayesian networks. *Environmental International* 80:8–18.

Black, D.G., Taylor, T.M., Kerr, H.J., Padhi, S., Montville, T.J., Davidson, P.M. 2008. Decontamination of fluid milk containing Bacillus spores using commercial household products. *Journal of Food Protection* 71(3):473–478.

Boelaert, F., Amore, G., Van der Stede, Y., Hugas, M. 2016. EU-wide monitoring of biological hazards along the food chain: Achievements, challenges and EFSA vision for the future. *Current Opinion in Food Science* 12:52–62; doi:10.1016/j.cofs.2016.08.004.

Brul, S., Bassett, J., Cook, P., Kathariou, S., McClure, P., Jasti, P.R., Betts, R. 2012. 'Omics' technologies in quantitative microbial risk assessment. *Trends in Food Science and Technology* 27(1):12–24.

Buchanan, R.L., Smith, J.L., Long, W. 2000. Microbial risk assessment: Dose-response relations and risk characterization. *International Journal of Food Microbiology* 58(3):159–172.

Byrne, B., Lyng, J., Dunne, G., Bolton, D.J. 2008. An assessment of the microbial quality of the air within a pork processing plant. *Food Control* 19(9):915–920.

Cassin, M.H., Lammerding, A.M., Todd, E.C., Ross, W., McColl, R.S. 1998. Quantitative risk assessment for *Escherichia coli* O157:H7 in ground beef hamburgers. *International Journal of Food Microbiology* 41(1):21–44.

Choi, B.K., Kang, D. 2013. *Modeling and Simulation of Discrete-Event Systems, Modeling and Simulation of Discrete-Event Systems.* Hoboken/New York: John Wiley & Sons Ltd.

Clopper, C.J., Pearson, E.S. 1934. The use of confidence or fiducial limits illustrated in the case of the binomial. *Biometrika* 26(4):404–413.

Codex Alimentarius Commission (CAC). 1999. *Principles and Guidelines for the Conduct of Microbiological Risk Assessment.* Rome: Codex Alimentarius Commission.

de Alba Aparicio, M., Tasja Buschhardt, A.S., Lars Valentin, O., Mesa-Varona, T.G., Plaza-Rodriguez, C., Filter, M. 2018. FSK-Lab – An open source food safety model integration tool. *Microbial Risk Analysis* 10:13–19.

Delignette-Muller, M.L., Cornu, M., Pouillot, R., Denis, J.-B. 2006. Use of Bayesian modelling in risk assessment: Application to growth of *Listeria monocytogenes* and food flora in cold-smoked salmon. *International Journal of Food Microbiology* 106(2):195–208.

EFSA (European Food Safety Authority). 2010. Application of systematic review methodology to food and feed safety assessments to support decision making. *EFSA Journal* 8(6):1637, 90 pp; doi:10.2903/j.efsa.2010.1637.

EFSA (European Food Safety Authority). 2013. Analysis of the baseline survey on the prevalence of *Listeria monocytogenes* in certain ready-to-eat foods in the EU, 2010–2011 part A: *Listeria monocytogenes* prevalence estimates. *EFSA Journal* 11(6):3241, 75 pp; doi:10.2903/j.efsa.2013.3241.

EFSA (European Food Safety Authority). 2014a. Update of the technical specifications for harmonised reporting of food-borne outbreaks through the European Union reporting system in accordance with Directive 2003/99/EC. *EFSA Journal* 12(3):3598, 228 pp; doi:10.2903/j.efsa.2014.3598.

EFSA (European Food Safety Authority). 2014b. Analysis of the baseline survey on the prevalence of *Listeria monocytogenes* in certain ready-to-eat foods in the EU, 2010–2011. Part B: Analysis of factors related to prevalence and exploring compliance. *EFSA Journal* 12(8):3810, 73 pp; doi:10.2903/j.efsa.2014.3810.

EFSA (European Food Safety Authority). 2014c. Guidance on expert knowledge elicitation in food and feed safety risk assessment. *EFSA Journal* 12(6):3734, 278 pp; doi:10.2903/j.efsa.2014.3734.

EFSA (European Food Safety Authority). 2015. An update on the risk of transmission of Ebola virus (EBOV) via the food chain - Part 2. *EFSA Journal* 13(1):4042, 17 pp; doi:10.2903/j.efsa2015.4042.

EFSA (European Food Safety Authority) and ECDC (European Centre for Disease Prevention and Control). 2019. The European Union summary report on antimicrobial resistance in zoonotic and indicator bacteria from humans, animals and food in 2017. *EFSA Journal* 17(2):5598, 278 pp; doi:10.2903/j.efsa.2019.5598.

EFSA and ECDC (European Food Safety Authority and European Centre for Disease Prevention and Control). 2016. The European Union summary report on trends and sources of zoonoses, zoonotic agents and food-borne outbreaks in 2015. *EFSA Journal* 14(12):4634, 231 pp; doi:10.2903/j.efsa.2016.4634.

EFSA and ECDC (European Food Safety Authority and European Centre for Disease Prevention and Control). 2018. The European Union summary report on trends and sources of zoonoses, zoonotic agents and food-borne outbreaks in 2017. *EFSA Journal* 16(12):5500, 262 pp; doi:10.2903/j.efsa.2018.5500.

EFSA and ECDC (European Food Safety Authority and European Centre for Disease Prevention and Control). 2019. The European Union one Health 2018 zoonoses report. *EFSA Journal* 17(12):5926, 276 pp; doi:10.2903/j.efsa.2019.5926.

EFSA (European Food Safety Authority), Hart, A., Maxim, L., Siegrist, M., Von Goetz, N., da Cruz, C., Merten, C., Mosbach-Schulz, O., Lahaniatis, M., Smith, A., Hardy, A. 2019. Guidance on communication of uncertainty in scientific assessments. *EFSA Journal* 17(1):5520, 73 pp; doi:10.2903/j.efsa.2019.5520.

EFSA BIOHAZ Panel (EFSA Panel on Biological Hazards). 2011. Scientific opinion on *Campylobacter* in broiler meat production: Control options and performance objectives and/or targets at different stages of the food chain. *EFSA Journal* 9(4):2105, 141 pp; doi:10.2903/j.efsa.2011.2105.

EFSA BIOHAZ Panel (EFSA Panel on Biological Hazards). 2013. Scientific opinion on the public health hazards to be covered by inspection of meat (bovine animals). *EFSA Journal* 11:3266, 261 pp; doi:10.2903/j.efsa.2013.3266.

EFSA BIOHAZ Panel (EFSA Panel on Biological Hazards). 2015. Scientific opinion on the development of a risk ranking toolbox for the EFSA BIOHAZ Panel. *EFSA Journal* 13(1):3939, 131 pp; doi:10.2903/j.efsa.2015.3939.

EFSA BIOHAZ Panel (EFSA Panel on Biological Hazards), Ricci, A., Allende, A., Bolton, D., Chemaly, M., Davies, R., Fernández Escámez, P.S., Girones, R., Herman, L., Koutsoumanis, K., Nørrung, B., Robertson, L., Ru, G., Sanaa, M., Simmons, M., Skandamis, P., Snary, E., Speybroeck, N., Ter Kuile, B., Threlfall, J., Wahlström, H., Takkinen, J., Wagner, M., Arcella, D., Da Silva Felicio, M.T., Georgiadis, M., Messens, W., Lindqvist, R. 2018a. Scientific opinion on the *Listeria monocytogenes* contamination of ready-to-eat foods and the risk for human health in the EU. *EFSA Journal* 16(1):5134, 173 pp; doi:10.2903/j.efsa.2018.5134.

EFSA BIOHAZ Panel (EFSA Panel on Biological Hazards), Koutsoumanis, K., Allende, A., Alvarez-Ordóñez, A., Bolton, D., Bover-Cid, S., Chemaly, M., Davies, R., De Cesare, A., Herman, L., Hilbert, F., Lindqvist, R., Nauta, M., Peixe, L., Ru, G., Simmons, M., Skandamis, P., Suffredini, E., Cacciò, S., Chalmers, R., Deplazes, P., Devleesschauwer, B., Innes, E., Romig, T., van der Giessen, J., Hempen, M., Van der Stede, Y., Robertson, L. 2018b. Scientific opinion on the public health risks associated with food-borne parasites. *EFSA Journal* 16(12):5495, 113 pp; doi:10.2903/j.efsa.2018.5495.

EFSA Scientific Committee, Benford, D., Halldorsson, T., Jeger, M.J., Knutsen, H.K., More, S., Naegeli, H., Noteborn, H., Ockleford, C., Ricci, A., Rychen, G., Schlatter, J.R., Silano, V., Solecki, R., Turck, D., Younes, M., Craig, P., Hart, A., Von Goetz, N., Koutsoumanis, K., Mortensen, A., Ossendorp, B., Martino, L., Merten, C., Hardy, A. 2018a. Guidance on uncertainty analysis in scientific assessments. *EFSA Journal* 16(1):5123, 39 pp; doi:10.2903/j.efsa.2018.5123.

EFSA Scientific Committee, Benford, D., Halldorsson, T., Jeger, M.J., Knutsen, H.K., More, S., Naegeli, H., Noteborn, H., Ockleford, C., Ricci, A., Rychen, G., Schlatter, J.R., Silano, V., Solecki, R., Turck, D., Younes, M., Craig, P., Hart, A., Von Goetz, N., Koutsoumanis, K., Mortensen, A., Ossendorp, B., Germini, A., Martino, L., Merten, C., Smith, A., Hardy, A. 2018b. Scientific opinion on the principles and methods behind EFSA's guidance on uncertainty analysis in scientific assessment. *EFSA Journal* 16(1):5122, 282 pp; doi:10.2903/j.efsa.2018.5122.

FAO (Food and Agriculture Organization). 2016. Fishery and aquaculture statistics. Global fisheries commodities production and trade 1976–2013 (FishstatJ). In: FAO Fisheries and Aquaculture Department [online or CD-ROM]. Rome. Updated 2016. http://www.fao.org/fishery/statistics/software/fishstatj/en (accessed March 29, 2019).

FAO (Food and Agriculture Organization). FAO/WHO GIFT. Global individual food consumption data tool. http://www.fao.org/gift-individual-food-consumption/en/ (accessed March 29, 2019).

FAO/WHO (Food and Agriculture Organization/World Health Organization). 2004. Risk assessment of *Listeria monocytogenes* in ready to eat foods— Technical report. Rome: Food and Agriculture Organization of the United Nations and World Health Organization Report No. Microbiological Risk Assessment Series 5. http://www.fao.org/docrep/010/y5394e/y5394e00.html (2004, accessed March 20, 2019).

FAO/WHO (Food and Agriculture Organization/World Health Organization). 2006. Food safety risk analysis. A guide for national food safety authorities. Rome: FAO. Food and Nutrition Paper 87.

Greiner, M., Smid, J., Havelaar, A.H., Müller-Graf, C. 2013. Graphical models and Bayesian domains in risk modelling: Application in microbiological risk assessment. *Preventive Veterinary Medicine* 110(1):4–11.

Haas, C.N., Rose, J.B., Gerba, C.P. 1999. *Quantitative Microbial Risk Assessment.* New York: John Wiley & Sons, Inc.

Haberbeck, L.U., Oliveira, R.C., Vivijs, B., Wenseleers, T., Aertsen, A., Michiels, C., Geeraerd, A.H. 2015. Variability in growth/no growth boundaries of 188 different *Escherichia coli* strains reveals that approximately 75% have a higher growth probability under low pH conditions than *E. coli* O157:H7 strain ATCC 43888. *Food Microbiology* 45(B):222–230.

Haberbeck, L.U., Plaza-Rodríguez, C., Desvignes, V., Dalgaard, V., Sanaa, M., Guillier, L., Nauta, M., Filter, M. 2018. Harmonized terms, concepts and metadata for microbiological risk assessment models: The basis for knowledge integration and exchange. *Microbial Risk Analysis* 10:3–12.

Hairer, E., Roche, M., Lubich, C. 1989. The numerical solution of differential-algebraic systems by Runge-Kutta methods. *Lecture Notes in Mathematics.* Berlin, Heidelberg: Springer; doi:10.1007/BFb0093947.

Havelaar, Arie H., Nauta, Maarten J., Jansen, Jaap T. 2004. Fine-tuning food safety objectives and risk assessment. *International Journal of Food Microbiology* 93(1):11–29.

Higgins, J.P.T., Green, S. (editors). 2011. *Cochrane Handbook for Systematic Reviews of Interventions* Version 5.1.0 [updated March 2011]. *The Cochrane Collaboration,* 2011. www.handbook.cochrane.org (accessed March 29, 2019).

Jaspers, S., De Troyer, E., Aerts, M. 2018. Machine learning techniques for the automation of literature reviews and systematic reviews in EFSA. *EFSA Supporting Publications* 2018:EN-1427. 83 pp.

Jofré, A., Garriga, M., Aymerich, T., Perez-Rodriguez, F., Valero, A., Carrasco, E., Bover-Cid, S. 2016. Closing gaps for performing a risk assessment on *Listeria monocytogenes* in ready-to-eat (RTE) foods: Activity 1, an extensive literature search and study selection with data extraction on L. *monocytogenes* in a wide range of RTE food. *EFSA Supporting Publications* 13(12):EN-1141, 184 pp; doi:10.2903/sp.efsa.2016.EN-1141.

Jongenburger, I., Bassett, J., Jackson, T., Zwietering, M.H., Jewell, K. 2012a. Impact of microbial distributions on food safety I. Factors influencing microbial distributions and modelling aspects. *Food Control* 26(2):601–609.

Jongenburger, I., Bassett, J., Jackson, T., Zwietering, M.H., Jewell, K. 2012b. Impact of microbial distributions on food safety II. Factors influencing microbial distributions and modelling aspects. *Food Control* 26:546–554.

Koutsoumanis, K.P., Aspridou, Z. 2016. Moving towards a risk-based food safety management. *Current Opinion in Food Science* 12:36–41.

Kusumaningrum, H.D., van Asselt, E.D., Beumer, R.R., Zwietering, M.H. 2004. A quantitative analysis of cross-contamination of *Salmonella* and *Campylobacter* spp. via domestic kitchen surfaces. *Journal of Food Protection* 67(9):1892–1903.

Lammerding, A. 2007. Using microbiological risk assessment (MRA) in food safety. Summary report of a workshop in Prague, Czech Republic, October 2006. ILSI Europe Report Series. Brussels: ILSI Europe. http://www.ilsi.org/Europe/Publications/R2007Usi_Mic.pdf (accessed March 29, 2019).

Lammerding, A.M., Paoli, G.M. 1997. Quantitative risk assessment: An emerging tool for emerging foodborne pathogens. *Emerging Infectious Diseases* 3(4):483–487.

Lunn, D.J., Thomas, A., Best, N., Spiegelhalter, D. 2000. WinBUGS — A Bayesian modelling framework: Concepts, structure, and extensibility. *Statistics and Computing* 10(4):325–337.

Mylius, S.D., Nauta, M., Havelaar, A.H. 2007. Cross-contamination during food preparation: A mechanistic model applied to chicken-borne Campylobacter. *Risk Analysis* 27(4):803–813.

National Research Council (NRC). 2009. Science and decisions: Advancing risk assessment. Washington, DC: National Academies Press. http://www.nap.edu/catalog.php?record_id=12209 (accessed March 29, 2019).

Nauta, H. 2000. Separation of uncertainty and variability in quantitative microbial risk assessment models. *International Journal of Food Microbiology* 57(1):9–18.

Nauta, M.J. 2005. Microbiological risk assessment models for partitioning and mixing during food handling. *International Journal of Food Microbiology* 100(1–3):311–322; doi:10.1016/j.ijfoodmicro.2004.10.027.

Nauta, M.J. 2008. The modular process risk model (MPRM): A structured approach to food chain exposure assessment. In: Schaffner, D.W., Doyle, M.P. (Eds.), *Microbial Risk Analysis of Foods*. Washington: American Society of Microbiology, 99–136.

Oakley, J.E., O'Hagan, A. 2004. Probabilistic sensitivity analysis of complex models: A Bayesian approach. *Journal of the Royal Statistical Society: Series B (Statistical Methodology)* 66(3):751–769.

OHAT/NTP. (Office of Health Assessment and Translation, Division of the National Toxicology Program, National Institute of Environmental Health Sciences). 2015. *Handbook for Conducting a Literature-Based Health Assessment Using OHAT Approach for Systematic Review and Evidence Integration*. US Department of Health and Human Services.

Pérez-Rodríguez, F., Carrasco, E., Bover-Cid, S., Jofré, A., Valero, A. 2017. Closing gaps for performing a risk assessment on *Listeria monocytogenes* in ready-to-eat (RTE) foods: Activity 2, a quantitative risk characterization on *L. monocytogenes* in RTE foods; starting from the retail stage. *EFSA Supporting Publications* 2017:EN-1252. 211 pp; doi:10.2903/sp.efsa.2017.EN-1252.

Pérez-Rodríguez, F., Valero, A., Carrasco, E., García, R.M., Zurera, G. 2008. Understanding and modelling bacterial transfer to foods: A review. *Trends in Food Science and Technology* 19(3):131–144.

Pérez-Rodríguez, F., Van Asselt, E.D., García-Gimeno, R.M., Zurera, G., Zwietering, M.H. 2007. Extracting additional risk managers information from a risk assessment of *Listeria monocytogenes* in deli meats. *Journal of Food Protection* 70(5):1137–1152.

Pérez-Rodríguez, F., Zwietering, M.H. 2012. Application of the central limit theorem in microbial risk assessment: High number of servings reduces the coefficient of variation of food-borne burden-of-illness. *International Journal of Food Microbiology* 153(3):413–419; doi:10.1016/j.ijfoodmicro.2011.12.005.

Plaza-Rodríguez, C., Haberbeck, L.U., Desvignes, V., Dalgaard, P., Sanaa, M., Nauta, M., Filter, M., Guillier, L. 2018. Towards transparent and consistent exchange of knowledge for improved microbiological food safety. *Current Opinion in Food Science* 19:129–137.

Plaza-Rodriguez, C., Thoens, C., Falenski, A., Weiser, A.A., Appel, B., Kaesbohrer, A., Filter, M. 2015. A strategy to establish food safety model repositories. *International Journal of Food Microbiology* 204:81–90.

Pouillot, R., Delignette-Muller, M.L. 2010. Evaluating variability and uncertainty separately in microbial quantitative risk assessment using two R packages. *International Journal of Food Microbiology* 142(3):330–340.

Pouillot, R., Hoelzer, K., Chen, Y.H., Dennis, S.B. 2015. *Listeria monocytogenes* dose response revisited incorporating adjustments for variability in strain virulence and host susceptibility. *Risk Analysis* 35(1):90–108.

R Development Core Team. 2009. R: A language and environment for statistical computing. R Foundation for Statistical Computing. Available at http://www.R-project.org.

Sales-Ortells, H., Fernandez-Cassi, X., Timoneda, N., Dürig, W., Girones, R., Medema, G. 2015. Health risks derived from consumption of lettuces irrigated with tertiary effluent containing Norovirus. *Food Research International* 68:70–77.

Schielke, Anika, Filter, Matthias, Appel, Bernd, Johne, Reimar. 2011. Thermal stability of hepatitis E virus assessed by a molecular biological approach. *Virology Journal* 8(1):487.

Smid, J.H., Swart, A.N., Havelaar, A.H., Pielaat, A. 2011. A practical framework for the construction of a biotracing model: Application to *Salmonella* in the pork slaughter chain. *Risk Analysis* 31(9):1434–1450.

Tenenhaus-Aziza, F., Ellouze, M. 2015. Software for predictive microbiology and risk assessment: A description and comparison of tools presented at the ICPMF8 Software Fair. *Food Microbiology* 45(B):290–299.

Tsafnat, G., Glasziou, P., Choong, M.K., Dunn, A., Galgani, F., Coiera, E. 2014. Systematic review automation technologies. *Systematic Reviews* 3:74. http://www.systematicreviewsjournal.com/content/3/1/74 (accessed March10, 2019).

Tuyl, F. 2011. Were Clopper & Pearson (1934) too careful? In: *Proceedings of the Fourth Annual ASEARC Conference*. Paramatta, Australia, 17–18.

van Asselt, E.D., Zwietering, M.H. 2006. A systematic approach to determine global thermal inactivation parameters for various food pathogens. *International Journal of Food Microbiology* 107(1):73–82.

van Gerwen, S.J., te Giffel, M.C., van't Riet, K., Beumer, R.R., Zwietering, M.H. 2000. Stepwise quantitative risk assessment as a tool for characterization of microbiological food safety. *Journal of Applied Microbiology* 88(6):938–951.

Vose, D. 2000. *Risk Analysis: A Quantitative Guide*. New York: John Wiley & Sons, Inc.

Voysey, P.A., Brown, M. 2000. Microbiological risk assessment: A new approach to food safety control. *International Journal of Food Microbiology* 58(3):173–179.

WHO/FAO (World Health Organization/Food and Agriculture Organization). 2008. Principles and guidelines for the conduct of microbiological risk management (MRM) CAC/GL 63. *Codex Aliment* 2007:1–19.

Zwietering, M.H. 2009. Quantitative risk assessment: Is more complex always better? Simple is not stupid and complex is not always more correct. *International Journal of Food Microbiology* 134(1–2):57–62.

10

Hazard Identification
Microbial Risks along the Food Chain

Sofia M. Santillana Farakos, Melinda M. Hayman and Michael B. Batz

Contents

10.1 OVERVIEW AND APPROACHES TO HAZARD IDENTIFICATION

As established by the Codex Alimentarius Commission (CAC, 1999), hazard identification is the first step in a risk assessment and is defined as the "identification of biological, chemical and physical agents capable of causing adverse health effects and which may be present in a particular food or group of foods". The scope of hazard identification can therefore range from a single hazard in a single commodity to multiple hazards in multiple commodities, though most tend to focus on a single pathogen–product pathway. Potential microbial hazards include bacteria and their toxins, viruses and parasites (Benford, 2001). Hazard identification not only identifies organisms but provides known information about the pathogens, the food products and the host (disease caused, susceptibility and pathogen–host interaction) (Bassett et al., 2012). It can thus be considered from the pathogen perspective (the agent), from the public health perspective (the illnesses and the population) and from the product perspective (food or water). Typically, in a hazard identification for a microbial risk assessment, more weight is given to the pathogen perspective, as this is the aspect that must be controlled to prevent the adverse effect (Benford, 2001). There are various phases involved in conducting a microbial hazard identification in order to be able to understand the disease process in humans, the microorganism of concern and the population burden of disease. These include confirming that the microorganism can in fact cause disease and characterizing the mechanisms, symptoms and outcomes associated with disease; obtaining information on the factors and characteristics that affect disease transmission (including identifying transmission routes); assessing virulence factors and the pathogen life cycle; and evaluating the incidence, prevalence and outbreaks, including characterizing the outcomes of exposure (hospitalizations and deaths) (Haas et al., 2014).

Microbial hazards for risk assessment can be identified in different ways. One can obtain data from published studies on the presence and characteristics of microorganisms throughout the farm to fork chain, clinical and epidemiological studies, surveillance and outbreak investigations, or foodborne disease reports (Bassett et al., 2012; Benford, 2001). These data can be collected from the scientific literature, food industry databases, government agencies and relevant international organizations as well as through consultation with experts (Bassett et al., 2012). A structured method for microbial hazard identification can use decision support tools, which also have the advantage that such tools allow the system to be automatically updated as new information becomes available (Bassett et al., 2012). Decision support models that combine self-defined quantitative and qualitative reasoning can be individually designed (Van Gerwen et al. (1997); Zwietering et al. (1993)). Alternatively, one can use existing tools, such as Risk Ranger, sQMRA (semi-quantitative risk-ranking framework prototype) or the U.S. Food and Drug Administration's (FDA) fresh produce risk-ranking tool and FDA-iRISK® (Bassett et al., 2012).

While the term *risk assessment* refers to a scientific process consisting of formal components that usually leads to a quantification of risk as established by CAC (1999), it can also refer to a more general, qualitative approach based on expert opinion and experience, which is the more typical approach used by primary producers and food manufacturers (Monaghan et al., 2017). Hazard identification is part of both processes, but the way it is approached and the types of information gathered may differ depending on the purpose. In the case of hazard identification for primary producers and food manufacturers, a team typically reviews the ingredients used in the product, the activities conducted at each step in the production process and the equipment used to make the product. The team also considers the storage methods, distribution, intended use and consumers of the product. Based on this discussion, the team develops a list of potential biological, chemical and physical hazards that may be introduced, increased or controlled at each step as described in the process flow diagram. Knowledge of any adverse health-related events historically associated with the product is also considered (Bernard & Scott, 2007).

From a quantitative perspective, hazard identification is the identification of known or potential health effects associated with a particular agent and consists of gathering information about the pathogen, its presence in foods and the adverse outcome (illness or death) associated with consumption of contaminated foods (Buchanan et al., 2004). The following are examples of quantitative and qualitative approaches to hazard identification in microbial risk assessment.

10.1.1 Hazard Identification Incorporating a Quantitative Approach: *Listeria monocytogenes* in Ready-to-Eat Foods

A joint Department of Health and Human Services Food and Drug Administration (DHHS/FDA) and United States Department of Agriculture Food Safety and Inspection Services (USDA/FSIS) 2003 Quantitative Assessment of Relative Risk to Public Health from Foodborne *Listeria monocytogenes* among Selected Categories of Ready-to-Eat Foods was developed in 2003 by the U.S. Food and Drug Administration (DHHS/ FDA) and the U.S. Department of Agriculture (USDA/FSIS) (DHHS/FDA and USDA/FSIS, 2003). In this risk assessment, the hazard identification step included qualitative and quantitative information. The hazard identification contained a description of the relationship between the pathogen, its presence in foods and the adverse outcome of its presence (illness or death). It included information on the characteristics of *L. monocytogenes*, a section on listeriosis, including quantitative information on incidence, pathogenesis and the distinct human health consequences of an infection (invasive versus non-invasive), as well as a description of the susceptible populations and the epidemiological patterns seen (sporadic and outbreak cases). The information as described in the hazard identification phase determined the need for the risk assessment. Based on the outcome of the Joint DHHS/FDA and USDA/FSIS 2003 Quantitative Risk Assessment (DHHS/FDA and USDA/FSIS, 2003), which included the relative risk of listeriosis that several categories of ready-to-eat (RTE) foods posed to the U.S. population, an action plan to reduce the risk of *L. monocytogenes* by FDA and the Centers for Disease Control and Prevention (CDC) was put in place and subsequently updated in 2008 (FDA, 2008). Additionally, FSIS conducted a complementary risk assessment to understand which interventions for RTE meat and poultry products would be most effective in preventing listeriosis (USDA, 2003). In 2013, an interagency (FDA/CFSAN, USDA/FSIS and CDC) risk assessment of *L. monocytogenes* in retail delicatessens was published (FDA, USDA and CDC, 2013). The need for this risk assessment was identified as a result of epidemiologic data showing a steady incidence of listeriosis in the United States despite a decline in the presence of *L. monocytogenes* in RTE meat and poultry products in processing plants (FDA, USDA and CDC, 2013). This 2013 risk assessment does not include a hazard identification step (as does the one reported for the Joint DHHS/FDA and USDA/FSIS 2003 Quantitative Risk Assessment 2003), but it is still quantitative, using data on published studies and surveys on contamination data as well as epidemiological data.

10.1.2 Qualitative Approach to Hazard Identification: Microbial Risks for Primary Producers of Leafy Vegetables

A qualitative approach to risk assessment proposed to be used for the primary production level has been published by Monaghan et al. (2017). In this approach to risk assessment, hazard identification is defined as the process of identifying microbial organisms that could lead to a foodborne illness outbreak from identified product types. For producers of leafy vegetables, a review of risks posed by food of non-animal origin revealed that the main hazards to consider in leafy salads are *Salmonella* and norovirus (EFSA, 2013). Data from published studies additionally showed that *Escherichia coli* O157, *Salmonella*, norovirus and *Cyclospora cayetanensis* are the main cause of foodborne illness outbreaks associated with leafy salads (Monaghan et al. (2017). The most probable route of contamination (i.e., risk factor) of produce with these microorganisms was identified to be direct or indirect fecal contamination from infected livestock or workers (Monaghan et al., 2017). The hazard identification step in this qualitative risk assessment resulted in the hazard being defined as "generic fecal hazard" in the product leafy vegetables (Monaghan et al., 2017).

As seen from the two examples provided, hazard identification can use different information and approaches (i.e., quantitative or qualitative), but regardless, the result is the identification of a hazard in a food commodity. In the next section, the data necessary to conduct a hazard identification for a microbial risk assessment are described in more detail, with more emphasis on what is typically needed when using a quantitative approach.

10.2 MICROORGANISMS IN FOODS

Qualitative data on the possible presence of certain microorganisms in foods generally can be obtained from food microbiology books (e.g., Doyle and Beuchat, 2007; ICMSF, 2005) as well as other peer-reviewed literature, including risk profiles (FAO/WHO, 2014). Quantitative data on the actual presence of a microorganism in a food are available through experimental surveys published in peer-reviewed journals (e.g., Calhoun et al., 2013; Gombas et al., 2003; Luchansky et al., 2017; Zhang et al., 2017), which may include data that are made available as a result of active government sampling programs (e.g., USDA, 2018; FDA, 2018a). Data from recalls and outbreak investigations are also useful (e.g., CDC, 2013; FDA, 2016). The major categories of microorganisms that can cause foodborne illness in humans are bacteria, viruses and protozoa (USDA and EPA, 2012). Mycotoxins produced by fungi can also cause illness but will not be considered in this chapter.

10.2.1 General Pathogen Characteristics

Once the hazard (microorganism) has been identified, information on the basic characteristics of the microorganism is gathered. This aids in developing the subsequent components of the hazard identification, as discussed later, as well as in identifying control measures for the hazard of interest. The relevant characteristics will vary depending on the type of microorganism and will comprise a combination of genotypic and phenotypic characteristics. For bacterial foodborne pathogens, the general pathogen characteristics of interest may include the following:

- Taxonomy, typically genus and species. For some pathogens, consideration of subtypes such as serotype or strain may be needed.
- Cell type (e.g., Gram positive or Gram negative), cell shape (e.g., rod, cocci or spiral) and ability to form an endospore.
- Factors that contribute to virulence, such as toxin production and antibiotic resistance.
- Growth requirements.

The correct identification and classification of the microorganism are needed to collect further information. In some genera, several of the species may be human pathogens, while in other genera, there is only one species of concern. For example, the genus *Vibrio* contains several species capable of causing food- and waterborne illness (*parahaemolyticus*, *vulnificus*, and *cholerae*). In contrast, the genus *Listeria* contains at least 17 species, with *L. monocytogenes* being the sole species implicated in foodborne disease to date (Ortiz et al., 2010), although *L. ivanovii* has been implicated in human illness that was possibly foodborne (Guillet et al., 2010). The serotype may be relevant for certain bacteria (e.g., pathogenic *E. coli* and *Salmonella enterica*), as some serotypes are more frequently associated with certain commodities and/or may have increased or reduced pathogenicity. This information could be used in many ways, for example in selection of strains for challenge studies.

The physiology of the bacterial cell is significant in several ways. The Gram reaction is usually one of the first steps in bacterial identification and identifies bacterial cells as either Gram positive or Gram negative. The stain differentiates the ability of cells to retain crystal violet dye due to the structure of the cell wall. Bacterial cells primarily occur as either rods or cocci, but some other types (such as curved or spiral) may also occur. Some foodborne pathogens, such as *Clostridium botulinum*, *C. perfringens* and *Bacillus cereus*, possess the ability to form an endospore. The structure and shape of the cell, as well as the capability to form an endo-spore, can affect the susceptibility of the cell to chemicals such as pre-servatives and sanitizers and impact the sensitivity of the organism to lethality treatments such as heat. Therefore, the type of cell and the ability to form endospores will impact the selection of methods for controlling the organism before, during and after processing of food. The structure of the cell wall can also impact the choice of antibiotic used to treat infection.

Factors that affect pathogenicity, such as toxin production and/or antibiotic resistance, may also be of interest for certain pathogens. Many foodborne pathogens cause illness when the cells multiply within the human body, but several pathogens cause foodborne illness through the formation of toxin. *C. botulinum*, *Staphylococcus aureus* and emetic strains of *B. cereus* are all examples of organisms that cause illness when they multiply within the food product and produce toxin (usually at high cell densities); consumption of the toxin rather than the cells causes illness. *C. perfringens* and diarrheal strains of *B. cereus* also cause illness via produc-tion of toxin; however, the cells need to grow to high levels in the food and to be consumed, and then, the toxin is produced *in vivo*.

Bacterial cells are also characterized by their growth (replication) requirements. Various tests have been traditionally used to identify each family, genus and species, but for risk assessment, two features of special interest will be growth temperature and oxygen requirement. These are important for the development of culture and detection methods and are also critical for establishing control strategies. Most foodborne pathogens are facultative anaerobes (i.e., they can grow with or without the presence of oxygen). Each organism also has an optimal growth temperature as well as a range of temperatures over which it will grow. Refrigeration is often used as a control for microbial growth, but some foodborne patho-gens (e.g., *L. monocytogenes* and *Yersinia enterocolitica*) can grow at <4°C, and other foodborne pathogens can grow at temperatures as low as 7°C (*B. cereus*, pathogenic *E. coli* and *Salmonella*).

The prevalence and expected load of the microorganism are impor-tant inputs for risk assessors. Thus, methodologies for detecting, identi-fying and quantifying the organism are needed. The method(s) should be validated for the intended use so that the organism can be reliably detected and accurately enumerated. Methods exist for the major bacterial foodborne pathogens and include government reference methods as well as methods developed or curated by standards organizations.

For microorganisms other than bacteria, factors to consider for the purposes of hazard identification may include:

- Taxonomy.
- Factors that affect pathogenicity.
- Life cycle and/or mode of replication in the host.
- Propagation method. In contrast to bacterial foodborne pathogens, which usually can be readily cultured in the laboratory, foodborne viruses and parasites cannot always be propagated in the laboratory. This makes detection and identification more difficult.

Information on foodborne pathogens and food/pathogen pairings is readily available from numerous resources, including text books, journal articles and various government reports. FDA's *Bad Bug Book*, 2nd Edition (FDA, 2012a) and ICMSF's *Microorganisms in Foods 6: Microbial Ecology of Food Commodities* (ICMSF, 2005) are excellent starting points. Table 10.1 provides examples of characteristics of interest for several bacterial pathogens (from Barach and Hayman, 2014; Lightfoot, 2003; NACMCF, 2010).

10.2.2 Sources of Microorganisms

The next step in hazard identification is to identify pathogen reservoirs. Some foods may become contaminated with zoonotic agents because they are derived from animals that are natural hosts or reservoirs, while other foods may become contaminated in the chain from farm to fork, often due to contamination at primary production. Understanding where the microorganism of interest may reside is crucial in understanding how food ingredients and food products become contaminated with foodborne pathogens and may also aid in strategies to manage the reservoir to prevent transmission of the disease. Some foodborne pathogens are considered wide-spread and have been isolated from many environments (e.g., *L. monocytogenes*), while others may have a more specific or narrow range. For example, *Vibrio parahaemolyticus* is associated with marine environments, with the geographic distribution and incidence determined by multiple interacting factors (Desmarchelier, 2003). The most important of these factors are salinity and temperature, with the organism found in tropical and temperate locations (Desmarchelier, 2003). The presence and prevalence of foodborne pathogens may also be impacted by seasonality and rainfall as well as other environmental factors.

Environments that harbor foodborne pathogens include soil, water, animals (e.g., intestine and hide), humans (e.g., intestine and skin), animal feed, plant material, air, dust and other fomites (such as surfaces within

Table 10.1 Characteristics of Bacterial Foodborne Pathogens[a]

Organism	Gram Reaction (Cell Type)	Cell Shape	Forms Endospore	Oxygen Requirement	Temperature Range for Growth (°C)[b]
Bacillus cereus	Positive	Rod	Yes	Facultative anaerobe	4–55
Campylobacter	Negative	Spiral	No	Microaerophilic	32–45
Clostridium botulinum	Positive	Rod	Yes	Anaerobic	3.3–45 (non-proteolytic) 10–48 (proteolytic)
Clostridium perfringens	Positive	Rod	Yes	Anaerobic	10–52
Pathogenic *Escherichia coli*	Negative	Rod	No	Facultative anaerobe	6.5–49.4
Listeria monocytogenes	Positive	Rod	No	Facultative anaerobe	–0.4–45
Salmonella enterica	Negative	Rod	No	Facultative anaerobe	5.2–46.2[c]
Shigella spp.	Negative	Rod	No	Facultative anaerobe	10–45
Staphylococcus aureus	Positive	Cocci	No	Facultative anaerobe	7–50
Vibrio parahaemolyticus	Negative	Curved rod	No	Facultative anaerobe	5–43

[a] Data from Barach and Hayman (2014); Lightfoot (2003); NACMCF (2010).
[b] Limits for growth when other conditions are near optimum.
[c] Depends on the strain; most serovars will not grow below 7°C.

279

food processing environments, utensils and clothing). Therefore, food has the potential to be contaminated at a variety of points throughout the supply chain, including during primary production, transportation and manufacture, at retail and within the home.

10.2.3 Exposure Routes to Microorganisms

Some organisms may have multiple routes of transmission, while others may have one route of transmission. Some are specific to the host and must undergo a precise sequence of steps, while others are flexible and opportunistic. Many foodborne pathogens are transmitted via the fecal-oral route. The pathogen is excreted in the stool of an infected host, which then contaminates food, water or fomites, and the new host becomes infected when the pathogen is taken up by mouth (through consuming contaminated food or water, touching the mouth after touching a contaminated surface, or encountering contaminated recreational water) (Haas et al., 2014).

Enteric bacterial pathogens such as *Shigella*, pathogenic *E. coli* and *Salmonella* multiply in the intestine of the host and are shed in the feces, and then can be further spread via the fecal-oral route (Haas et al., 2014). In some instances, this is a rather direct route; for example, humans are the primary carriers of *Shigella* (i.e., human stool→food or water→human). Foodborne viruses that are pathogenic to humans are also usually solely carried and transmitted by humans. Norovirus and hepatitis A are shed in feces, and transmission occurs when an infected individual practices poor toilet hygiene and then handles food, which is then consumed by others. Fecal-oral transmission can also occur when an infected person handles and contaminates an object, which is subsequently touched by another individual, and then the microorganism is ingested. Direct person-to-person transmission of some foodborne pathogens is also possible.

In other situations, the transmission route is much more complex, and it may be difficult to elucidate the source. Pathogens from human or animal excreta can enter water sources, including the ocean and estuaries, rivers, lakes and ground water (Haas et al., 2014). Pathogens, such as pathogenic *E. coli* and *Salmonella*, can be transmitted from livestock or wild animals to irrigation water, to crops, to a manufacturing facility, and then to humans. Parasites differ from bacteria and typically have a complex life cycle. Many species require an intermediate host to complete their life cycle and form the structures that are infectious to humans (e.g., tapeworms [*Taenia* spp.]) require transit through a mammalian intermediate host before infecting humans (FDA, 2012a).

Cross contamination via fomites can occur in the processing plant, at home and in retail establishments. *L. monocytogenes* and *Salmonella* have been recognized as environmental pathogens (FDA, 2015a). Published studies have reported that certain strains of *L. monocytogenes* can persist in wet manufacturing environments for years (Nucera et al. (2010); Olsen et al. (2005); Ortiz et al. (2010)). Outbreaks have been attributed to cross contamination during packing, as in the 2011 U.S. cantaloupe outbreak (FDA, 2012b), as well as manufacturing (FDA, 2015b). Contamination at retail, such as slicing of luncheon meats in delis, is another significant concern (USDA, FDA and CDC, 2013). *Salmonella* has been recognized to persist for long periods of time in facilities manufacturing low–water activity foods. For example, two outbreaks of salmonellosis caused by the same strain of *S. agona* occurred 10 years apart (1998 and 2008). In both outbreaks, the implicated product was breakfast cereal, which was produced in the same manufacturing facility. Whole genome sequencing (WGS) showed that the 1998 and 2008 isolates were highly related, which supports the hypothesis that the strain persisted in the facility over the 10-year period (FDA, 2017). Food manufacturers need to be aware of whether these hazards are applicable to their process and implement effective programs to control and monitor these pathogens so that they are not transmitted to RTE foods.

WGS of bacterial foodborne pathogens has been increasingly employed in food microbiology. This technology has many applications and is being increasingly utilized by regulatory agencies in many ways, including foodborne outbreak investigation and characterizing isolates. WGS can be used to identify whether patients are infected by the same strain and therefore are part of a common outbreak. Regulatory agencies such as the FDA are sequencing foodborne pathogens isolated from foods and processing environments, including *L. monocytogenes*, *Salmonella* and pathogenic *E. coli*; the sequences and metadata are then deposited into a database (FDA, 2018b). When an isolate is recovered from a patient, it can be cross-referenced with isolates in the database to determine whether there is a match. This technology can be used to elucidate transmission routes. For example, in 2012, a *Salmonella* outbreak affected 425 people in the United States. Traditional epidemiology traced the illnesses to a tuna "scrape" product (tuna backmeat, which is specifically scraped off the bones and looks like a ground product) imported from India. The *Salmonella* isolated from product and patient samples, as well as previously collected *Salmonella* Bareilly isolates, were whole genome sequenced. It was determined that the outbreak isolates were highly related to an isolate recovered from shrimp that had come from southwest India. The shrimp processing plant was about 5 miles from the plant involved in the outbreak, and the shrimp isolate had been collected several years before the outbreak. This was a significant finding, because it was an indicator that the pairing of genomic information with geographic information might have the potential to be a powerful tool for traceback investigations (FDA, 2017).

WGS can also be used within a processing plant to trace transmission routes. Isolates obtained from the processing environment from routine environmental monitoring or investigational testing can be characterized using WGS. The firm can use the information to determine whether strains are persisting in the environment over time, which could indicate deficiencies in control programs (e.g., sanitation or hygienic zoning). The firm can also use the information to trace sources of contamination and movement of isolates through the production environment. For example, a study conducted in a meat processing facility demonstrated that *L. monocytogenes* isolated from the slaughter line was highly related to isolates found in the RTE packing room, indicating spread of the organism from the slaughter line into RTE areas (Nastasijevic et al., 2017). This type of information can be used to investigate how the organism is being spread from the raw to the RTE areas.

10.2.4 Survival and Growth of Microorganisms in Foods

Bacteria can replicate within foods, while other microorganisms (e.g., viruses and protozoa) do not. Preventing the growth of pathogens before, during and after the manufacture of foods is an important component of food safety, since the greater the number of pathogens, the greater the risk of illness. The characteristics of the food are termed *intrinsic parameters*; the characteristics of the food storage environment are *extrinsic parameters*. Interaction between microorganisms, food and the environment not only affects whether microorganisms grow or survive in foods but can also impact risk by affecting the pathogenicity of the cells.

Examples of intrinsic parameters are the nutrient profile of the food, pH, salt concentration, water activity, redox potential and the presence of antimicrobial compounds (e.g., preservatives) (Jay, 2000). The formulation of the food and determining the appropriate criteria for these characteristics are crucial for food safety in numerous food products.

Extrinsic parameters include the temperature of storage, the relative humidity, the presence and concentration of gases, and the presence and activities of other microorganisms (Jay, 2000). These parameters can be applied to prevent or slow the growth of foodborne pathogens in foods. Examples include freezing, modified atmosphere packaging and fermentation. The limits of bacterial foodborne growth due to both intrinsic and extrinsic parameters have been widely studied and are available from various sources. Sources of information include the Pathogen Modeling Program Online (USDA), ComBase (USDA/University of Tasmania) and publications such as "Parameters for Determining Inoculated Pack/Challenge Study Protocols" (NACMCF, 2010).

The "Hurdle Effect" is achieved by using combinations of parameters to control microbial growth (Leistner, 1992). It is possible to prevent the growth of foodborne pathogens using milder conditions, for example using a combination of low pH and low water activity, which individually do not prevent growth. An example is a formulated meat product, which may use a combination of refrigeration, low pH, salt and reduced water activity to prevent the outgrowth of *L. monocytogenes*.

10.2.5 Food Production and Beyond

The information now needs to be tied together to characterize the hazard in the food(s) of interest. Questions to consider include what the characteristics of the food are, what other microorganisms are present, how the food is produced, transported and stored, what the shelf life is, how the food is used and how these factors will impact the hazard in question.

Food manufacturing achieves several outcomes. One is to produce a food that is desirable and palatable to the consumer. It is also of paramount importance to produce a product that is safe. Food manufacturing should reduce risk. A thorough understanding of the hazards known to be associated with the food allows manufacturers to take the appropriate steps at each stage of the process to reduce or eliminate risks of illness from foods. These steps could include a microbial kill step during the production of the food (e.g., pasteurization or retorting), formulation, appropriate packaging and preventing recontamination of foods once they are processed. Other considerations include establishing the shelf life of the food and understanding how the food will be stored and used by the consumer.

However, if not done properly, food manufacturing can increase the risk. There have been many examples of outbreaks that have occurred when manufacturers have not adequately controlled the risk. Lack of control could include poor sanitation, poor employee practices, not identifying the pertinent hazards, not applying a kill step or applying it incorrectly, and recontamination/spreading contamination. The potential for these to occur and the impact on the hazard need to be understood as part of the hazard identification process. It has been increasingly recognized that preventing recontamination of RTE foods by pathogens such as *L. monocytogenes* and *Salmonella* is vital for producing safe foods. In addition to adopting robust controls to prevent contamination, such as sanitation programs, manufacturers implement environmental monitoring programs that verify the effectiveness of the controls. An environmental monitoring program typically includes root cause analysis and corrective actions to address pathogen or pathogen indicator findings.

10.3 HUMAN ADVERSE HEALTH OUTCOMES

In the fields of toxicology and environmental health related to chemical hazards, a big focus of the hazard identification step in risk assessment is to determine whether there is enough evidence to consider a chemical substance to be the cause of an adverse health effect like cancer. However, in microbial risk assessment, the hazard is usually already identified as being capable of causing human illness prior to the initiation of the risk assessment. Microbial pathogens are often isolated from the individual(s) who exhibit(s) the adverse health effect(s), which can provide positive evidence for a cause-and-effect relationship. This is because for microbial hazards, the cause-and-effect relationship can often be measured over short periods of time (acute illness), and the likelihood of an adverse effect exhibited in a population to be associated with a pathogen/food combination can be relatively large. Because the link between pathogen and disease is well established, it does not require detailed evaluation, but the information is collected and presented to provide insight and a framework of reference around the assessment (Lammerding & Fazil, 2000).

A microbial pathogen infection is the process by which a microorganism multiplies or grows in a host (Haas et al., 2014). Infection can result in either an asymptomatic state, where no illness is observed, or symptomatic disease (clinical illness). The capacity of a microbial pathogen to cause disease is known as virulence (Haas et al., 2014). Once infection begins after exposure to a pathogen, there are various possible adverse health outcomes, including acute illness, chronic illness or death. The type of adverse health outcome is dependent not only on the pathogen (determined by pathogenicity and virulence) but also on the host (affected by factors such as the immune system and age).

Assessing the public health significance of a hazard requires characterizing the symptoms, severities and likelihoods of outcomes that can be caused by a given pathogen. Different acute manifestations and chronic sequelae of illness can result from the same pathogen in different individuals, and similar adverse health effects can result from different pathogens. For instance, *Salmonella*, *E. coli* O157:H7 and *Campylobacter* can all lead to diarrhea as an acute illness. However, chronic outcomes, such as reactive arthritis (from *Salmonella*) and Guillain–Barré syndrome (from *Campylobacter*) may occur in some individuals. Examples of long-term sequelae associated with major foodborne pathogens are shown in Table 10.2 (adapted from Batz et al., 2013). Some of these symptoms may particularly affect sensitive populations, which depending on the pathogen, includes pregnant women, the elderly, infants and immunocompromised individuals (Haas et al., 2014).

One of the goals of hazard identification is to define and describe these adverse health outcomes for the microorganism of interest, including a description of the pathogenesis of the microorganism and the consequences of the disease given a sensitive population.

10.4 THE POPULATION BURDEN OF DISEASE

One of the key goals of hazard identification is to assess – at least qualitatively and if possible quantitatively – the magnitude and scope of the burden of disease on the population that can be attributed to the pathogens in question in the foods in question. Whereas quantitative microbial risk assessments are a "bottom-up" approach that uses predictive microbiological modeling to estimate the burden of illness due to a pathogen–product pathway, the hazard identification phase often uses "top down" approaches that are based on analyzing data on human illness.

10.4.1 Public Health Surveillance

The primary data source for assessing the magnitude and scope of disease burden is national public health surveillance systems, which facilitate the ongoing, systematic collection, management, analysis and interpretation of health data (Thacker et al., 2012). Published summaries of these data, often at the national or regional level, can be critical to hazard identification efforts.

Table 10.2 Examples of Long-Term Sequelae for Major Foodborne Pathogens

Campylobacter	Chronic diarrhea, Guillain–Barré syndrome (peripheral neuropathies resulting in muscle weakness, ascending paralysis, respiratory failure, permanent paraplegia and walking difficulties), irritable bowel syndrome, inflammatory bowel disease, reactive arthritis
Cryptosporidium	Chronic diarrhea, reactive arthritis, cognitive, developmental and fitness deficits in small children
Escherichia coli O157:H7	HUS (hemolytic-uremic syndrome), chronic kidney disease, renal failure and other post-HUS sequelae (e.g., hypertension, pancreatitis, seizures, hemiplegia), irritable bowel syndrome, dyspepsia, reactive arthritis
Giardia lamblia	Chronic diarrhea, irritable bowel syndrome, inflammatory bowel disease, reactive arthritis, severe malabsorption leading to cognitive, developmental and fitness deficits in small children
Listeria monocytogenes	Chronic neurologic manifestations (e.g., cranial nerve palsies, epilepsy, impaired executive and cognitive function, memory loss, vision and hearing loss), congenital neurologic sequelae (cerebral palsy, epilepsy, hearing loss, cognitive deficits, chronic lung disease)
Norovirus	Irritable bowel syndrome
Salmonella enterica (nontyphoidal)	Chronic diarrhea, irritable bowel syndrome, dyspepsia, inflammatory bowel disease, reactive arthritis
Shigella	HUS, chronic kidney disease, renal failure and other post-HUS sequelae (e.g., hypertension, pancreatitis, seizures, hemiplegia), irritable bowel syndrome, dyspepsia, inflammatory bowel disease, reactive arthritis
Toxoplasma gondii	Myocarditis, encephalitis, meningitis, irreversible ocular impairment, congenital toxoplasmosis (neurological manifestations including hearing impairment, cognitive deficits, learning disabilities, epilepsy, palsies and growth retardation), psychiatric sequelae (schizophrenia, depression)
Vibrio vulnificus	Sepsis, secondary lesions with necrotizing fasciitis or vasculitis necessitating debridement or amputation, altered mental status, hypotension, pneumonia, endometritis
Yersinia enterocolitica	Appendicitis-like mesenteric lymphadenitis, chronic diarrhea, autoimmune thyroid disease (Graves' disease), reactive arthritis

Adapted from Batz et al. (2013)

There are many kinds of surveillance systems, but for the purposes of foodborne disease, the three most important distinctions are between passive surveillance, active surveillance and outbreak surveillance. Passive surveillance systems rely on hospitals, clinics or other health care units to report cases of illness to the relevant health jurisdiction (Nsubuga et al., 2006). In many countries, there are lists of "nationally notifiable" diseases with mandatory reporting requirements and associated surveillance programs. In active surveillance, staff members monitor or solicit reporting from health care providers or laboratories and may follow up with individual patients, which vastly increases the timeliness, accuracy and completeness of data. Because this is so resource-intensive, however, these systems often use a sentinel surveillance model with limited population coverage (Ford et al., 2015). The rapid detection and investigation of foodborne disease outbreaks, usually defined as the occurrence of two or more cases of a similar illness resulting from consumption of the same food, is a goal of public health. Data from these investigations form the basis of outbreak surveillance. Unlike other surveillance systems, outbreak surveillance routinely links cases of human illness to contaminated food vehicles, contributing factors, and the settings in which food was prepared and consumed.

In the United States, passive surveillance for some foodborne pathogens is conducted by the National Notifiable Disease Surveillance System (NNDSS), which collects data on over 100 diseases and conditions (Adams et al., 2017). Based on these data, CDC publishes annual summaries, which include tables listing the number of cases and rates of each condition by region, month, age, gender, race and ethnicity, as well as maps and figures to provide additional context (e.g., Adams et al., 2015; Adams et al., 2017; Adams et al., 2016). In contrast, the Foodborne Diseases Active Surveillance Network, or FoodNet, is active surveillance; it monitors the incidence of laboratory-confirmed illnesses caused by nine foodborne pathogens across 10 catchment areas covering about 15% of the U.S. population (Scallan et al., 2011). FoodNet publishes cases of infection, incidence and trends on an annual basis (Marder et al., 2017; CDC, 2017a). Surveillance for foodborne outbreaks in the United States is provided by CDC's National Outbreak Reporting System (NORS).

CDC publishes annual summaries of reported outbreaks(Gould et al., 2013; CDC, 2017b), and makes these data searchable and downloadable online (CDC, 2013).

Similar surveillance systems are in place elsewhere in the world and are used to produce government reports on the incidence and risk factors associated with foodborne disease. In the European Union (EU), for example, member states are mandated to publish annual reports describing surveillance of zoonotic infections in humans and animals as well as the occurrence of zoonotic agents in food and feeding stuffs (DTU Food, 2017). These reports are based on data from active, passive and outbreak surveillance and are available online (EFSA, 2018). Based on these and other data, the European Centre for Disease Prevention and Control (ECDC) and the European Food Safety Authority (EFSA) coordinate annual summary reports for the EU on the trends and sources of zoonoses, zoonotic agents and foodborne outbreaks for 37 countries on over two dozen pathogens (EFSA and ECDC, 2017). Similarly, Australia's OzFoodNet publishes annual reports on eight foodborne pathogens (Ozfoodnet Working Group, 2015). Canada's FoodNet program also publishes annual reports based on active surveillance of human illness due to 10 foodborne pathogens as well as results of sampling programs in retail food, animals and water sources (PHAC,2017). Japan's National Institute of Infectious Diseases publishes data and reports on notifiable diseases and foodborne outbreaks on a regular basis (NIID, 2018).

10.4.2 Estimating the Burden of Foodborne Disease

Disease surveillance systems provide insight into the numbers of cases of illness that have been reported to authorities, but these illnesses reflect only a small slice of the overall burden of disease in the population. Most illnesses go unreported (Figure 10.1). Although circumstances vary by disease severity, few who become ill with acute gastroenteritis symptoms

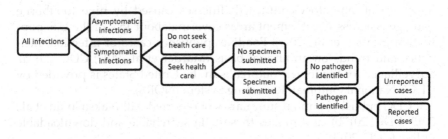

Figure 10.1 Outcome tree of underestimation of disease burden based on reported cases of illness. Adapted from Gibbons et al. (2014).

seek medical care, and only a fraction of those cases that seek medical care are asked by physicians to provide specimens for identification of the causative organism by a clinical laboratory. In turn, laboratories may fail to identify the causative organism. Clinics and laboratories may also fail to report lab-confirmed cases to local, state or federal authorities, particularly under passive surveillance.

To estimate the true incidence of foodborne diseases for a particular pathogen, the number of reported cases can be multiplied by factors to account for under-ascertainment of community infections by health care providers, under-diagnosis within health care settings, and under-reporting to surveillance systems (Gibbons et al., 2014). Of these estimates, some percentage may be assumed or estimated to be due to foodborne exposure.

Epidemiological studies are also used to measure the incidence of disease when disease surveillance is either unavailable or deemed insufficient for the need. For example, in the United Kingdom, a prospective cohort study approach is used to estimate the burden of foodborne disease (Tam et al., 2012). In Japan, the burden of foodborne campylobacteriosis is based on extrapolating from national hospital patient surveys (Kumagai et al., 2015). Norovirus is rarely estimated based on surveillance; rather, in the United States, it is often based on an estimate of the overall burden of acute gastroenteritis, based on U.S. health survey data, multiplied by a fraction assumed to be caused by norovirus (Scallan et al., 2011).

Burden of foodborne disease estimates have been developed for a number of countries, including Australia (Kirk et al., 2014), Canada (Thomas et al., 2013), France (Vaillant et al., 2005), Japan (Kumagai et al., 2015), the Netherlands (Mangen et al., 2015), New Zealand (Lake et al., 2010), South Korea (Park et al., 2015), the United Kingdom (Tam et al., 2012) and the United States (Scallan et al., 2011). Pilot studies have also been conducted in Albania, Thailand and Uganda (Lake et al., 2015). Most of these studies also estimate the number of annual hospitalizations and deaths due to foodborne illness.

In a very large study, the World Health Organization estimated the global burden of foodborne diseases, including rates of foodborne illnesses, hospitalizations and deaths, for six regions, as shown in Table 10.3 (Kirk et al., 2015). These estimates are based on data synthesis from national studies, systematic reviews of published literature, and primary data collection efforts.

Numbers of illnesses, hospitalizations and deaths are useful summary measures but do not fully account for differences in symptoms

Table 10.3 World Health Organization Estimates of the Median Rates of Foodborne Illness per 100,000 Persons for Select Pathogens, by Region, in the Year 2010 (Kirk et al., 2015)

	African Region	Region of the Americas	Eastern Mediterranean Region	European Region	South-East Asian Region	Western Pacific Region	Global
Campylobacter	2,221	1,389	1,873	522	1,152	876	1,390
Cryptosporidium	205	114	346	21	78	32	125
Entamoeba histolytica	796	212	737	0	256	229	407
E. coli, Shiga-toxin producing	5	16	65	18	19	3	17
E. coli, enteropathogenic	454	189	430	8	594	166	346
E. coli, enterotoxigenic	982	1,281	4,971	6	1,075	555	1,257
Giardia	809	309	670	54	159	354	410
Hepatitis A	232	12	237	11	494	51	199
Listeria monocytogenes	0.1	0.3	0.1	0.2	0.1	0.2	0.2
Norovirus	1,749	2,491	2,276	1,652	841	1,305	1,814
Salmonella enterica, non-typhoidal	896	1002	1610	186	906	898	1,140
Salmonella typhi	108	10	73	1	250	77	110
Shigella	523	278	627	3	1084	689	741
Vibrio cholerae	43	0.02	9	0.03	17	0.2	11
Total[a]	10,304	7,937	16,865	2,506	8,068	6,491	8,369

[a] Total median estimates include diarrheal and enteric invasive diseases caused by agents not shown in this table.

or severities or reflect the long-term health sequelae often associated with foodborne illness. Integrated measures of disease burden, such as Disability Adjusted Life Years (DALYs), Quality Adjusted Life Years (QALYs) and monetary costs of illness, are used to quantify and combine these various health impacts into a single number (Mangen et al., 2010). Many estimates of foodborne disease burden have published total and per-case estimates using these measures (Havelaar et al., 2015; Hoffmann et al., 2012; Kumagai et al., 2015; Mangen et al., 2015; Minor et al., 2015).

10.4.3 Attributing Illnesses to Foods

Unfortunately, national estimates of the burden of foodborne disease do not often answer the question of the role of specific foods in these illnesses. Additional studies are needed to quantify the percentage of foodborne illnesses caused by a specific pathogen that can be attributed to a food or category of foods (Batz et al., 2005).

The most common approaches used for foodborne illness source attribution include mathematical modeling studies that compare microbial subtypes isolated from human patients with those obtained from food and animal sources (De Knegt et al., 2015; Sheppard et al., 2009); prospective epidemiological studies such as case-control studies, some of which incorporate microbial subtyping and additional modeling (Macdonald et al., 2015; Rosner et al., 2017); analysis of aggregated foodborne outbreak data (Greig & Ravel, 2009; Painter et al., 2013); and structured elicitations of expert judgment to fill data gaps or synthesize data from multiple sources (Hald et al., 2016; Hoffmann et al., 2007).

For example, Figure 10.2 shows the estimated percentage of *E. coli* O157 cases in the United States that were attributed to categories of foods based on statistical modeling of foodborne outbreaks that occurred between 1998 and 2013, with 90% credibility intervals (IFSAC, 2017). The number of annual illnesses estimated due to *E. coli* O157 using data from Scallan et al. (2011) can be multiplied by these percentages to estimate the number of illnesses due to a pathogen–product pathway.

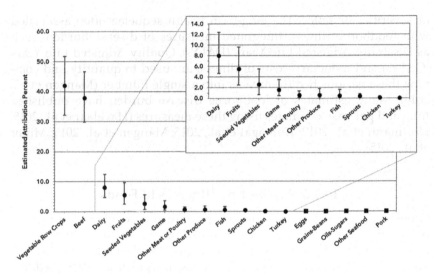

Figure 10.2 Estimated percentage of *E. coli* O157 illnesses (with 90% credibility intervals) for 2013, in descending order, attributed to each of 17 food categories, based on statistical modeling of foodborne outbreaks from 1998 to 2013. Adapted from IFSAC (2017).

10.5 CONCLUSIONS

In conclusion, this chapter provides an overview of hazard identification, different approaches with examples, and the key steps, including detailed information on the types of data needed and resources to obtain the data.

REFERENCES

Adams, D., Fullerton, K., Jajosky, R., Sharp, P., Onweh, D., Schley, A., … Kugeler, K. (2015). Summary of notifiable infectious diseases and conditions – United States, 2013. *MMWR Morb Mortal Wkly Rep*, *62*(53), 1–122. doi:10.15585/mmwr.mm6253a1.

Adams, D. A., Thomas, K. R., Jajosky, R. A., Foster, L., Baroi, G., Sharp, P., … Nationally Notifiable Infectious Conditions Group. (2017). Summary of notifiable infectious diseases and conditions – United States, 2015. *MMWR Morb Mortal Wkly Rep*, *64*(53), 1–143. doi:10.15585/mmwr.mm6453a1.

Adams, D. A., Thomas, K. R., Jajosky, R. A., Foster, L., Sharp, P., Onweh, D. H., … Nationally Notifiable Infectious Conditions Group. (2016). Summary of notifiable infectious diseases and conditions – United States, 2014. *MMWR Morb Mortal Wkly Rep, 63*(54), 1–152. doi:10.15585/mmwr.mm6354a1.

Barach, J. T., & Hayman, M. M. (2014). *HACCP A Systematic Approach to Food Safety* (5th ed.). Washington, DC: Grocery Manufacturers Association.

Bassett, J., Nauta, M., Lindqvist, R., & Zwietering, M. H. (2012). *Tools for Microbiological Risk Assessment*. Brussels, Belgium: ILSI Europe.

Batz, M. B., Doyle, M. P., Morris, G., Jr., Painter, J., Singh, R., Tauxe, R. V., … Food Attribution Working Group. (2005). Attributing illness to food. *Emerg Infect Dis, 11*(7), 993–999. doi:10.3201/eid1107.040634.

Batz, M. B., Henke, E., & Kowalcyk, B. (2013). Long-term consequences of foodborne infections. *Infect Dis Clin North Am, 27*(3), 599–616. doi:10.1016/j.idc.2013.05.003.

Benford, D. (2001). *Principles of Risk Assessment of Food and Drinking Water Related to Human Health*. Brussels, Belgium: ILSI Europe.

Bernard, D., & Scott, V. N. (2007). Hazard analysis and critical control point system: Use in controlling microbiological hazards. In: Doyle, M. P. and Beuchat, L. R. (Eds.), *Food Microbiology: Fundamentals and Frontiers* (3rd ed.). Washington, DC: American Society of Microbiology.

Buchanan, R. L., Dennis, S., & Miliotis, M. (2004). Initiating and managing risk assessments within a risk analysis framework: FDA/CFSAN's practical approach. *J Food Prot, 67*(9), 2058–2062.

Calhoun, S., Post, L., Warren, B., Thompson, S., & Bontempo, A. R. (2013). Prevalence and concentration of *Salmonella* on raw shelled peanuts in the United States. *J Food Prot, 76*(4), 575–579. doi:10.4315/0362-028x.jfp-12-322.

Codex Alimentarius Commission (CAC). (1999). *Principles and Guidelines for the Conduct of Microbiological Risk Assessment*. Retrieved from http://www.fao.org/docrep/004/y1579e/y1579e05.htm.

De Knegt, L. V., Pires, S. M., & Hald, T. (2015). Using surveillance and monitoring data of different origins in a *Salmonella* source attribution model: A European Union example with challenges and proposed solutions. *Epidemiol Infect, 143*(6), 1148–1165. doi:10.1017/S0950268814000429.

Desmarchelier, P. M. (2003). Pathogenic vibrios. In: A. D. Hocking (Ed.), *Foodborne Microorganisms of Public Health Significance* (pp. 333–358). Waterloo, DC: Australian Institute of Food Science and Technology Incorporated.

Doyle, M. P., & Beuchat, L. R. (2007). *Food Microbiology: Fundamentals and Frontiers* (3rd ed.). Washington, D.C.: American Society of Microbiology.

European Food Safety Authority (EFSA). (2013). *Scientific Opinion on the Risk Posed by Pathogens in Food of Non-Animal Origin, Part 1*. Retrieved from https://www.efsa.europa.eu/en/efsajournal/pub/3025.

European Food Safety Authority (EFSA). (2018). *Biological Hazards Reports: National Zoonoses Country Reports*. Retrieved from https://www.efsa.europa.eu/en/biological-hazards-data/reports.

European Food Safety Authority (EFSA) and European Centre for Disease Prevention and Control (ECDC). (2017). *The European Union Summary Report on Trends and Sources of Zoonoses, Zoonotic Agents and Food-Borne Outbreaks in 2016*. Retrieved from https://www.efsa.europa.eu/en/efsajournal/pub/5077.

Food and Agriculture Organization of the United Nations and World Health Organization (FAO/WHO). (2014). *Ranking of Low Moisture Foods in Support of Microbiological Risk Management*. Retrieved from http://ucfoodsafety.ucdavis.edu/files/209893.pdf.

Food Safety Inspection Service, USFDA, & Centers for Disease Control (FSIS/FDA/CDC). (2013). *Interagency Risk Assessment: Listeria monocytogenes in Retail Delicatessens*.

Ford, L., Miller, M., Cawthorne, A., Fearnley, E., & Kirk, M. (2015). Approaches to the surveillance of foodborne disease: A review of the evidence. *Foodborne Pathog Dis, 12*(12), 927–936. doi:10.1089/fpd.2015.2013.

Gibbons, C. L., Mangen, M. J., Plass, D., Havelaar, A. H., Brooke, R. J., Kramarz, P., ... Burden of Communicable diseases in Europe Consortium. (2014). Measuring underreporting and under-ascertainment in infectious disease datasets: A comparison of methods. *BMC Public Health, 14*, 147. doi:10.1186/1471-2458-14-147.

Gombas, D. E., Chen, Y., Clavero, R. S., & Scott, V. N. (2003). Survey of *Listeria monocytogenes* in ready-to-eat foods. *J Food Prot, 66*(4), 559–569.

Gould, L. H., Walsh, K. A., Vieira, A. R., Herman, K., Williams, I. T., Hall, A. J., & Cole, D. (2013). Surveillance for foodborne disease outbreaks – United States, 1998–2008. *MMWR Surveill Summ, 62*(2), 1–34.

Greig, J. D., & Ravel, A. (2009). Analysis of foodborne outbreak data reported internationally for source attribution. *Int J Food Microbiol, 130*(2), 77–87. doi:10.1016/j.ijfoodmicro.2008.12.031.

Guillet, C., Join-Lambert, O., Le Monnier, A., Leclercq, A., Mechai, F., Mamzer-Bruneel, M. F., ... Lecuit, M. (2010). Human listeriosis caused by *Listeria ivanovii*. *Emerg Infect Dis, 16*(1), 136–138. doi:10.3201/eid1601.091155.

Haas, C. N., Rose, J. B., & Gerba, C. P. (2014). Conducting the hazard identification (HAZ ID). In: Haas, C. N. (Ed.), *Quantitative Microbial Risk Assessment* (pp. 91–127). Hoboken, NJ: John Wiley & Sons, Inc.

Hald, T., Aspinall, W., Devleesschauwer, B., Cooke, R., Corrigan, T., Havelaar, A. H., ... Hoffmann, S. (2016). World Health Organization estimates of the relative contributions of food to the burden of disease due to selected foodborne hazards: A structured expert elicitation. *PLOS ONE, 11*(1), e0145839. doi:10.1371/journal.pone.0145839.

Havelaar, A. H., Kirk, M. D., Torgerson, P. R., Gibb, H. J., Hald, T., Lake, R. J., ... World Health Organization Foodborne Disease Burden Epidemiology Reference Group. (2015). World Health Organization global estimates and regional comparisons of the burden of foodborne disease in 2010. *PLOS Med, 12*(12), e1001923. doi:10.1371/journal.pmed.1001923.

Hoffmann, S., Batz, M. B., & Morris, J. G., Jr. (2012). Annual cost of illness and quality-adjusted life year losses in the United States due to 14 foodborne pathogens. *J Food Prot, 75*(7), 1292–1302. doi:10.4315/0362-028X.JFP-11-417.

Hoffmann, S., Fischbeck, P., Krupnick, A., & McWilliams, M. (2007). Using expert elicitation to link foodborne illnesses in the United States to foods. *J Food Prot, 70*(5), 1220–1229.

Interagency Food Safety Analytics Collaboration (IFSAC) (CDC/FDA/USDA. FSIS). (2017). *Foodborne Illness Source Attribution Estimates for 2013 for Salmonella, Escherichia coli O157, Listeria monocytogenes, and Campylobacter Using Multi-Year Outbreak Surveillance Data, United States.* Retrieved from https://www.cdc.gov/foodsafety/ifsac/publications.html.

International Commission on Microbiological Specifications for Foods (ICMSF). (2005). *Microorganisms Foods 6: Microbial Ecology of Food Commodities* (2nd ed.). Boston, MA: Springer.

Jay, J. M. (2000). *Modern Food Microbiology* (6th ed.). Gaithersburg, MD: Aspen Publishers.

Kirk, M., Ford, L., Glass, K., & Hall, G. (2014). Foodborne illness, Australia, circa 2000 and circa 2010. *Emerg Infect Dis, 20*(11), 1857–1864. doi:10.3201/eid2011.131315.

Kirk, M. D., Pires, S. M., Black, R. E., Caipo, M., Crump, J. A., Devleesschauwer, B., ... Angulo, F. J. (2015). World Health Organization estimates of the global and regional disease burden of 22 foodborne bacterial, protozoal, and viral diseases, 2010: A data synthesis. *PLOS Med, 12*(12), e1001921. doi:10.1371/journal.pmed.1001921.

Kumagai, Y., Gilmour, S., Ota, E., Momose, Y., Onishi, T., Bilano, V. L., ... Shibuya, K. (2015). Estimating the burden of foodborne diseases in Japan. *Bull World Health Organ, 93*(8), 540–549C. doi:10.2471/BLT.14.148056.

Lake, R. J., Cressey, P. J., Campbell, D. M., & Oakley, E. (2010). Risk ranking for foodborne microbial hazards in New Zealand: Burden of disease estimates. *Risk Anal, 30*(5), 743–752. doi:10.1111/j.1539-6924.2009.01269.x.

Lake, R. J., Devleesschauwer, B., Nasinyama, G., Havelaar, A. H., Kuchenmüller, T., Haagsma, J. A., ... Speybroeck, N. (2015). National studies as a component of the World Health Organization initiative to estimate the global and regional burden of foodborne disease. *PLOS ONE, 10*(12), e0140319. doi:10.1371/journal.pone.0140319.

Lammerding, A. M., & Fazil, A. (2000). Hazard identification and exposure assessment for microbial food safety risk assessment. *Int J Food Microbiol, 58*(3), 147–157. doi:10.1016/s0168-1605(00)00269-5.

Leistner, L. (1992). Food preservation by combined methods. *Food Res Int, 25*(2), 151–158. doi:10.1016/0963-9969(92)90158-2.

Lightfoot, D. (2003). Shigella. In: A. D. Hocking (Ed.), *Foodborne Microorganisms of Public Health Significance* (pp. 543–552). Washington, DC: Food and Agriculture Organization.

Luchansky, J. B., Chen, Y., Porto-Fett, A. C. S., Pouillot, R., Shoyer, B. A., Johnson-DeRycke, R., … Dennis, S. (2017). Survey for *Listeria monocytogenes* in and on ready-to-eat foods from retail establishments in the United States (2010 through 2013): Assessing potential changes of pathogen prevalence and levels in a decade. *J Food Prot, 80*(6), 903–921. doi:10.4315/0362-028x.Jfp-16-420.

MacDonald, E., White, R., Mexia, R., Bruun, T., Kapperud, G., Lange, H., … Vold, L. (2015). Risk factors for sporadic domestically acquired *Campylobacter* infections in Norway 2010–2011: A national prospective case-control study. *PLOS ONE, 10*(10), e0139636. doi:10.1371/journal.pone.0139636.

Mangen, M. J., Batz, M. B., Kasbohrer, A., Hald, T., Morris, J. G., Jr., Taylor, M., & Havelaar, A. H. (2010). Integrated approaches for the public health prioritization of foodborne and zoonotic pathogens. *Risk Anal, 30*(5), 782–797. doi:10.1111/j.1539-6924.2009.01291.x.

Mangen, M. J., Bouwknegt, M., Friesema, I. H., Haagsma, J. A., Kortbeek, L. M., Tariq, L., … Havelaar, A. H. (2015). Cost-of-illness and disease burden of food-related pathogens in the Netherlands, 2011. *Int J Food Microbiol, 196*, 84–93. doi:10.1016/j.ijfoodmicro.2014.11.022.

Marder, E. P., Cieslak, P. R., Cronquist, A. B., Dunn, J., Lathrop, S., Rabatsky-Ehr, T., … Geissler, A. L. (2017). Incidence and trends of infections with pathogens transmitted commonly through food and the effect of increasing use of culture-independent diagnostic tests on surveillance – Foodborne Diseases Active Surveillance Network, 10 U. S. Sites, 2013–2016. *MMWR Morb Mortal Wkly Rep, 66*(15), 397–403. doi:10.15585/mmwr.mm6615a1.

Minor, T., Lasher, A., Klontz, K., Brown, B., Nardinelli, C., & Zorn, D. (2015). The per case and total annual costs of foodborne illness in the United States. *Risk Anal, 35*(6), 1125–1139. doi:10.1111/risa.12316.

Monaghan, J. M., Augustin, J. C., Bassett, J., Betts, R., Pourkomailian, B., & Zwietering, M. H. (2017). Risk assessment or assessment of risk? Developing an evidence-based approach for primary producers of leafy vegetables to assess and manage microbial risks. *J Food Prot, 80*(5), 725–733. doi:10.4315/0362-028x.Jfp-16-237.

Nastasijevic, I., Milanov, D., Velebit, B., Djordjevic, V., Swift, C., Painset, A., & Lakicevic, B. (2017). Tracking of *Listeria monocytogenes* in meat establishment using whole genome sequencing as a food safety management tool: A proof of concept. *Int J Food Microbiol, 257*, 157–164. doi:10.1016/j.ijfoodmicro.2017.06.015.

National Advisory Committee on Microbiological Criteria of Foods (NACMCF). (2010). Parameters for determining inoculated pack/challenge study protocols. *J Food Prot, 73*(1), 140–202.

National Food Institute (DTU Food). (2017). *Annual Report on Zoonoses in Denmark 2016*. Retrieved from https://www.food.dtu.dk/Publikationer/Sygdoms fremkaldende-mikroorganismer/Zoonoser-aarlige-rapporter.

National Institute of Infectious Diseases (NIID). (2018). Retrieved from https://www.niid.go.jp/niid/en/.

Nsubuga, P., White, M. E., Thacker, S. B., Anderson, M. A., Blount, S. B., Broome, C. V., ... Trostle, M. (2006). Public health surveillance: A tool for targeting and monitoring interventions. In: D. T. Jamison, J. G. Breman, A. R. Measham, G. Alleyne, M. Claeson, D. B. Evans, P. Jha, A. Mills, & P. Musgrove (Eds.), *Disease Control Priorities in Developing Countries.* Washington, DC: The International Bank for Reconstruction and Development/The World Bank.

Nucera, D., Lomonaco, S., Bianchi, D. M., Decastelli, L., Grassi, M. A., Bottero, M. T., & Civera, T. (2010). A five year surveillance report on PFGE types of *Listeria monocytogenes* isolated in Italy from food and food related environments. *Int J Food Microbiol, 140*(2), 271–276. doi:10.1016/j.ijfoodmicro.2010.04.016.

Olsen, S. J., Patrick, M., Hunter, S. B., Reddy, V., Kornstein, L., MacKenzie, W. R., ... Mead, P. (2005). Multistate outbreak of *Listeria monocytogenes* infection linked to delicatessen turkey meat. *Clin Infect Dis, 40*(7), 962–967. doi:10.1086/428575.

Ortiz, S., López, V., Villatoro, D., López, P., Dávila, J. C., & Martínez-Suárez, J. V. (2010). A 3-year surveillance of the genetic diversity and persistence of *Listeria monocytogenes* in an Iberian pig slaughterhouse and processing plant. *Foodborne Pathog Dis, 10*(10), 1177–1184.

OzFoodNet Working Group. (2015). Monitoring the incidence and causes of diseases potentially transmitted by food in Australia: Annual report of the OzFoodNet network, 2011. *Commun Dis Intell Q Rep, 39*(2), E236–264.

Painter, J. A., Hoekstra, R. M., Ayers, T., Tauxe, R. V., Braden, C. R., Angulo, F. J., & Griffin, P. M. (2013). Attribution of foodborne illnesses, hospitalizations, and deaths to food commodities by using outbreak data, United States, 1998–2008. *Emerg Infect Dis, 19*(3), 407–415. doi:10.3201/eid1903.111866.

Park, M. S., Kim, Y. S., Lee, S. H., Kim, S. H., Park, K. H., & Bahk, G. J. (2015). Estimating the burden of foodborne disease, South Korea, 2008–2012. *Foodborne Pathog Dis, 12*(3), 207–213. doi:10.1089/fpd.2014.1858.

Public Health Agency of Canada (PHAC). (2017). *FoodNet Canada Short Report 2015.* Ottawa: PHAC.

Rosner, B. M., Schielke, A., Didelot, X., Kops, F., Breidenbach, J., Willrich, N., ... Stark, K. (2017). A combined case-control and molecular source attribution study of human *Campylobacter* infections in Germany, 2011–2014. *Sci Rep, 7*(1), 5139. doi:10.1038/s41598-017-05227-x.

Scallan, E., Hoekstra, R. M., Angulo, F. J., Tauxe, R. V., Widdowson, M. A., Roy, S. L., ... Griffin, P. M. (2011). Foodborne illness acquired in the United States-- Major pathogens. *Emerg Infect Dis, 17*(1), 7–15.

Sheppard, S. K., Dallas, J. F., Strachan, N. J., MacRae, M., McCarthy, N. D., Wilson, D. J., ... Forbes, K. J. (2009). *Campylobacter* genotyping to determine the source of human infection. *Clin Infect Dis, 48*(8), 1072–1078.

Tam, C. C., Rodrigues, L. C., Viviani, L., Dodds, J. P., Evans, M. R., Hunter, P. R., ... Committee, I. I. D. S. E. (2012). Longitudinal study of infectious intestinal disease in the UK (IID2 study): Incidence in the community and presenting to general practice. *Gut, 61*(1), 69–77. doi:10.1136/gut.2011.238386.

Thacker, S. B., Qualters, J. R., Lee, L. M., & Centers for Disease, C., & Prevention. (2012). Public health surveillance in the United States: Evolution and challenges. *MMWR Suppl, 61*(3), 3–9.

Thomas, M. K., Murray, R., Flockhart, L., Pintar, K., Pollari, F., Fazil, A., ... Marshall, B. (2013). Estimates of the burden of foodborne illness in Canada for 30 specified pathogens and unspecified agents, circa 2006. *Foodborne Pathog Dis, 10*(7), 639–648. doi:10.1089/fpd.2012.1389.

United States Centers for Disease Control (CDC). (2013). *Foodborne Outbreak Online Database (FOOD Tool) Data (1998–2015).* Retrieved from https://wwwn.cdc.gov/foodborneoutbreaks/.

United States Centers for Disease Control (CDC). (2017a). *FoodNet 2015 Surveillance Report (Final Data).* Retrieved from https://www.cdc.gov/foodnet/reports/annual-reports-2015.html.

United States Centers for Disease Control (CDC). (2017b). *Surveillance for Foodborne Disease Outbreaks, United States, 2015: Annual Report.* Retrieved from https://www.cdc.gov/foodsafety/pdfs/2015FoodBorneOutbreaks_508.pdf.

United States Department of Agriculture Food Safety and Inspection Service (USDA/FSIS). (2003). *FSIS Risk Assessment for Listeria monocytogenes in Deli Meats.* Retrieved from https://www.fsis.usda.gov/shared/PDF/Lm_Deli_Risk_Assess_Final_2003.pdf.

United States Department of Agriculture Food Safety and Inspection Service (USDA/FSIS). (2018). *Data Collection and Reports.* Retrieved from https://www.fsis.usda.gov/wps/portal/fsis/topics/data-collection-and-reports/microbiology.

United States Department of Agriculture Food Safety and Inspection Service. (USDA/FSIS) and United States Environmental Protection Agency (EPA). (2012). *Microbial Risk Assessment Guideline: Pathogenic Microorganisms with Focus on Food and Water.* Retrieved from https://www.fsis.usda.gov/shared/PDF/Microbial_Risk_Assessment_Guideline_2012-001.pdf.

United States Food and Drug Administration (FDA). (2008). *Current FDA Activities Related to the Listeria monocytogenes Action Plan.* Retrieved from https://www.fda.gov/Food/FoodScienceResearch/RiskSafetyAssessment/ucm208995.htm.

United States Food and Drug Administration (FDA). (2012a). *Bad Bug Book: Foodborne Pathogenic Microorganisms and Natural Toxins Handbook* (2nd ed.). Silver Spring, Maryland.

United States Food and Drug Administration (FDA). (2012b). *Information on the Recalled Jensen Farms Whole Cantaloupes.* Retrieved from http://wayback.archive-it.org/7993/20171114155043/https://www.fda.gov/Food/RecallsOutbreaksEmergencies/Outbreaks/ucm272372.htm#final.

United States Food and Drug Administration (FDA). (2015a). *Current Good Manufacturing Practice, Hazard Analysis, and Risk-Based Preventive Controls for Human Food. Final Rule.* Retrieved from https://www.federalregister.gov/documents/2015/09/17/2015-21920/current-good-manufacturing-practice-hazard-analysis-and-risk-based-preventive-controls-for-human.

United States Food and Drug Administration (FDA). (2015b). *FDA Investigates Listeria monocytogenes in Ice Cream Products from Blue Bell Creameries*. Retrieved from http://wayback.archive-it.org/7993/20171114154904/https://www.fda.gov/Food/RecallsOutbreaksEmergencies/Outbreaks/ucm438104.htm.

United States Food and Drug Administration (FDA). (2016). *Recalls, Market Withdrawals & Safety Alerts (Archive for Recalls)*. Retrieved from http://www.fda.gov/Safety/Recalls/ArchiveRecalls/default.htm.

United States Food and Drug Administration (FDA). (2017). *Examples of How FDA Has Used Whole Genome Sequencing of Foodborne Pathogens for Regulatory Purposes*. Retrieved from https://www.fda.gov/Food/FoodScienceResearch/WholeGenomeSequencingProgramWGS/ucm422075.htm.

United States Food and Drug Administration (FDA). (2018a). *Microbiological Surveillance Sampling*. Retrieved from https://www.fda.gov/Food/ComplianceEnforcement/Sampling/ucm473112.htm.

United States Food and Drug Administration (FDA). (2018b). *Whole Genome Sequencing (WGS) Program*. Retrieved from https://www.fda.gov/Food/FoodScienceResearch/WholeGenomeSequencingProgramWGS/.

United States Food and Drug Administration (DHHS/FDA) and Food Safety Inspection Service United States Department of Agriculture (FSIS/USDA). (2003). *Joint DHHS/FDA and USDA/FSIS 2003 Quantitative Assessment of Relative Risk to Public Health from Foodborne Listeria monocytogenes Among Selected Categories of Ready-to-Eat Foods*. Retrieved from https://www.fda.gov/Food/FoodScienceResearch/RiskSafetyAssessment/ucm183966.htm.

Vaillant, V., de Valk, H., Baron, E., Ancelle, T., Colin, P., Delmas, M. C., ... Desenclos, J. C. (2005). Foodborne infections in France. *Foodborne Pathog Dis, 2*(3), 221–232. doi:10.1089/fpd.2005.2.221.

van Gerwen, S. J. C., de Wit, J. C., Notermans, S., & Zwietering, M. H. (1997). An identification procedure for foodborne microbial hazards. *Int J Food Microbiol, 38*(1), 1–15. doi:10.1016/s0168-1605(97)00077-9.

Zhang, G., Hu, L., Melka, D., Wang, H., Laasri, A., Brown, E. W., ... Hammack, T. S. (2017). Prevalence of *Salmonella* in cashews, hazelnuts, macadamia nuts, pecans, pine nuts, and walnuts in the United States. *J Food Prot, 80*(3), 459–466. doi:10.4315/0362-028x.jfp-16-396.

Zwietering, M. H., Wijtzes, T., Wit, d. J. C., & Riet, v. t. K. (1993). A decision support system for prediction of the microbial spoilage in foods. *J Food Prot, 55*, 973–979. doi:urn:nbn:nl:ui:32-20354.

11

Predictive Microbiology Tools for Exposure Assessment

Jean Carlos Correia Peres Costa, Daniel Angelo Longhi,
Letícia Ungaretti Haberbeck, and Gláucia Maria Falcão de Aragão

Contents

11.1 PREDICTIVE MICROBIOLOGY FOR QUANTITATIVE MICROBIOLOGICAL RISK ASSESSMENT

In the food chain from farm to fork, microorganisms can encounter a series of food processing and preservation steps. The area of predictive microbiology makes use of a number of mathematical models to describe the behaviour of bacterial populations in these steps. Predictive microbiology models are of great importance for quantitative microbiological risk assessments (QMRA), as QMRA should explicitly consider the dynamics of microbiological growth, survival and death in foods (Codex 1999).

In recent years, predictive microbiology has proved to be a promising tool to estimate the changes in microbial concentration in foods over the farm-to-fork chain (Messens et al. 2018). This occurred mainly due to successfully validated models and active software development in the area (Koutsoumanis et al. 2016; Mejlholm et al. 2015; Mejlholm et al. 2010). Predictive microbial models can help to understand the microbial behaviour in food systems depending on environmental conditions, being a powerful tool to evaluate the microbial exposure in the exposure assessments step within a QMRA (Koutsoumanis et al. 2016).

The application of predictive models in exposure assessments is not always straightforward, especially for QMRA aiming to evaluate the status of public health concerning a specific hazard and/or food product. While QMRA studies assess probabilities and therefore need to use stochastic models, preferably second-order Monte Carlo models, predictive growth models are generally deterministic models, i.e. they are developed and validated to produce point estimates as outputs (Nauta 2002). This deficiency highlighted the need for the development of models expressing populations of microorganisms in terms of probability and drove the start of the so-called "stochastic predictive microbiology" (Koutsoumanis et al. 2016; Nicolai and Van Impe 1996). In the last 20 years, a number of stochastic predictive modelling approaches, aimed at quantifying and integrating different types of variability, have been reported (Augustin et al. 2011; Couvert et al. 2010; Delignette-Muller et al. 2006; Koutsoumanis et al. 2007; Koutsoumanis et al. 2010; Mejlholm et al. 2015; Membré et al. 2005; Pouillot et al. 2003). Besides, transfer, mixing, partitioning and some growth/no growth boundary models apply a probabilistic modelling approach.

The farm-to-fork food chain has a large variety of processes that require different models. The variety of models that can be used is large, and a description of the food chain may be difficult (Nauta 2008). Nauta (2001) introduced the use of modular process risk models (MPRMs) as a tool for QMRA. This approach splits the food pathway into processing steps that describe one of six basic processes: growth, inactivation, partitioning, mixing, removal and cross-contamination. In theory, once the modelling of these basic processes is established, any food pathway can be modelled when it is described as a sequence of consecutive basic processes. Traditionally, models developed in the area of predictive microbiology describe microbial growth and inactivation. These models, together with cross-contamination (transfer), mixing, partitioning and removal, are discussed in this chapter.

11.2 PREDICTIVE MICROBIOLOGY MODEL TYPES

The literature on predictive microbiology presents several classifications of mathematical models. Predictive models can be classified as (i) primary, secondary and tertiary, (ii) kinetic or probabilistic, and (iii) empirical or mechanistic (McDonald and Sun 1999). These proposed classifications are based on modelling levels, the way of obtaining the experimental data and the construction form of the mathematical model, respectively.

The classification of models by modelling level was proposed by Whiting and Buchanan (1993). Primary models measure the response of the microorganism with time to a single set of conditions, in which each population versus time curve can be described by a set of specific values for each of the parameters in the model (e.g. lag phase, growth rate and D-value). Secondary models describe the response of one or more parameters of a primary model to changes in one or more of the environmental conditions. The environmental conditions can be intrinsic to the food product, such as pH, water activity (a_w) and organic acids concentration, and/or extrinsic, such as temperature, pressure and air composition of the packaging, among others. Tertiary models are applications of one or more secondary models to generate systems for providing predictions to non-modellers (e.g. user-friendly or applications software and expert systems) (Whiting and Buchanan 1993).

Primary and secondary models are based on experimental data at constant conditions. However, environmental conditions (e.g. temperature) can change during distribution and storage. Thus, dynamic models have been developed to predict the behaviour of microorganisms under conditions that vary with time, especially under non-isothermal conditions (Haberbeck et al. 2012; Longhi et al. 2013; Silva et al. 2017). Dynamic models can be categorized as tertiary models, which are widely used to describe microbial growth from information obtained with the primary and secondary models for varying environmental conditions. Some of these tertiary models are available in user-friendly software, such as ComBase, MicroHibro, PMP and FSSP, among others. These software may include algorithms for calculating changes in environmental conditions, comparing microbial behaviour under different conditions or constructing growth curves of more than one microorganism simultaneously.

Baty and Delignette-Muller (2004) presented the classification of the microbial growth models by their mathematical form of construction. The models can be characterized as sigmoid models, models with an adjustment function or compartmental models. Sigmoid models (e.g. logistic and Gompertz models) were historically used to describe the increase in the logarithm of the bacterial cell density with time. Models with an adjustment function (e.g. Baranyi and Roberts model) are less empirical and based on differential equations. Compartmental models (e.g. logistic model with delay and Buchanan three-phase linear model) were developed in order to model the lag phase (Baty and Delignette-Muller 2004). These models will be discussed later.

McMeekin and Ross (2002) reported the classification of the models by the construction form. Predictive models can be classified as mechanistic, which are models that present physical, chemical and/or biological explanations for their parameters, or as empirical, which are models without explanations for their parameters and are usually proposed based only on the observation of the format of common mathematical functions (McMeekin and Ross 2002). Most of the models used in predictive microbiology are not purely mechanistic, and some of them are simply empirical models with simple mathematical function adjustments. For Zwietering and Den Besten (2011), the use of models for the description of microbial growth kinetics does not assume that the mechanism has been fully understood, and the acceptable performance of model fit is not a guarantee that its mechanism is right. The development of fully mechanistic microbial growth models has been limited by the inability to provide quantitative values for all model parameters (McMeekin et al. 1993). The investigation of correlations between results of different experiments is important because biological principles can be found and mechanisms can be inferred. On the other hand, Corradini and Peleg (2005) stated that a completely different approach consists in abandoning the attempt to find a universal growth model and relying only on the microorganism's pattern of growth and the environmental conditions. The format of the model can be chosen through convenient mathematical considerations using the principle of parsimony.

11.3 PRIMARY MODELS: GROWTH, INTERACTION AND INACTIVATION MODELS

11.3.1 Growth Models

Growth models have been a major area of development in predictive microbiology over the past 25 years. Traditionally, these models rely on the generation of kinetic data, under defined environmental conditions, allowing the description of microbial curves in food. These microbial curves present four phases: lag, exponential growth, stationary and decline (see Figure 11.1). In practice, microbial growth models in foods assume a sigmoidal growth function, i.e. the decline phase is ignored (Amézquita et al. 2011; Buchanan et al. 1997).

11.3.1.1 Gompertz and Logistic Models

Gompertz and logistic models were introduced by Gibson et al. (1987) and have been successfully used to describe nonlinear microbial responses,

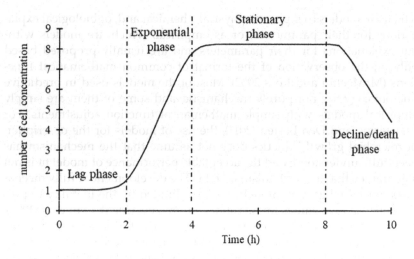

Figure 11.1 Typical microbial growth curve as a function of time representing four phases: lag, exponential growth, stationary and decline.

making it possible to express the growth in log colony forming units (cfu)/ (g or mL) as a function of time. These models are the most common sigmoidal functions used to fit to microbial growth data, because they consist of three phases, similar to the microbial growth curve (Pérez-Rodríguez and Valero 2013).

The Gompertz equation (Equation 11.1) is a function based on the limitation of space and/or nutrients as well as the production of toxic metabolites, where the growth rate is not constant. Typically, growth rate would increase to a maximum, and then it would decrease. In this way, the maximum growth rate is determined at the point of inflection in the microbial curve (Devlieghere et al. 2009; McKellar and Lu 2003).

$$\log N(t) = A + C \exp\left\{-\exp\left[-B(t - M)\right]\right\} \tag{11.1}$$

where $N(t)$ is the microbial concentration (cfu/[g or mL]) at time t (h), A (log cfu/[g or mL]) is the lower asymptotic value as t decreases to zero, C (log cfu/[g or mL]) is the difference between the upper and lower asymptotic value, M (h) is the time when the absolute growth rate is maximum, and B (h^{-1}) is the relative growth rate at M.

The regression empirical parameters of the Gompertz model can be replaced by parameters with biological meaning (e.g. lag time, maximum growth rate, initial and maximum microbial population) through adequate mathematical expressions, as proposed by Zwietering et al. (1990). Thus, the lag time duration (λ) (h) (Equation 11.2) and the maximum growth rate (μ) (log [cfu/g or mL/h]) (Equation 11.3) can be calculated:

$$\lambda = M - \frac{1}{B} \tag{11.2}$$

$$\mu = \frac{B \cdot C}{e} \tag{11.3}$$

where e is equal to 2.7182.

In order to simplify the fitting process, the reparametrized Gompertz model (Equation 11.4) proposed by Zwietering et al. (1990) can be written:

$$y = D \exp\left\{-\exp\left[\frac{\mu\, e}{A}(\lambda - t) + 1\right]\right\} \tag{11.4}$$

where y is the logarithm of the relative population size ($y = \ln[N/N_0]$), N_0 is the microbial concentration at time zero (cfu/[g or mL]), D is the maximum increase of population attained in the stationary phase. Table 11.1 summarizes all variables presented in the different equations of this chapter.

Several studies have used the deterministic Gompertz models to describe the counting of microbial growth in foods (Dabadé et al. 2015; Kim et al. 2018; Slongo et al. 2009; Tarlak et al. 2018) and growth curves based on turbidity data (Chatterjee et al. 2015; Mytilinaios et al. 2012; Perni et al. 2005). However, some authors (Dalgaard et al. 1994; Membré et al. 1999) reported some limitations associated with the use of the Gompertz model, for instance, overestimated growth rate compared with the usual definition of the maximum growth rate (Pérez-Rodríguez and Valero 2013). An additional limitation is that experimental data are required over the whole growth range in order to get a good model fit (McMeekin et al. 2013; Peleg 1997).

Another widely applied primary growth model to describe the microbial growth is the logistic model (Equation 11.5), also proposed by Gibson et al. (1987). The difference between the logistic and Gompertz models is that the former is described with a symmetric growth pattern.

$$\log N(t) = A + \frac{C}{1 + \exp\left[-B\left(t - M\right)\right]} \tag{11.5}$$

Table 11.1 Summary of All Variables Presented in the Different Equations

Symbol	Description	Unit
A	Lower asymptotic value as time decreases to zero	log cfu/(g or mL)
$A(t)$	An adjustment function for the modified Baranyi and Roberts model	h
a_f and A_f	Accuracy factor	unitless
B	Relative growth rate at M	h^{-1}
b	Regression coefficient	$(h^{-0.5}°C^{-1})$
b_f and B_f	Bias factor	unitless
C	Difference between the upper and lower asymptotic values	log cfu/(g or mL)
c	Proportionality constant	$°C^{-1}$
D	Maximum increase of population attained in the stationary phase	log cfu/(g or mL)
D_{ref}	The decimal reduction time at a reference temperature T_{ref} (°C)	s or min
D-value	The decimal reduction time	s or min
E_a	Activation energy of the system	J/mol
H	Removal function	log cfu/(g or mL)
h_0	Physiological state of the cells at $t = t_0$	unitless
K	First-order rate constant	s^{-1} or min^{-1}
k_{max}	Maximum specific decay rate	min^{-1} or h^{-1}
M	Time when the absolute growth rate is maximum	h
m	Curvature factor	unitless
n	Number of observations	unitless
$N(t)$, N_t or N	Microbial concentration at time t	log cfu/(g or mL)
N_0	Microbial concentration at time zero	log cfu/(g or mL)
N_{max}	Maximum microbial concentration	log cfu/(g or mL)
N_{res}	Residual microbial concentration	log cfu/(g or mL)
N_{cri}	Maximum critical concentration	log cfu/(g or mL)

(*Continued*)

Table 11.1 (Continued)   Summary of All Variables Presented in the Different Equations

Symbol	Description	Unit
N_{out}	Microbial concentration at the end of the process step	log cfu/(g or mL)
N_{in}	Microbial concentration at the beginning of the process step	log cfu/(g or mL)
Q	Measure of the physiological state of cells at time t	unitless
R	Universal gas constant	J/mol.K
R^2	Coefficient of determination	unitless
r_0	Fraction of the growing initial subpopulation	unitless
RMSE	Root-mean-square error	unitless
$S(t)$	Momentary ("instantaneous") survival ratio at time t	unitless
T	Time	h, min or s
t	Temperature	°C
t_L	Time before inactivation	h, min or s
t_{lag}	Time when the lag phase ends	h
υ	Rate of increase of the limiting substrate, generally assumed to be equal to μ_{max}	h^{-1}
t_{max}	Time when N_{max} is reached	h
T_{max}	Temperature in the upper part of the range over which growth is not possible	°C
T_{min}	Minimum theoretical temperature at which growth is detected	°C
T_{ref}	Reference temperature at which the shelf-life is known	°C
y	Logarithm of the relative population size ($y = \ln[N/N_0]$)	Unitless
Y	Natural logarithm of cell concentration at time t ($Y = \ln [N]$	ln cfu/(g or mL)

(*Continued*)

Table 11.1 (Continued)   Summary of All Variables Presented in the Different Equations

Symbol	Description	Unit
y_0	Natural logarithm of the cell concentration at $t = t_0$	log cfu/(g or mL)
y_{max}	Natural logarithm of the maximum cell concentration	log cfu/(g or mL)
$y_{observed}$	Data observed experimentally	according to the data
$y_{predicted}$	Data predicted by the model	according to the data
z-value	The temperature increase required to reduce the D-value by a factor of 10	°C
α	Scale parameter	s or min
α_{AB}	The competition factor parameter of Population A on Population B	unitless
α_{BA}	The competition factor parameter of Population B on Population A	unitless
β	Shape parameter	unitless
λ	Lag time duration	h
μ_{max}	Maximum specific growth rate	h^{-1}
μ	Specific growth rate	log cfu/(g or mL)/h
μ_G	Maximum specific growth	h^{-1}
μ_D	Maximum specific death	h^{-1}
$\bar{y}_{observed}$	Arithmetic mean of the observed data	according to the data

The parameters λ and μ can be defined by Equations 11.6 and 11.7, respectively.

$$\lambda = M - \frac{2}{B} \qquad (11.6)$$

$$\mu = \frac{B \cdot C}{4} \qquad (11.7)$$

To correct a failure in the logistic equation, several modifications were proposed, accounting for asymmetric growth curves and improving its accuracy (Augustin and Carlier 2000a; Fujikawa et al. 2003; Zwietering et al. 1990). Rosso et al. (1996) proposed a modification to the logistic model (Equation 11.8) with delay ($\lambda > 0$) and without delay ($\lambda = 0$).

$$\begin{cases} \text{If} \quad t < \lambda \qquad Log\left(N_t\right) = Log\left(N_0\right) \\ \\ \text{If} \quad t \geq \lambda \qquad Log\left(N_t\right) = Log\left(\dfrac{N_{max}}{1+\left(\dfrac{N_{max}}{N_0}-1\right)\cdot\left(\exp\left(-\mu_{max}\cdot\left(t-\lambda\right)\right)\right)} \right) \end{cases} \qquad (11.8)$$

where N_{max} is the maximum microbial concentration (cfu/[g or mL]), and μ_{max} is the maximum specific growth rate (h^{-1}).

The application of the logistic model with and without delay has provided good accuracy to describe deterministic microbial growth curves (Bover-Cid et al. 2019; Hereu et al. 2014; Wang et al. 2012). Ancelet et al. (2012) applied the logistic model with delay in a farm-to-fork QMRA model to determine the growth of *Bacillus cereus* in courgette purée.

11.3.1.2 Baranyi and Roberts Model

Baranyi and Roberts (1994) developed a mechanistic model for bacterial growth (Equation 11.9), including an exponential linear growth phase and a lag phase calculated by an adjustment function $A(t)$ (Equation 11.10). In this model, the lag phase extension depends on the cell's physiologic state and its adaptation to the new environment. For instance, lag phase is longer if the cells are not adapted to the new environment. Once the cells have adjusted to the new environment, they grow exponentially until they reach the stationary phase, limited by restrictions dictated by the growth medium (Baranyi and Roberts 1994; Pérez-Rodríguez 2013; McKellar and Lu 2003).

$$Y = y_0 + \mu_{max}\cdot A\left(t\right) - \frac{1}{m}\cdot\ln\left(1 + \frac{e^{m\cdot\mu_{max}\cdot A(t)} - 1}{e^{m\left(y_{max}-y_0\right)}}\right) \qquad (11.9)$$

$$A(t) = t + \frac{\ln\left(e^{-\mu_{max}t} + e^{-h_0} - e^{-v_t - h_0}\right)}{\mu_{max}} \qquad (11.10)$$

where m is a curvature factor and $A(t)$ represents an adjustment function for the model.

The Baranyi and Roberts model has been used extensively to describe the growth of pathogenic and spoilage bacteria (Carrasco et al. 2006; Silva et al. 2018; Tarlak et al. 2018), including the increasing colony diameter of heat-resistant fungi (Tremarin et al. 2015). The model has shown advantages when compared with other growth models: (i) it is available on two

311

user-friendly software programs: DMFit and MicroFit; (ii) it has good fitting capacity; and (iii) most of the model parameters are biologically interpretable. Another advantage of the Baranyi and Roberts model is that it is available as differential equations that allow microbial growth to be modelled in a dynamic environment, generally resulting from non-isothermal conditions (Costa et al. 2016; Gospavic et al. 2008; McKellar and Lu 2003; Xanthiakos et al. 2006).

11.3.1.3 Buchanan Three-Phase Linear Model

The three-phase linear model (Equation 11.11) proposed by Buchanan et al. (1997) was developed to describe the three bacterial growth phases (lag, exponential and stationary) by applying straight lines. The authors elaborated a biological basis for justifying the suitability of the linear model. During the lag phase, it is assumed that the cells do not divide due to adaptation in the new environment. For this reason, the growth rate is zero. During the exponential phase, it is assumed that the growth rate is a constant, in which the logarithm of the cellular concentration increases linearly with time. Once the stationary phase is reached, there is no increase in the cells, and the growth rate returns to zero.

$$N(t) \begin{cases} \text{For } t \leq t_{lag}, \log N_t = \log N_0 & - \text{lag phase} \\ \text{For } t_{lag} < t < t_{max}, \log N_t = \log N_0 + \mu_{max}\left(t - t_{lag}\right) & - \text{exponential growth phase} \\ \text{For } t \geq t_{max}, \log N_t = \log N_{max} & - \text{stationary phase} \end{cases} \quad (11.11)$$

where t_{lag} is the time when the lag phase ends (h) and t_{max} is the time when the N_{max} is reached (h).

Buchanan et al. (1997) compared this model with the Gompertz and Baranyi and Robert models. The curves predicted by the three models presented good fit to the data, and the growth kinetic parameters were similar. The three-phase linear model was more robust than the other models, especially when there were few experimental data. The QMRA developed by Koutsoumanis et al. (2010) to evaluate the growth of *Listeria monocytogenes* in pasteurized milk from production to the time of consumption used the three-phase linear model to calculate the pathogen's growth.

11.3.2 Interaction Models

The growth of microorganisms in foods is a complex system where different microbial populations can coexist and interact. The interaction between two different microorganisms can be direct or indirect, and the effect of the interaction may be positive, neutral or negative. Direct interactions occur when two populations use *quorum sensing* or physical contact, while indirect interactions occur through a change of the environment. Indirect interactions can be specific, due for example to the production of bacteriocin and organic acids and competition for nutrients (Casla et al. 1996; Greer and Dilts 1995; Fredrickson, 1977). Table 11.2 describes the five types of indirect interaction between two populations (Fredrickson 1977; Sieuwerts et al. 2008).

Many studies have observed that natural food microbiota can inhibit or reduce the growth of pathogenic bacteria (Buchanan and Bagi 1999; Brillet et al. 2004; Hwang and Sheen 2011; Lardeux et al. 2015). Nevertheless, microbial interaction is often not considered in predictive microbiology and QMRA due to its complexity (Malakar et al. 2003). Interaction models are usually intended to quantify how much the growth of one population is reduced by the growth of other populations (Cornu et al. 2011; Pérez-Rodríguez and Valero 2013). Two model approaches can be used to describe the interaction between microorganisms. One approach considers the Jameson effect, which describes the simultaneous stopping of growth of all bacterial species at the time when the dominant bacterial

Table 11.2 The Five Types of Microbial Indirect Interactions: The Effect of Coexistence of Two Populations Is Shown with an Example of the Interaction

Interaction Types	Effect on Population A	Effect on Population B	Example
Amensalism	Neutral	Negative	Bacteriocin produced by Population A.
Competition	Negative	Negative	Nutrient competition.
Commensalism	Neutral	Positive	Production of lactic acid by lactic acid bacteria (Population A) is metabolized by yeast and fungi species (Population B) (Mounier et al. 2005).
Mutualism	Positive	Positive	Exchange of growth factors.
Parasitism	Negative	Positive	Bacteriophages (Population B) attach to bacteria (Population A).

population reaches its stationary phase (Giménez and Dalgaard 2004; Jameson 1962; Mellefont et al. 2008). The other approach considers the Lotka–Volterra competition model, which describes the dynamics of two competing bacterial populations in food products by incorporating an additional reduction term in the population growth rate that is proportional to the population density of another competing population (Powell et al. 2004; Valenti et al. 2013; Vereecken et al. 2000).

11.3.2.1 Jameson Effect Model

The Jameson effect model is based on Equations 11.12 through 11.15, which assume that the growth of the pathogen halts when the dominant microbial population reaches its N_{max} (Cornu et al. 2011; Jameson 1962).

$$\frac{dN_A}{dt} = N_A \cdot \mu_{maxA} \cdot \left(1 - \frac{N_A}{N_{maxA}}\right) \cdot \left(1 - \frac{N_B}{N_{maxB}}\right) \cdot \left(\frac{Q_A}{1+Q_A}\right) \qquad (11.12)$$

$$\frac{dN_B}{dt} = N_B \cdot \mu_{maxB} \cdot \left(1 - \frac{N_B}{N_{maxB}}\right) \cdot \left(1 - \frac{N_A}{N_{maxA}}\right) \cdot \left(\frac{Q_B}{1+Q_B}\right) \qquad (11.13)$$

$$\frac{dQ_A}{dt} = Q_{A\,t-1} \cdot \mu_{maxA} \qquad (11.14)$$

$$\frac{dQ_B}{dt} = Q_{B\,t-1} \cdot \mu_{maxB} \qquad (11.15)$$

where subscript A or B represents the different microbial populations and Q is a measure of the physiological state of cells at time t.

The value of Q at $t = 0$ (Q_0) can be calculated for both microorganisms (Equation 11.16).

$$Q_0 = \frac{1}{e^{(\mu_{max} \cdot \lambda)} - 1} \qquad (11.16)$$

where e is Euler's number and λ is the lag time estimated from co-culture experiments.

Different variants of the Jameson model have been proposed to evaluate the interaction between two microbial populations (Cornu et al. 2011; Giménez and Dalgaard 2004; Møller et al. 2013). Costa et al. (2019) proposed a modification of the Jameson effect model that includes two parameters (N_{criA} and N_{criB}) describing the maximum critical concentration that one population should reach to inhibit the growth of the other population. Mejlholm et al. (2015) developed a stochastic model for *Listeria monocytogenes* growth in naturally contaminated lightly preserved seafood applying the Jameson effect. The quantitative assessment of the exposure to *L. monocytogenes* from cold-smoked salmon consumption developed by Pouillot et al. (2007) considered the competitive bacterial growth between the pathogen and the background flora using the Jameson effect model principle.

11.3.2.2 Lotka–Volterra Model

The classic Lotka–Volterra (Equations 11.17 through 11.20), also called the predator–prey model, describes the interaction of two competing bacteria in a co-culture. This model is based on the logistic growth model and includes two empirical parameters reflecting the degree of interaction between microbial species (α_{AB} and α_{BA}) (Giuffrida et al. 2007; Fujikawa et al. 2014; Lotka 1956). Depending on the empirical parameter value for Population A (α_A), the growth of Population B (α_B) can be affected in three different ways: i) if $0 < \alpha_{AB} < 1$, Population B grows with reduced μ_{max} after Population A reaches N_{max}; ii) if $\alpha_{AB} = 1$, Population B stops growing when the Population A reaches its N_{max}; and iii) if $\alpha_{AB} > 1$, Population B declines when Population A reaches its N_{max}.

$$\frac{dN_A}{dt} = N_A \cdot \mu_{maxA} \cdot \left(1 - \frac{N_A + \alpha_{AB} \cdot N_B}{N_{maxA}}\right) \cdot \left(\frac{Q_A}{1 + Q_A}\right) \qquad (11.17)$$

$$\frac{dN_B}{dt} = N_B \cdot \mu_{maxB} \cdot \left(1 - \frac{N_B + \alpha_{AB} \cdot N_A}{N_{maxB}}\right) \cdot \left(\frac{Q_B}{1 + Q_B}\right) \qquad (11.18)$$

$$\frac{dQ_A}{dt} = Q_{At-1} \cdot \mu_{maxA} \qquad (11.19)$$

$$\frac{dQ_B}{dt} = Q_{Bt-1} \cdot \mu_{maxB} \qquad (11.20)$$

where α_{AB} and α_{BA} are, respectively, the competition factor parameters of Population A on Population B and vice versa.

A number of mathematical models have been developed to predict microbial interaction, mainly between lactic acid bacteria and *L. monocytogenes* in various food matrices (Blanco-Lizarazo et al. 2016; Mejlholm and Dalgaard 2007; Mejlholm and Dalgaard 2015; Quinto et al. 2016;

Ye et al. 2014; Østergaard et al. 2014) and between natural microbiota and pathogenic bacteria (Buchanan and Bagi 1999; Guillier et al. 2008; Koseki et al. 2011; Le Marc et al. 2009).

11.3.3 Phoenix Phenomenon Model

The experimental observation that some bacteria, after exposition to stress conditions (osmotic or thermal, for example) at constant temperature, present a kinetic growth characterized by a decrease in the initial cell counts followed by an exponential increase in the level of the count is known as the Phoenix phenomenon (Amézquita et al. 2005). This term was probably applied because the Phoenix, in ancient times, symbolized immortality through "death and resurrection" (Shoemaker and Pierson 1977). Colle et al. (1961) and Shoemaker and Pierson (1977) described this microbiological phenomenon in *Clostridium perfringens*, Kelly et al. (2003) in *Campylobacter jejuni*, and several authors in *Salmonella* (Airoldi and Zottola 1988; Aspridou et al. 2018; Zhou et al. 2011; Paganini et al. submitted).

The Phoenix phenomenon was observed by Shoemaker and Pierson (1977) in log count curves for *C. perfringens* after thermal stress. The authors explain this behaviour considering three phases: injury, recovery and growth. Mellefont et al. (2005), working with *Salmonella* Typhimurium M48 under osmotic stress, thought that the phenomenon was due to an initial inactivation of a portion of the population followed by growth. However, they considered that the Phoenix phenomenon represented the responses of different subpopulations, in which some cells were dying but both recovery of culturability and exponential growth were occurring simultaneously. By studying single cells of *Salmonella enterica* serotype Agona, Aspridou et al. (2018) concluded that the Phoenix phenomenon occurred under severe osmotic stress (5.7% and 6.75% NaCl) as a result of simultaneous growth, survival and death of cells.

The modified Baranyi and Roberts (1994) model was suggested by Zhou et al. (2011) to model the Phoenix phenomenon (Equations 11.21 and 11.22) based on the assumption that the log count curve of the total population was the sum of a dying and a surviving-then-growing subpopulation.

$$\ln(N) = \ln\left\{\exp\left[(1-r_0)\ln(N_0) - \mu_D t\right] + \frac{\exp\left[r_0\ln(N_0) + \mu_G A(t)\right]}{\left[1 + \dfrac{e^{\mu_G A(t)} - 1}{e^{\ln(N_{max}) - r_0\ln(N_0)}}\right]}\right\} \quad (11.21)$$

where μ_G and μ_D are the maximum specific growth and death rates (h^{-1}), respectively, r_0 is the fraction of the growing initial subpopulation ($0 \leq r_0 < 1$).

$$A(t) = t - \lambda + \frac{\ln\left[1 - e^{-\mu_G t} + e^{-\mu_G(t-\lambda)}\right]}{\mu_G} \tag{11.22}$$

with λ being the lag time of the growing subpopulation (therefore, as per definition, of the whole population).

Paganini et al. (submitted) applied a modified Baranyi and Roberts model similar to Zhou et al. (2011) to describe curves of four serotypes of *Salmonella enterica* Typhimurium under osmotic stress (a$_w$ 0.95) for different inoculum conditions. The predictive ability of the model was assessed through statistical indexes, with good results ($R^2 > 0.973$ and RMSE < 0.288).

The modified Baranyi and Roberts model can be applied to studies about the Phoenix phenomenon, which show the different cellular responses and the complexity in the behaviour of microbial populations in conditions close to the boundary of growth.

11.3.4 Inactivation Models

The kinetics of microbial inactivation has been receiving attention since the 1920s and constitutes one of the earliest forms of predictive microbiology. Inactivation models describe the decrease in a microbial population over time when it is exposed to a lethal process or agent, such as thermal treatments, non-thermal technologies and drying/dehydration processes. The distinction between inactivation and survival models is not always clear, but survival processes are usually associated with slowly declining patterns, while inactivation refers to a lethal process showing a rapid decrease of the microbial population (Pérez-Rodríguez 2013; Peleg 2006).

Traditionally, microbial inactivation has been assumed to show first-order kinetics (McKellar and Lu 2003; Peleg 2003). However, non-log-linear microbial inactivation models have been frequently presented in the literature and are used to describe the kinetics of inactivation of a wide variety of microorganisms with the most varied behaviour (Bevilacqua et al. 2015; Peleg and Cole 1998). The models most commonly used to describe microbial inactivation in foods are presented in Figure 11.2, and a brief description of the various models is provided next.

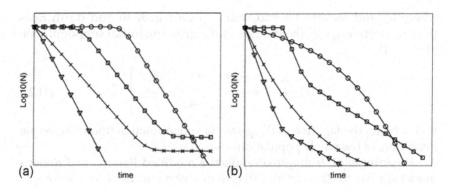

Figure 11.2 Representation of eight different shapes of inactivation curves. (a) Linear (∇, shape I), linear with tailing (×, shape II), sigmoidal-like ($\square$, shape III), linear with a preceding shoulder ($\bigcirc$, shape IV). (b) Biphasic (∇, shape V), concave (×, shape VI), biphasic with a shoulder ($\square$, shape VII), and convex ($\bigcirc$, shape VIII). (From Geeraerd et al. 2005 with permission of the authors.)

11.3.4.1 Bigelow Model

The Bigelow or linear model (Equation 11.23) was first proposed to describe the inactivation of microorganisms and enzymes in the canning industry, in particular to establish the D-value (decimal reduction time) as a function of temperature (Bigelow and Esty 1920; Bigelow 1921). This model assumes that the inactivation kinetics is first-order and that all cells or spores of a microbial population have identical resistance to lethal treatments (Whiting and Buchanan 1997).

$$\ln(N) = \ln N_0 - kt \tag{11.23}$$

where k is the first-order rate constant (s^{-1} or min^{-1}).

By rearranging Equation 11.23, Equation 11.24 can be obtained.

$$\log \frac{N}{N_0} = \log S(t) = -\frac{t}{D_{value}} \tag{11.24}$$

where D-value is the decimal reduction time (D-value = 2.303/k, units in minutes or seconds) and $S(t)$ is the momentary ("instantaneous") survival ratio at time t (Peleg 2006).

The D-value is a measure of the thermal resistance of a microorganism at a given temperature required to destroy one log cycle (90%) of the target microorganism. When the log D-values are plotted versus the treatment exposure temperature, the reciprocal of the slope is equal to the z-value, which is the temperature increase required to reduce the D-value by a factor of 10 in order to increase the rate of destruction by a factor of 10 (Bigelow and Esty 1920).

11.3.4.2 Weibull Model

The Weibull model (Equation 11.25) has been used to describe the inactivation of microorganisms due to its mathematical simplicity and great flexibility (Peleg and Cole 1998). This model is based on the different distribution of resistance or sensitivity between the individuals in a microbial population (Van Boekel 2002; Van Derlinden et al. 2012).

$$\log S(t) = -\frac{1}{2.303}\left(\frac{t}{\alpha}\right)^{\beta}$$

(11.25)

where α and β are parameters related to the scale and shape of the inactivation curve, respectively.

The shape parameter accounts for upward concavity of a survival curve ($\beta < 1$), a linear survival curve ($\beta = 1$), and downward concavity ($\beta > 1$). Therefore, when $\beta = 1$, no biological variation is assumed (each cell is equally susceptible to being destroyed), and the survivor curve is linear with first-order kinetics. Although the Weibull model is of an empirical nature, a link can be made with physiological effects. Curves with upward concavity ($\beta < 1$) forming a tail indicate that the population of microorganisms contains members that die rapidly. However, as the destruction process occurs, the survivors are most resistant, leading to a longer inactivation time, while curves with downward concavity ($\beta > 1$) indicate that the population progressively decreases, and the time required to destroy the same fraction of microorganisms decreases over time (Aragao et al. 2007; Peleg 2006). This model has successfully described survival and inactivation curves of B. pumilus, B. coagulans, L. monocytogenes, C. jejuni, S. enterica and Yersinia enterocolitica (Chun et al. 2009; Couvert et al. 2005; Haberbeck et al. 2012; Patil et al. 2009; Virto et al. 2005).

11.3.4.3 Shoulder/Tail Model (Geeraerd Model)

The shoulder/tail model (Equation 11.26) developed by Geeraerd et al. (2000) is based on the physiological state of cells and the residual population density on the basis of the existence of shoulder (or lag time) and tail regions. The shoulder describes the initial segment of the inactivation curve, in which the microbial population remains at similar levels, and the tail region corresponds to the final segment of the inactivation curve, representing the residual population more resistant to the lethal treatment.

$$N(t) = \left[(N_0 - N_{res}) \exp(-k_{max}t) \frac{\exp(-k_{max}t_L)}{1 + \exp((-k_{max}t_L) - 1) \exp(-k_{max}t)} + N_{res} \right] \quad (11.26)$$

where k_{max} is the maximum specific decay rate (min^{-1} or h^{-1}), t_L is the time before inactivation (min^{-1} or h^{-1}) and N_{res} is the residual microbial population concentration (cfu/g or mL).

Geeraerd et al. (2005) created GInaFiT, a freeware add-in for Microsoft Excel. This tool includes nine different types of primary inactivation models, including shoulder/tail models.

11.4 SECONDARY MODELS

11.4.1 Ratkowsky or Square Root Model

The Ratkowsky model (Ratkowsky et al. 1982), known as the square root model (Equation 11.27), is one of the most popular models to describe the effect of temperature on maximum specific growth rate. This model has the advantage of homogenizing the data variance and providing a linear response in the region of temperatures below the optimum growth temperature (McMeekin et al. 2013).

$$\sqrt{\mu_{max}} = b \cdot (T - T_{min}) \quad (11.27)$$

where b represents the regression coefficient (h$^{-0.5}$°C^{-1}), T is the temperature (°C) and T_{min} is the minimum theoretical temperature at which growth is detected (°C).

Ratkowsky et al. (1983) extended the square root model to describe the effect on the growth rate in the whole region ranging from the minimum to the maximum temperature for growth (Equation 11.28).

$$\sqrt{\mu} = b \cdot (T - T_{min}) \cdot \left(1 - \exp(c(T - T_{max}))\right) \quad (11.28)$$

where, in the expanded model, the parameters c and T_{max} are the proportionality constant (°C^{-1}) and the temperature (°C) in the upper part of the range over which growth is not possible, respectively.

Other adaptations include the effect of alternative environmental factors, such as pH (Adams et al. 1991; Wijtzes et al. 2001), a_w (McMeekin et al. 1987; Miles et al. 1997), carbon dioxide and phenol (Dalgaard et al. 1997; Giménez and Dalgaard 2004) and lactic acid (Ross et al. 2003). In some cases, the square root model has been applied also to describe lag time (Mataragas et al. 2006; Sant'Ana et al. 2012). For the quantitative assessment of the exposure to *L. monocytogenes* from cold-smoked salmon consumption, Pouillot et al. (2007) applied the square root model to describe the effect of temperature on the specific growth rate.

Square root models are widely applied in predictive microbiology due to their simplicity and easy interpretability of the model parameters (e.g. T_{min}, pH_{max} and a_{wmin}) (Ross and Dalgaard 2004). Another advantage is that these models can be easily adapted to encompass the whole biokinetic range of environmental factors, making them more attractive to predictive microbiology practitioners (Pérez-Rodríguez and Valero, 2013).

11.4.2 Arrhenius-Type Model

The Arrhenius-type model is considered fundamental in different scientific fields. This model is derived empirically, based on thermodynamics, to describe chemical reaction kinetics and/or biological processes (Labuza and Riboh, 1982). In predictive microbiology, this model is used to describe the relationship between the maximum specific growth rate (μ_{max}) of a microorganism and the growth temperature (Equation 11.29) (Gonzales-Barron 2012; McMeekin et al. 2013).

$$\ln(\mu) = \mu_0 \cdot \left(\frac{\Delta E_a}{R \cdot T} \right) \tag{11.29}$$

where μ_0 is a constant, T is the absolute temperature (K), R is the universal gas constant (8.314 J/mol.K), and E_a is the so-called activation energy of the system (J/mol).

An Arrhenius-type model was used to predict mould growth (Longhi et al. 2014; Silva et al. 2010), bacterial growth (Silva et al. 2018; Huang et al. 2011; Koutsoumanis et al. 2000) and inactivation (Amos et al. 2001; Cerf et al. 1996). This model was extended by Davey (1989) to represent the relationship between the maximum growth rate or the death rate of microorganisms under additional processing conditions, such as pH and/or a_w. The model has also been used to calculate the relative rate of spoilage (RRS), which is defined as the shelf-life (determined by sensory evaluation) at a reference temperature (T_{ref}) divided by the shelf-life observed at the actual storage temperature (Equation 11.30) (Dalgaard 2002).

$$RRS = \frac{\text{Shelf - life at } T_{ref}}{\text{Shelf - life at } T} = \exp\left[\frac{E_a}{R} \cdot \left(\frac{1}{T} - \frac{1}{T_{ref}}\right)\right]$$ (11.30)

where T_{ref} is a reference temperature (°C) at which the shelf-life is known.

RRS models are interesting because they enable shelf-life to be predicted at different temperatures and for products where the specific spoilage organisms or the type of reaction responsible for spoilage are not known (McKellar and Lu 2003).

11.4.3 Polynomial or Response Surface Models

Polynomial or response surface models are the most commonly used models to describe the relationship between environmental conditions and microbial growth parameters as well as to determine optimal process conditions (Dalcanton et al. 2018; Devlieghere et al. 2000). Generally, second-order polynomial equations (Equation 11.31) are used, including three terms: first-order, second-order (quadratic) and interaction terms (Pérez-Rodríguez and Valero 2013).

$$y_i = \beta_0 + \sum_{j=1}^{k} \beta_j X_j + \sum_{j=1}^{k} \beta_{jj} X_j^2 + \sum_{j \neq l}^{k} \beta_{jl} X_j X_l + \varepsilon$$ (11.31)

where y_i is the dependent variable (e.g. growth rate or lag phase), β_0, β_j, β_{jj} and β_{jl} are the estimated regression coefficients, X_j and X_l are the independent variables (environmental factors), and ε is the error term.

Polynomial models are characterized by a high number of parameters, which increases exponentially when the number of factors included in the model is increased. Response surface models were evaluated to predict the growth parameters of *Leuconostoc mesenteroides*, cultivated under different combinations of temperature, pH, NaCl and NaNO$_2$, and the optimal composition of culture media for production of biosurfactants by the probiotic bacteria *Lactococcus lactis* and *Streptococcus thermophilus* (Rodrigues et al. 2006; Zurera-Cosano et al. 2006).

11.4.4 Bigelow Model

A secondary model equivalent to Bigelow (1921) was developed to describe the effect of temperature on the microbial inactivation rate (Equation 11.32).

$$k_{max} = \frac{\ln 10}{D_{ref}} \cdot \exp\left(\frac{\ln 10}{z} \cdot (T - T_{ref}) \right) \qquad (11.32)$$

where k_{max} (min^{-1} or h^{-1}) is the maximum inactivation rate, D_{ref} (min or h) is the decimal reduction time (D-value) at a reference temperature T_{ref} (°C). Different Bigelow-type model approaches have been developed, including additional physicochemical factors as well as temperature (e.g. pH, a_w, high pressure, etc.) (Adekunte et al. 2010; Gaillard et al. 1998; Mafart and Leguerinal 1998).

11.5 TRANSFER MODELS

Cross-contamination is one of the most important factors linked to food-borne outbreaks and food spoilage. According to Pérez-Rodríguez et al. (2008), the cross-contamination phenomenon is defined as a general term that refers to the transfer, direct or indirect, of microorganisms from a contaminated product to a non-contaminated product or due to the environmental conditions where the foods are processed, air, poor hygiene of handlers and contaminated equipment.

In general, the microbial transfer can be classified into three specific types: i) air-to-food transfer (AF); ii) surface-to-food in fluids (SFF); and iii) surface-to-food contact (SFC). The last is the most frequent type, given its high incidence in the domestic consumption phase (Pérez-Rodríguez et al. 2008). Although microbial transference phenomena have always been understood as an important cause of food contamination, together with the pronounced need to incorporate them into risk assessment studies, transfer models have only in recent years been implemented in predictive microbiology. Møller et al. (2012) developed a model to describe the transfer and survival of S. typhimurium during the grinding of pork. The model satisfactorily predicted the observed concentrations of S. typhimurium during grinding of meat pork. In another study, Møller et al. (2016) evaluated the robustness of the model obtained before to predict the transfer and survival of Salmonella spp. and L. monocytogenes during the grinding of meat pork and beef, using two different grinders and different sizes and different numbers of pieces of meat to be ground. The parameters obtained under different conditions may not be applied to describe cross-contamination. However, the risk estimates showed that the risk of food-borne disease can be reduced when meat is ground in a grinder made of stainless steel using a well-sharpened knife in a cooling room with temperature below 4 °C. Through the application of transfer models, it is possible to identify routes and risk factors associated with microbial transfer using a quantitative approach and to predict the number of microorganisms transferred from one surface to another (Den Aantrekker et al. 2003; Reij and Den Aantrekker 2004). Cross-contamination models are discussed further in Chapter 12.

11.6 MIXING, PARTITIONING AND OTHERS

Mixing, partitioning and removal are among the main types of unit operations in the food industry. These processes will influence the microbial status of the product in terms of both likelihood of contamination and numbers. Thus, descriptive data of these processes should be collected.

11.6.1 Mixing

Mixing is one of the commonest unit operations in the food processing industries with a primary objective of achieving a homogeneous mixture. Generally, this means attaining a nearly uniform distribution of the ingredients for improving food quality, such as texture and colour development. Food mixtures involve many ingredients, including liquids, powders and gases. Some important ingredients are contained only in minor quantities, which should be dispersed evenly and efficiently in the final mixture (Levine and Boehmer 1997). Nowadays, a large majority of processed ready-to-eat products are indeed multiphase dispersions, such as solids dispersed in liquids (e.g. canned foods), emulsions (e.g. soups, margarines and spreads) or bubbly dispersions (e.g. meringue, ice cream and sponge cake, among others) (Niranjan 2009).

In a mixing process, food units are combined into a new large unit, as shown in Figure 11.3. This process affects the food matrix and causes physical or chemical change in the materials being processed. As a consequence, the microorganisms can be redistributed, resulting in an increase in prevalence and a decrease in the mean concentration in contaminated food units (Bassett et al., 2012; Nauta 2005). The consequence of the mixing process for public health risk is uncertain, because it depends on the mean individual risk, which will be affected in various ways according to the total number of microbial cells reallocated and the pathogenicity of the foodborne pathogen of concern (Augustin et al. 2017). This basic mixing process is commonly encountered in many food production scenarios, such as ground beef manufacturing (Smith et al. 2013), milk collection

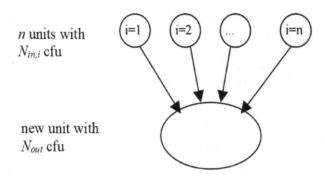

n units with $N_{in,i}$ cfu

new unit with N_{out} cfu

Figure 11.3 Mixing: n small units, containing $N_{in,i}$ cells (particles and spores) in the units i ($I = 1 \dots n$) are put together to form a new larger unit with N_{out} cells. The objective is to describe the probability distribution of N_{out} given a distribution of the $N_{in,i}$. (From Nauta 2005 with permission of the authors.)

from multiple farm tanks (Albert et al. 2005), shredded lettuce processing (Danyluk and Schaffner 2011) and processed food manufacturing (Daelman et al. 2013).

There is little published information on the effect of mixing operations on microorganisms. It is unlikely that the shearing conditions or temperature in a mixer would reduce the number of microorganisms, and hence, mixing does not have a preservative effect. In some instances, especially where the temperature of the food is allowed to rise during mixing, there may be an increase in number of microbial contaminants, caused in part by the greater availability of nutrients as a result of the mixing action (Fellows 2009). Although these processes do not result in an increase or decrease in the number of microorganisms in the total amount of food produced, they change the distribution of microorganisms among food items. This has an impact on the variability of doses between servings and therefore may have an impact on the risk assessment (Bassett et al. 2012).

The modelling approach for mixing suggested by Nauta (2005, 2008) assumes that if the numbers of cells on or in all small units are known, summation can be used to model the effect of mixing on the number of cells per unit: if n units are combined, with unit i containing $N_{in,i}$ cells ($i = 1 \dots n$), the larger unit (N_{out}) will contain the sum of all, as in Equation 11.33:

$$N_{in,i}\text{cells} = N_{out} = \sum_{n} N_{in,i} \tag{11.33}$$

Hence, the total number of cells in the system remains the same. The fraction of contaminated units (prevalence) will increase with unit size. Assuming random homogeneous mixing and equally sized small units, the increased prevalence after mixing of n small units can be estimated as shown in Equation 11.34.

$$P_{out} = 1 - (1 - P_{in})^n \qquad (11.34)$$

Nauta (2005) discusses complications related to the efficient and correct modelling of mixing processes and some practical solutions for food chain risk assessment modelling. In model simulations, mixing can be modelled by application of the Dirichlet distribution, a counterpart of the multinomial distribution as applied to partitioning (Nauta 2008). Mathematical methodologies relevant to modelling mixing and partitioning are presented in Nauta (2001, 2008) and Nauta et al. (2001).

11.6.2 Partitioning

Partitioning (fractionation) is a process used in the food industry when a large volume (e.g. an industrial batch) is split up into several small units (e.g. consumer packages), as shown in Figure 11.4; for example, when milk from different cows is placed in the same milk tank and distributed in bottles, or cuts of different carcasses are joined for grinding and then distributed in packages. Here, the redistribution of microorganisms could

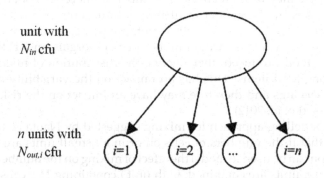

unit with
N_{in} cfu

n units with
$N_{out,i}$ cfu

$i=1$ $i=2$ ... $i=n$

Figure 11.4 Partitioning: a major unit containing N_{in} cells (particles and spores) is split up into n smaller units i ($I = 1 \ldots n$) that contain $N_{out,i}$ cells. The objective is to describe the distribution of the $N_{out,i}$ over the small units given the values of N_{in} and n (From Nauta 2005 with permission of the authors).

result in a lower prevalence and a higher concentration in contaminated food units (Bassett et al. 2012).

For the partitioning model, it is possible to describe the reallocation of the N cells (or spores, particles or colony forming units [cfu]) present in a large unit over n small units. Basically, the N_{in} cells are distributed into n portions $N_{out,i}$ ($I = 1 \dots n$). The challenge is to find the appropriate model to describe this distribution, which ultimately results in a distribution of N_{out} resulting from a distribution of N_{in}. As the number of units in the food chain increases with portioning, and the total number of cells in the system remains equal, the prevalence is likely to decrease. In the literature, several partitioning models have been proposed and discussed by Nauta (2005).

11.6.3 Removal

The removal process during food processing can be classified into two categories (Nauta 2008). First, removal can be considered as a process whereby some units (or parts of units) are selectively removed from the production. Examples are the rejection of carcasses by veterinary inspectors in the slaughterhouse or the visual discarding of vegetables or fruits due to, for instance, injuries. Thus, in this category, the removal of units is not a random process, as removal is performed because there is a presumed relation between microbial contamination and visual appearance. Second, removal can occur during the process of washing, peeling, cutting, derinding and filtering, for instance. In this category, the removal process in principle aims at all units, yielding a similar model as an inactivation model, with a decrease in the number of cells that can be described by Equation 11.35.

$$\log\left(N_{out}\right) = \log\left(N_{in}\right) - h\left(.\right) \tag{11.35}$$

where N_{out} and N_{in} are the microbial concentration (cfu/[g or mL]) at the end and the beginning of the process step, respectively, and $h(.)$ is a positive "removal" function. For instance, if the removal process is a washing process, $h(.)$ can be assumed to be constant, or variable for a process step. More complex models can be developed if the removal mechanisms are incorporated in the model (Nauta 2008).

The basic process of removal of *Y. enterocolitica* was considered during the derinding of pork belly cuts to produce minced meat (Van Damme et al. 2017). The number of *Y. enterocolitica* after derinding was described through a binomial distribution with the number of *Y. enterocolitica* per belly cut after cutting (process occurring before derinding) and the proportion of the pathogen remaining on the belly cut after derinding as parameters.

11.7 GROWTH PROBABILITY MODELS (GROWTH/NO GROWTH)

Growth probability models, also known as growth/no growth, are employed to predict the probability that growth of a microorganism of concern could occur in a food product as a function of intrinsic and extrinsic factors. Growth is not always the only feature of interest for food producers and scientists; on many occasions, the possibility of growth is the most important issue of concern (Carrasco et al. 2012). This approach is often used to predict suitable combinations of hurdles making microbial growth highly unlikely in a specific food product during, for instance, the storage time (Masana and Baranyi 2000). These models were first applied concerning the prediction of the probability of formation of staphylococcal enterotoxin or botulinum toxin within a specific period of time under conditions of storage and product composition (Genigeorgis 1981; Gibson et al. 1987). Knowledge of these growth/no growth limits under different environmental conditions is important since it enables better quality and safety management of foods (McMeekin et al. 2000).

Studies on growth probability test combinations of hurdles used for food preservation under different levels. Typically, a number of repetitions are tested for each combination. After a certain time of incubation, each repetition is examined to determine whether growth occurred or not; 100% growth is considered when growth is confirmed for all repetitions and 0% when no growth is observed for all repetitions. The classical graphical representation of growth probabilities is illustrated in Figure 11.5. The hypothetical data in Figure 11.5 show the combined effect of pH and temperature on the probability of growth of a microorganism over a certain time period. In this example, from a food safety point of view, a food product with a pH equal to 4.5 should be held at temperatures lower than 20 °C to prevent microbial growth.

The transition from the growth zone, where the probability of growth is 100%, to the no growth zone, where the probability of growth is 0%, can occur gradually, as depicted in Figure 11.5 and in Haberbeck et al. (2015), Belessi et al. (2011), Mertens et al. (2011) and Vermeulen et al. (2009).

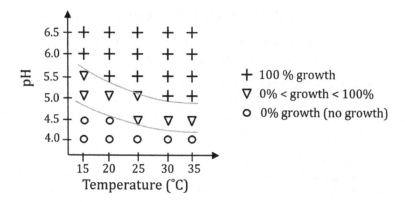

Figure 11.5 Example of a classical graphical representation for growth probabilities according to two environmental factors: in this example, temperature and pH.

Consequently, a growth boundary zone where the probability of growth can be between 0% and 100% exists. For instance, if 5 out of 10 biological repetitions grew, the estimated growth probability equals 50%. This behaviour reflects the heterogeneity in phenotypic response within a population (Sumner and Avery 2002). The observation that some cells of a microbial population are sensitive to an environmental stress, such as low pH and temperature, whereas others are resistant is a readily observed manifestation of heterogeneity (Avery 2006). In other cases, this transition can occur abruptly, and in this case, the zone with different growth probabilities would not be present in the graphical representation, as in López et al. 2007; McKellar and Lu 2001; Presser et al. 1998 and Skandamis et al. 2007. Importantly, the growth/no growth zone is time dependent. Thus, the longer the incubation period prior to the verification of growth, the wider the growth zone may become, since the cells would have more chances to initiate growth. Consequently, there is a greater chance that the transition zone will occur abruptly.

Various approaches have been suggested to model the growth/no growth boundary: empirical deterministic approaches (Augustin and Carlier 2000b; Le Marc et al. 2002; Membré et al. 2001), logistic regression (Haberbeck et al. 2015; Kakagianni et al. 2018) and artificial neural networks (Hajmeer and Basheer 2002, 2003). Koutsoumanis and Angelidis (2007), Koutsoumanis et al. (2010) and Sanaa et al. (2004) are examples of studies applying growth/no growth models and concepts in QMRA.

11.8 MODEL GENERATION PROCESS

11.8.1 Data Generation: Experimental Design, Data Acquisition and Data Process

The generation of experimental data to describe and to predict microbial behaviour in foods should be structured in some basic steps, starting with the experimental design, followed by the data collection and finally, the data processing.

11.8.1.1 Experimental Design

A premise that must be considered in the experimental design is about the factors that influence the microbial behaviour. All intrinsic (e.g. pH, water activity and salt concentration) and extrinsic (e.g. temperature and gas concentration) factors related to the food that may vary during the process under study must be carefully checked to incorporate such variations into the predictive model. For refrigerated perishable foods, for instance, the impact of temperature variation on microbial behaviour is commonly analysed, since temperature is a factor that has a great influence on microbial growth, and it can vary greatly throughout the food production chain from manufacturing to consumption. Factors with very small or no variation during the production chain are generally considered constant and are not incorporated into the experimental design. The incorporation of many factors into the experimental design increases the complexity of experiment execution and the proposition of a predictive mathematical model.

The definition of the main factors to be investigated allows one to proceed to the definition of the levels to be considered in the study. The levels depend directly on the range of expected values in each factor to be reached by the product throughout the processes. In the case of storage of perishable products, for instance, the Brazilian legislation allows refrigeration up to 10 °C (Brasil 1984), while in the United Kingdom it is recommended that temperatures should be ≤5 °C (FSA 2015) and in the United States, ≤4.4 °C (USA FDA 2014). However, in practice, if some problem occurs in the cold chain (during transportation of the product or exposure for sale in supermarkets), the product temperature can exceed the recommended maximum. If experiments are designed and performed to describe the microbial behaviour only in the range from 4 °C to 10 °C (e.g. 4 °C, 6 °C, 8 °C and 10 °C), the predictions by the mathematical model will be valid only within that range of values. Thus, the extent of the model predictions at temperatures above or below the tested range are treated as extrapolation and should be avoided (as depicted in Figure 11.6). In this context, many studies involving the prediction of microbial behaviour in perishable foods consider higher than actual temperature ranges to avoid extrapolation, e.g. the growth of *Pseudomonas* spp. in poultry from 2 °C to 20 °C (Gospavic et al. 2008), *L. monocytogenes* in pasteurized milk from 1.5 °C to 16 °C (Xanthiakos et al. 2006), *S. enteritidis* SE86 on homemade mayonnaise from 7 °C to 37 °C (Elias et al. 2016) and *Lactobacillus viridescens* in vacuum-packed sliced ham from 4 °C to 30 °C (Silva et al. 2017).

The appropriate levels for the study can be defined by different strategies of experimental design. Traditional statistical designs, e.g. full and factional factorial designs, in many cases are not suitable in predictive microbiology, since the responses (dependent variables) usually do not present linear and/or parabolic dependence in relation to the factors (as described by the quadratic equation of the response surface). The responses usually present other nonlinear behaviours (secondary models to describe different nonlinear responses were covered in Section 11.4). In this context, the optimal experimental design is an interesting alternative, for which some applications in predictive microbiology have been proposed since the late 1990s (Balsa-Canto et al. 2008; Cunha et al. 1997; Longhi et al. 2018; Stamati et al. 2016; Van Derlinden et al. 2008; Versyck et al. 1999). If appropriate criteria and parameters are used in the design and the experiments are correctly performed, the optimization results in time and resource savings, ensuring parameter estimation with great accuracy. One advantage of the optimal experimental design is the

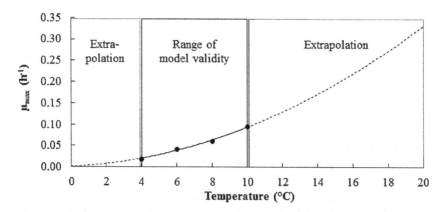

Figure 11.6 Graphical representation of the range of validity of a mathematical model (temperature between 4 °C and 10 °C) and extrapolation.

execution of some experiments with optimal variation in the levels within the desired range instead of performing several experiments under fixed levels within the desired range. A graphical representation of four kinetic growth curves under fixed levels and one optimally designed experiment, both in the range from 4 °C to 10 °C, is shown in Figure 11.7.

11.8.1.2 Data Collection

A widely used technique for obtaining experimental data in food microbiology is plate counting, which is used to determine the number of viable cells in a sample. The method is usually associated with a high cost of analysis and a long time to obtain the responses and often presents moderate uncertainty in the responses. Modern methods, many of them with sophisticated equipment, have been proposed in recent years to obtain microbiological experimental data. These methods are based on molecular biology techniques (e.g. polymerase chain reaction [PCR], restriction fragment length polymorphism [RFLP] and DNA microarray assay), immunological techniques (e.g. enzyme-linked immunosorbent assay [ELISA]), and biophysical and biochemical principles with the application of biosensors (e.g. bioluminescence sensors, bio-analytical sensors utilizing enzymes, electrical impedometry and flow cytometry) (Mandal et al., 2011). They often depend on high initial investment for equipment acquisition and intensive laboratory work to validate the responses obtained.

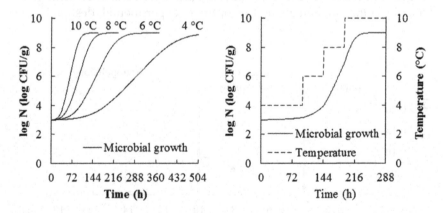

Figure 11.7 Graphical representation of kinetic growth curves under fixed levels (isothermal experiments at 4 °C, 6 °C, 8 °C and 10 °C) and under optimal variation of the levels (non-isothermal experiment between 4 °C and 10 °C) with the respective temperature profiles.

The U.S. Food and Drug Administration Bacterial Analytical Manual recommends counting in the range from 25 to 250 cfu (based on 1 mL of sample) on one plate (Maturin and Peeler 2001). As the microbiological concentration in a food can be higher than this range, dilutions of the sample are necessary, which may propagate experimental errors and generate uncertainty in the experimental measurements. The number of samples analysed at each time in the experiment is a decisive aspect to minimize this problem; it is recommended that at least two samples should be collected and analysed to generate a duplicate of the experimental data. Then, the result is expressed by the average of the measurements at that time, also allowing the measurement error to be calculated. Depending on the characteristics of the experiment and the available resources, three data points can be obtained, providing triplicate results, which increases confidence in the experimental data. Another important recommendation in data collection is the repetition of the experiment, aiming to verify its reproducibility. The results of the initial experiment and the repetitions under the same experimental conditions should be as close as possible to prove reproducibility. The treatment of experimental data for replicates and repetitions will be presented in the next section.

When performing experiments, monitoring and recording data on intrinsic and extrinsic factors is extremely important. Preferably, this should be done by appropriate calibrated equipment connected to information storage equipment (usually computers). This monitoring ensures that the experiment occurred at the designed levels and that the mathematical model incorporates the correct information.

In the case of kinetic experiments, time is the independent variable and deserves special attention. The establishment of the times required to obtain experimental kinetic data is essential, and for this, the different phases of the microbial behaviour should be remembered. For microbial growth, the lag, exponential and stationary phases (in most cases) can be observed (Figure 11.1), while for microbial death, the inactivation (at constant or variable rate), shoulder and/or tail phases can be observed (Figure 11.2). The total experimental time should be long enough to ensure that all phases of microbial behaviour are detected, but it should be short enough to optimize the response and avoid wasting time. In practice, the experimental time is difficult to determine and usually depends on the researcher's experience and ability to predict the responses to be obtained.

The number of experimental data points and the time for collecting each experimental data point should be delineated in order to optimize the cost of the experiments and later, to ensure that the parameters of the microbial behaviour can be estimated with accuracy. For the parameter estimation of the mathematical models, the minimum number of experimental data points required is equal to the number of parameters of the model plus one. Since most mathematical models have four parameters, the minimum number of experimental data to be collected is five, but for an accurate estimation of the parameters, these five data must be strategically collected to represent all phases of microbial behaviour and phase transitions (e.g. Experiment A in Figure 11.8), which is a very difficult task even for experienced researchers. In turn, less experienced researchers tend to perform experiments with some additional data that could have been avoided (e.g. Experiment C in Figure 11.8). Therefore, a general recommendation is the collection of at least eight experimental data points for a sufficient representation of the complete kinetic curve (e.g. Experiment B in Figure 11.8), allowing the three phases of the microbial behaviour, in addition to the phase transitions, to be clearly identified, guaranteeing an accurate estimation of the parameters of the mathematical model.

11.8.1.3 Data Processing

The initial recommendation for data processing is the organization of information in spreadsheets, especially in software that will be used later to fit the mathematical models. Among the software available for this task, there are some free software options (no need to pay for a licence),

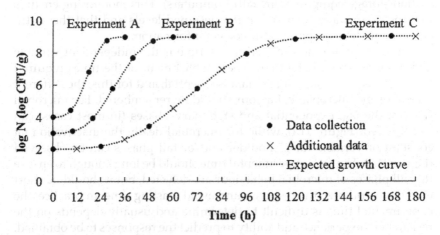

Figure 11.8 Representation of complete growth kinetics model (lag, exponential and stationary phases) with minimal (Experiment A), sufficient (Experiment B) and additional (Experiment C) data collection and experimental time.

such as R, Scilab and Open Office. The feeding of the experimental data into the spreadsheet should be performed in an organized way to carry out the first data processing.

In the case of kinetic studies, it is recommended to dedicate one column to the experimental time and successive columns to the responses obtained from the microbial behaviour. As mentioned in the data collection section, the mean values of each point of the experiment (replicates) should be calculated as well as the mean errors of each point. Software tools can be used to generate the spread of the experimental data, which contributes to a better presentation and interpretation of the microbial behaviour in relation to the data presentation in tables. In each set of experimental data, it is recommended to check whether there is any very discrepant kinetic data, a so-called outlier. In general, this discrepant data is the result of some experimental error and can be discarded. Figure 11.9 shows a schematic representation of kinetic growth curves

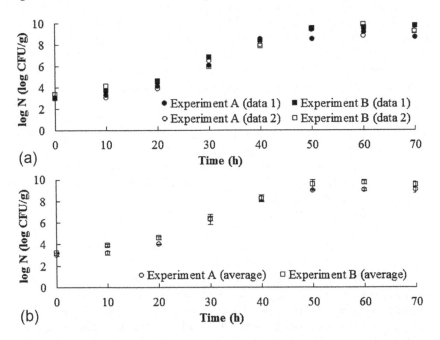

Figure 11.9 (a) Representation of microbial growth kinetics with duplicates of experimental data (data 1 and 2) and repetitions of experiments (Experiments A and B). (b) treatment of experimental data with average experimental data and error bars.

335

with experimental data collected in duplicate (Data 1 and 2) and with repeated experiments (Experiments A and B).

11.8.2 Model Fitting and Goodness-of-Fit Indexes

11.8.2.1 Model Fitting

The logarithmic transformation of the experimental data is a statistical strategy for the linearization of the exponential phase of the microbial curve. Similarly, the square root or logarithm transformation of the maximum specific growth rate values in the square root and Arrhenius models, respectively, have the same purpose of linearization of the curves. The transformation usually also helps to reduce the variation among the data, which is very important when analysing the impact of the residuals in the step of fitting the models to the data and in the estimation of more precise parameters.

Let us analyse a hypothetical example of fitting a model to the growth data of a microorganism in a food with initial and final concentrations of 10^3 and 10^9 cfu/g, respectively. If the mean relative residuals of the fit are equal to 1% in both sets of data, the absolute residuals are of a very different order of magnitude: 10^1 and 10^7 cfu/g in the measurements of the initial and final concentrations, respectively. As the procedure of fitting the model to the experimental data is based on the minimization of the residuals between the observed data and those predicted by the model, the residual of the final concentration data has much more impact than the residual of the initial concentration data. With the logarithmic transformation of the data, the initial and final concentrations become 3 and 9 log cfu/g, respectively, and the absolute residuals of the measurements become equal (0.00436 log cfu/g).

The fitting of a mathematical model to the experimental data is carried out using software, for which there are several options available today. One can choose to use specific predictive microbiology software available on the internet (e.g. ComBase and GInaFiT, among others, as presented by Plaza-Rodríguez et al. 2017 and Tenenhaus-Aziza and Ellouze 2015) or program the fitting procedure manually using commercial or free software (e.g. MATLAB® or R). The chosen software should not be considered as a decisive aspect, since it is only auxiliary in the procedure of fitting the model to the data. The mathematical method of fitting the model to the data is also not usually decisive, and different methods of fitting should lead to solutions equivalent to the estimated parameters. The most decisive criterion is to select the model that best describes the experimental data, and this can be done by analysing statistical indexes of goodness-of-fit of the model to the data.

11.8.2.2 Goodness-of-Fit Indexes

The main statistical indices used to verify the goodness of fit in predictive microbiology are the coefficient of determination (R^2), root-mean-square error (RMSE), bias and accuracy factors, bias and accuracy discrepancy. The R^2 is the proportion of the variance in the dependent variable that is predictable from the independent variables, and its calculation is defined by Equation 11.36. The RMSE measures the differences between the data predicted by a model and the experimental data observed, and its calculation is defined by Equation 11.37.

$$R^2 = \frac{\sum_{i=1}^{n} \left(y_{predicted,i} - y_{observed,i} \right)^2}{\sum_{i=1}^{n} \left(y_{observed,i} - \bar{y}_{observed} \right)^2} \tag{11.36}$$

$$RMSE = \sqrt{\frac{\sum_{i=i}^{n} \left(y_{predicted,i} - y_{observed,i} \right)^2}{n}} \tag{11.37}$$

where $y_{predicted}$ is the data predicted by the model, $y_{observed}$ is the observed data, $\bar{y}_{observed}$ is the arithmetic mean of the observed data and n is the number of observations.

The accuracy factor (a_f) and bias factor (b_f) were proposed by Ross (1996) to evaluate the performance of models in predictive microbiology. According to the author, the indices assess the level of confidence one can have in the predictions of the model and whether the model displays any bias that could lead to fail-dangerous predictions. The accuracy factor is defined by Equation 11.38), and the bias factor is defined by Equation 11.39. In addition to being quantitative, the bias factor is also qualitative, since a bias factor greater than one indicates that the model is "fail-dangerous" because it predicts higher data than those observed, and a bias factor lower than one indicates that a model is "fail-safe" because it predicts lower data than those observed.

$$a_f = 10^{\left(\frac{\sum_{i=1}^{n} \left| \log\left(y_{predicted,i} / y_{observed,i} \right) \right|}{n} \right)} \tag{11.38}$$

$$b_f = 10^{\left(\frac{\sum_{i=1}^{n} \log\left(y_{predicted,i}/y_{observed,i}\right)}{n}\right)} \tag{11.39}$$

Baranyi et al. (1999) proposed a different formulation for a new accuracy factor (A_f) and bias factor (B_f), as defined by Equations 11.40 and 11.41, respectively. According to the authors, the advantage of the modified definition is that it is consistent with the least squares algorithm of fitting models to observed values. The new factors allow the calculation of "*per cent* discrepancy" (%D_f) and "per cent bias" (%B_f) between the model and observations, as defined by Equations 11.42 and 11.43, respectively. Equation 11.44 indicates whether the overall bias is negative or positive, in which if %B_f > 0, on average, the model predicts higher data than the observations, and if % B_f < 0, on average, the model predicts lower data than the observations.

$$A_f = \exp\left(\sqrt{\frac{\sum_{i=1}^{n}\left(\ln y_{predicted,i} - \ln y_{observed,i}\right)^2}{n}}\right) \tag{11.40}$$

$$B_f = \exp\left(\frac{\sum_{i=1}^{n}\left(\ln y_{predicted,i} - \ln y_{observed,i}\right)}{n}\right) \tag{11.41}$$

$$\%D_f = \left(A_f - 1\right) \cdot 100\% \tag{11.42}$$

$$\mathrm{sgn}\left(\ln B_f\right) = \begin{cases} +1, & \text{if } \ln B_f > 0 \\ 0, & \text{if } \ln B_f = 0 \\ -1, & \text{if } \ln B_f < 0 \end{cases} \tag{11.43}$$

$$\%B_f = \mathrm{sgn}\left(\ln B_f\right) \cdot \left(\exp\left(\ln B_f\right) - 1\right) \cdot 100\% \tag{11.44}$$

Confidence on the estimated parameters is another important criterion of analysis. Many models have great goodness of fit but parameters with greater uncertainty. This characteristic can be verified by analysing the confidence intervals of the parameters.

11.8.3 Model Validation

Model validation is a procedure that aims to assess the performance of fitted models and to determine whether they can be used to aid decision-making (Haberbeck et al. 2018). It consists of comparing model predictions with independent experimental/observational data though indices of model performance such as bias factor, accuracy factor and the acceptable simulation zone approach in combination with graphical methods (Mejlholm et al. 2010; Oscar 2005; Østergaard et al. 2014).

Although some predictive models were constructed in real foods, the vast majority were constructed from experiments performed in laboratory culture media. Ideally, for both cases, the validation process should include comparisons with the behaviour of microorganisms in real foods or during real food processes. However, due often to cost but also to other factors, validation can be done in model systems or by using previously published data (Brocklehurst 2004).

11.9 CONCLUSIONS

Modelling the microbial behaviour on a QMRA through the entire farm-to-fork chain is a complex process, but the increasing number of validated predictive models in the literature facilitates this task. The aim of this chapter was to give an overview of predictive models and of the process to generate these models. Deterministic predictive models have been developed and successfully validated for more than 20 years. However, more research into so-called "stochastic predictive microbiology" should improve the use of predictive models in QMRA. Finally, a generation process and validation of robust predictive models are of great importance, since the final value of the model is significantly influenced by the experimenter's choices. In addition, these predictive models can be used in food processing industries as powerful tools to support and develop food safety standards.

REFERENCES

Adams, M. R., Little, C. L., and Easter, M. C. 1991. Modelling the effect of pH, acidulant and temperature on the growth rate of *Yersinia enterocolitica*. *J. Appl. Bacteriol.* 71(1): 65–71.

Adekunte, A., Valdramidis, V. P., Tiwari, B. K., Slone, N., Cullen, P. J., Donnell, C. P. O., and Scannel, A. 2010. Resistance of *Cronobacter sakazakii* in reconstituted powdered infant formula during ultrasound at controlled temperatures: A quantitative approach on microbial responses. *Int. J. Food Microbiol.* 142(1–2): 53–59.

Airoldi, A. A., and Zottola, E. A. 1988. Growth and survival of *Salmonella typhimurium* at low temperature in nutrient deficient media. *J. Food Sci.* 53(5): 1511–1513.

Albert, I., Pouillot, R., and Denis, J. B. 2005. Stochastically modeling *Listeria monocytogenes* growth in farm tank milk. *Risk Anal.* 25(5): 1171–1185.

Amézquita, A., Kan-King-Yu, D., and Le Marc, Y. 2011. Modelling microbiological shelf life of foods and beverages. In: *Food and Beverage Stability and Shelf Life*, eds. D. Kilcast and P. Subramaniam, 405–458. Cambridge: Woodhead Publishing.

Amézquita, A., Weller, C. L., Wang, L., Thippareddi, H., and Burson, D. E. 2005. Development of an integrated model for heat transfer and dynamic growth of *Clostridium perfringens* during the cooling of cooked boneless ham. *Int. J. Food Microb.* 101(2): 123–144.

Amos, S. A., Davey, K. R., and Thomas, C. J. 2001. A comparison of predictive models for the combined effect of UV dose and solids concentration on disinfection kinetics of *Escherichia coli* for potable water production. *Process Saf. Environ. Prot.* 79(3): 174–182.

Ancelet, C. S. R., Carlin, F., Nguyen-thé, C., and Albert, I. 2012. Inferring an augmented Bayesian network to confront a complex quantitative microbial risk assessment model with durability studies: Application to *Bacillus cereus* on a courgette purée production chain. *Risk Anal.* 33: 877–892.

Aragão, G. M. F., Corradini, M. G., Normand, M. D., and Peleg, M. 2007. Evaluation of the Weibull and log normal distribution functions as survival models of *Escherichia coli* under isothermal and non isothermal conditions. *Int. J. Food Microbiol.* 119(3): 243–257.

Aspridou, Z., Akritidou, T., and Koutsoumanis, K. P. 2018. Simultaneous growth, survival and death: The trimodal behavior of *Salmonella* cells under osmotic stress giving rise to "Phoenix phenomenon". *Int. J. Food Microb.* 285: 103–109.

Augustin, J. C., and Carlier, V. 2000a. Mathematical modelling of the growth rate and lag time for *Listeria monocytogenes*. *Int. J. Food Microbiol.* 25(1): 29–51.

Augustin, J. C., and Carlier, V. 2000b. Modelling the growth of *Listeria monocytogenes* with a multiplicative type model including interactions between environmental factors. *Int. J. Food Microbiol.* 56(1): 53–70.

Augustin, J. C., Bergis, H., Midelet-Bourdin, G., Cornu, M., Couvert, O., Denis, C., Huchet, V., Lemonnier, S., Pinon, A., Vialette, M., Zuliani, V., and Stahl, V. 2011. Design of challenge testing experiments to assess the variability of *Listeria monocytogenes* growth in foods. *Food Microbiol.* 28(4): 746–754.

Augustin, J. C., Ellouze, M., and Guillier, L. 2017. Microbial risk assessment: Integrating and quantifying the impacts of food processing operations on food safety. In: *Quantitative Microbiology in Food Processing Modeling the Microbial Ecology*, ed. A. S. Sant'Ana, 583–596. Campinas, Brazil: Wiley.

Avery, S. V. 2006. Microbial cell individuality and the underlying sources of heterogeneity. *Nat. Rev. Microbiol.* 4(8): 577–587.

Balsa-Canto, E., Alonso, A. A., and Banga, J. R. 2008. Computing optimal dynamic experiments for model calibration in predictive microbiology. *J. Food Process Eng.* 31(2): 186–206.

Baranyi, J., and Roberts, T. A. 1994. A dynamic approach to predicting bacterial growth in food. *Int. J. Food Microbiol.* 23(3–4): 277–294.

Baranyi, J., Pin, C., and Roos, T. 1999. Validating and comparing predictive models. *J. Food Microb.* 48(3): 159–166.

Bassett, J., Nauta, M., Linqvist, R., and Zwietering, M. 2012. *Tools for Microbial Risk Assessment.* International Life Sciences Institute. http://ilsi.org/publication/tools-for-microbiological-risk-assessment/.

Baty, F., and Delignette-Muller, M. 2004. Estimating the bacterial lag time: Which model, which precision? *Int. J. Food Microbiol.* 91(3): 261–277.

Belessi, C. E. A., Gounadaki, A. S., Schvartzman, S., Jordan, K., and Skandamis, P. N. 2011. Evaluation of growth/no growth interface of *Listeria monocytogenes* growing on stainless steel surfaces, detached from biofilms or in suspension, in response to pH and NaCl. *Int. J. Food Microbiol.* 145 Supplement 1: S53–S60.

Bevilacqua, A., Speranza, B., Sinigaglia, M., and Corbo, M. R. 2015. A focus on the death kinetics in predictive microbiology: Benefits and limits of the most important models and some tools dealing with their application in foods. *Foods* 4(4): 565–580.

Bigelow, W. D. 1921. The logarithmic nature of thermal death time curves. *J. Infect. Dis.* 29(5): 528–236.

Bigelow, W. D., and Esty, J. R. 1920. The thermal death point in relation to time of typical thermophilic organisms. *J. Infect. Dis.* 27(6): 602–617.

Blanco-Lizarazo, C. M., Sotelo-Díaz, I., and Llorente-Bousquets, A. 2016. In vitro modelling of simultaneous interactions of *Listeria monocytogenes, Lactobacillus sakei*, and *Staphylococcus carnosus. Food Sci. Biotechnol.* 25(1): 341–348.

Bover-Cid, S., Serra-Castelló, C., Dalgaard, P., Garriga, M., and Jofré, A. 2019. "New insights on *Listeria monocytogenes* growth in pressurised cooked ham: A piezo-stimulation effect enhanced by organic acids during storage. *Int. J. Food Microbiol.* 290: 150–158.

Brasil. 1984. Ministério da Agricultura e Ministério da Saúde. Resolução CISA/MA/MS n° 10, de 31 de julho de 1984. http://www.anvisa.gov.br/anvisalegis/resol/10_84.htm.

Brillet, A., Pilet, M. F., Prevost, H., Bouttefroy, A., and Leroi, F. 2004. Biodiversity of *Listeria monocytogenes* sensitivity to bacteriocin-producing *Carnobacterium* strains and application in sterile cold-smoked salmon. *J. Appl. Microbiol.* 97(5): 1029–1037.

Brocklehurst, T. 2004. Challenge of food and the environment. In: *Modeling Microbial Responses in Foods*, eds. R. C. McKellar, and X. Lu, 197–232. Boca Raton: CRC Press.

Buchanan, R. L., and Bagi, L. K. 1999. Microbial competition: Effect of *Pseudomonas fluorescens* on the growth of *Listeria monocytogenes. Food Microbiol.* 16(5): 523–529.

341

Buchanan, R. L., Whiting, R. C., and Damert, W. C. 1997. When is simple good enough: A comparison of the Gompertz, Baranyi, and Three-phase linear models for fitting bacterial growth curves. *Food Microbiol.* 4(4): 313–326.

Carrasco, E., García-Gimeno, R., Seselovsky, R., Valero, A., Pérez, F., Zurera, G., and Todd, E. 2006. Predictive model of *Listeria monocytogenes*' Growth rate under different temperatures and acids. *Food Sci. Technol. Int.* 12(1): 47–56.

Carrasco, E., del Rosal, S., Racero, J. C., and García-Gimeno, R. M. 2012. A review on growth/no growth *Salmonella* models. *Food Res. Int.* 47(1): 90–99.

Casla, D., Requena, T., and Gómez, R. 1996. Antimicrobial activity of lactic acid bacteria isolated from goat's milk and artisanal cheeses: Characteristics of a bacteriocin produced by *Lactobacillus curvatus* IFPL105. *J. Appl. Bacteriol.* 81(1): 35–41.

Cerf, O., Davey, K. R., and Sadoudi, A. K. 1996. Thermal inactivation of bacteria - A new predictive model for the combined effect of three environmental factors: Temperature, pH and water activity. *Food Res. Int.* 29(3–4): 219–226.

Chatterjee, T., Chatterjee, B. K., Majumdar, D., and Chakrabarti, P. 2015. Antibacterial effect of silver nanoparticles and the modeling of bacterial growth kinetics using a modified Gompertz model. *BBA Gen. Subj.* 1850(2): 299–306.

Chun, H., Kim, J., Chung, K., Won, M., and Song, K. B. 2009. Inactivation kinetics of *Listeria monocytogenes, Salmonella enterica* serovar Typhimurium, and *Campylobacter jejuni* in ready-to-eat sliced ham using UV-C irradiation. *Meat Sci.* 83(4): 599–603.

Codex Alimentarius Commission. 1999. Principles and Guidelines for the Conduct of Microbiological Risk Assessment. In Joint FAO/WHO *Food Standards Programme* (Ed.), CAC/GL-30, Rome.

Collee, J. G., Knowlden, J. A., and Hobbs, B. C. 1961. Studies of the growth, sporulation and carriage of *Clostridium welchii* with special reference to food poisoning strains. *J. Appl. Bacteriol.* 26: 326–339.

Cornu, M., Billoir, E., Bergis, H., Beaufort, A., and Zuliani, V. 2011. Modeling microbial competition in food: Application to the behavior of *Listeria monocytogenes* and lactic acid flora in pork meat products. *Food Microbiol.* 28(4): 639–647.

Corradini, M., and Peleg, M. 2005. Estimating non-isothermal bacterial growth in foods from isothermal experimental data. *J. Appl. Microbiol.* 99(1): 187–200.

Costa, J. C. C. P., Tremarin, A., Longhi, D. A., Silva, A. P. R., Carciofi, B. A. M., Laurindo, J. B., and Aragão, G. M. F. 2016. Modeling of the growth of *Lactobacillus viridescens* under non-isothermal conditions. *Proc. Food Sci.* 7: 29–32.

Costa, J. C. C. P., Bover-Cid, S., Bolívar, A., Zurera, G., and Pérez-Rodríguez, F. 2019. Modelling the interaction of the sakacin-producing *Lactobacillus sakei* CTC494 and *Listeria monocytogenes* in filleted gilthead sea bream (*Sparus aurata*) under modified atmosphere packaging at isothermal and non-isothermal conditions. *Int. J. Food Microb.* 297: 72–84.

Couvert, O., Gaillard, S., Savy, N., Mafart, P., and Leguérine, I. 2005. Survival curves of heated bacterial spores: Effect of environmental factors on Weibull parameters. *Int. J. Food Microbiol.* 1(1): 73–81.

Couvert, O., Pinon, A., Bergis, H., Bourdichon, F., Carlin, F., Cornu, M., Denis, C., Gnanou Besse, N., Guillier, L., Jamet, E., Mettler, E., Stahl, V., Thuault, D., Zuliani, V., and Augustin, J. C. 2010. Validation of a stochastic modelling approach for *Listeria monocytogenes* growth in refrigerated foods. *Int. J. Food Microbiol.* 144(2): 236–242.

Cunha, L. M., Oliveira, F. A. R., Brandão, T. R. S., and Oliveira, J. C. 1997. Optimal experimental design for estimating parameters of the Bigelow model. *J. Food Eng.* 33(1–2): 111–128.

Dabadé, D. S., Azokpota, P., Nout, M. J. R., Hounhouigan, D. J., Zwietering, M. H., and Den Besten, H. M. W. 2015. Prediction of spoilage of tropical shrimp (*Penaeus notialis*) under dynamic temperature regimes. *Int. J. Food Microbiol.* 2010: 121–130.

Daelman, J., Membré, J. M., Jacxsens, L., Vermeulen, A., Devlieghere, F., and Uyttendaele, M. 2013. A quantitative microbiological exposure assessment model for *Bacillus cereus* in REPFEDs. *Int. J. Food Microbiol.* 166(3): 433–449.

Dalcanton, F., Carrasco, E., Pérez-Rodríguez, F., Posada-Izquierdo, G. D., Aragão, G. M. F., and García-Gimeno, R. M. 2018. Modeling the combined effects of temperature, pH, and sodium chloride and sodium lactate concentrations on the growth rate of *Lactobacillus plantarum* ATCC 8014. *J. Food Qual.* 2018: 1–10.

Dalgaard, P. 2002. Modelling and prediction the shelf-life of seafood. In: *Safety and Quality Issues in Fish Processing*, ed. H. A. Bremmer, 191–219. Cambridge: Woodhead Publishing.

Dalgaard, P., Ross, T., Kamperman, L., Neumeyer, K., and McMeekin, T. A. 1994. Estimation of bacterial growth rates from turbidimetric and viable count data. *Int. J. Food Microbiol.* 23(3–4): 391–404.

Dalgaard, P., Mejlholm, O., and Huss, H. H. 1997. Application of an iterative approach for development of a microbial model predicting the shelf-life of packed fish. *Int. J. Food Microbiol.* 38(2–3): 169–179.

Danyluk, M. D., and Schaffner, D. W. 2011. Quantitative assessment of the microbial risk of leafy greens from farm to consumption: Preliminary framework, data, and risk estimates. *J. Food Prot.* 74(5): 700–708.

Davey, K. R. 1989. A predictive model for combined temperature and water activity on microbial growth during the growth phase. *J. Appl. Bacteriol.* 67(5): 483–488.

Delignette-Muller, M. L., Cornu, M., Pouillot, R., and Denis, J. B. 2006. Use of Bayesian modelling in risk assessment: Application to growth of *Listeria monocytogenes* and food flora in cold-smoked salmon. *Int. J. Food Microbiol.* 106(2): 195–208.

Den Aantrekker, E. D., Boom, R. M., Zwietering, M. H., and Van Schothorst, M. 2003. Quantifying recontamination through factory environments - A review. *Int. J. Food Microbiol.* 25(2): 117–130.

Devlieghere, F., Geeraerd, A. H., Versyck, K. J., Bernaert, H., Van Impe, J. F., and Debevere, J. 2000. Shelf life of modified atmosphere packed cooked meat products: Addition of Na-lactate as a fourth shelf life determinative factor in a model and product validation. *Int. J. Food Microbiol.* 30(1–2): 93–106.

343

Devlieghere, F., Francois, K., Vermeulen, A., and Debevere, J. 2009. Predicitve microbiology. In: *Predictive Modeling and Risk Assessment*, eds. R. Costa and K. Kristbergsson, 29–53. Boston: Springer.

Elias, S. O., Alvarenga, V. O., Longhi, D. A., Sant'Ana, A. S., and Tondo, E. C. 2016. Modeling growth kinetic parameters of *Salmonella* Enteritidis SE86 on homemade mayonnaise under isothermal and nonisothermal conditions. *Foodborne Pathog. Dis.* 13(8): 462–467.

Fellows, P. J. 2009. *Food Processing Technology: Principles and Practice.* Cambridge: Woodhead Publishing.

Food Standards Agency [FSA]. 2015. Chilling. Available: https://www.food.gov. uk/northern-ireland/nutritionni/niyoungpeople/survivorform/dontget-sick/chilling.

Fredrickson, A. G. 1977. Behavior of mixed cultures of microorganisms. *Ann. Rev. Microbiol.* 31: 63–87.

Fujikawa, H., Kai, A., and Morozumi, S. 2003. A new logistic model for bacterial growth. *J. Food Hyg. Soc.* 44(3): 155–160.

Fujikawa, H., Munakata, K., and Sakha, M. Z. 2014. Development of a competition model for microbial growth in mixed culture. *Biocontrol Sci.* 19(2): 61–71.

Gaillard, S., Leguerinel, I., and Mafart, P. 1998. Model for combined effects of temperature, pH and water activity on thermal inactivation of *Bacillus cereus* spores. *J. Food Sci.* 63(5):887–889.

Geeraerd, A. H., Herremans, C. H., and Van Impe, J. F. 2000. Structural model requirements to describe microbial inactivation during a mild heat treatment. *Int. J. Food Microbiol.* 59(3): 185–209.

Geeraerd, A. H., Valdramidis, V. P., and Van Impe, J. F. 2005. GInaFiT, a freeware tool to assess non-log-linear microbial survivor curves. *Int. J. Food Microbiol.* 102(1): 95–105.

Genigeorgis, C. A. 1981. Factors affecting the probability of growth of pathogenic microorganisms in foods. *J. Am. Vet. Med. Assoc.* 179(12): 1410–1417.

Gibson, A. M., Bratchell, N., and Roberts, T. A. 1987. The effect of sodium chloride and temperature on the rate and extent of growth of *Clostridium botulinum* type A in pasteurized pork slurry. *J. App. Bacteriol.* 62(6): 479–490.

Giménez, B., and Dalgaard, P. 2004. Modelling and predicting the simultaneous growth of *Listeria monocytogenes* and spoilage micro-organisms in cold-smoked salmon. *J. Appl. Microbiol.* 96(1): 96–109.

Giuffrida, A., Ziino, G., Valenti, D., Donato, G., and Panebianco, A. 2007. Application of an interspecific competition model to predict the growth of *Aeromonas hydrophila* on fish surfaces during the refrigerated storage. *Arch. Lebensmittelhyg.* 58: 136–141.

Gonzales-Barron, U. A. 2012. Predictive microbial modelling. In: *Handbook of Food Safety Engineering*, ed. D. W. Sun, 108–140. Oxford: Wiley-Blackwell.

Gospavic, R., Kreyenschmidt, J., Bruckner, S., Popov, V., and Haque, N. 2008. Mathematical modelling for predicting the growth of *Pseudomonas* spp. in poultry under variable temperature conditions. *Int. J. Food Microbiol.* 127(3): 290–297.

Greer, G. G., and Dilts, B. D. 1995. Lactic acid inhibition of the growth of spoilage bacteria and cold tolerant pathogens on pork. *Int. J. Food Microbiol.* 25(2): 141–151.

Guillier, L., Stahl, V., Hezard, B., Notz, E., and Briandet, R. 2008. Modelling the competitive growth between *Listeria monocytogenes* and biofilm microflora of smear cheese wooden shelves. *Int. J. Food Microbiol.* 128(1): 51–57.

Haberbeck, L. U., Dannenhauer, C., Salomão, B. D. C. M., and Aragão, G. M. F. 2012. Estimation of the thermochemical nonisothermal inactivation behaviour of *Bacillus coagulans* spores in nutrient broth with oregano essential oil. *J. Food Process. Preserv.* 37: 962–968.

Haberbeck, L. U., Riehl, C. A. S., Salomão, B. D. C. M., and Aragão, G. M. F. 2012. *Bacillus coagulans* spore inactivation through the application of oregano essential oil and heat. *LWT Food Sci. Technol.* 46(1): 267–273.

Haberbeck, L. U., Oliveira, R. C., Vivijs, B., Wenseleers, T., Aertsen, A., Michiels, C., and Geeraerd, A. H. 2015. Variability in growth/no growth boundaries of 188 different *Escherichia coli* strains reveals that approximately 75% have a higher growth probability under low pH conditions than *E. coli* O157:H7 strain ATCC 43888. *Food Microbiol.* 45(B): 222–230.

Haberbeck, L. U., Plaza-Rodríguez, C., Desvignes, V., Dalgaard, P., Sanaa, M., Guillier, L., Nauta, M., and Filter, M. 2018. Harmonized terms, concepts and metadata for microbiological risk assessment models: The basis for knowledge integration and exchange. *Microb. Risk Anal.* 10: 3–12.

Hajmeer, M. N., and Basheer, I. 2002. A probabilistic neural network approach for modeling and classification of bacterial growth/no-growth data. *J. Microbiol. Methods* 51(2): 217–226.

Hajmeer, M. N., and Basheer, I. A. 2003. A hybrid Bayesian-neural network approach for probabilistic modeling of bacterial growth/no-growth interface. *Int. J. Food Microbiol.* 82(3): 233–243.

Hereu, A., Dalgaard, P., Garriga, M., Aymerich, T., and Bover-Cid, S. 2014. Analysing and modelling the growth behaviour of *Listeria monocytogenes* on RTE cooked meat products after a high pressure treatment at 400 MPa. *Int. J. Food Microbiol.* 186: 84–94.

Huang, L., Hwang, A., and Phillips, J. 2011. Effect of temperature on microbial growth rate-mathematical analysis: The Arrhenius and Eyring-Polanyi connections. *J. Food Sci.* 76(8): E553–E560.

Hwang, C. A., and Sheen, S. 2011. Growth characteristics of *Listeria monocytogenes* as affected by a native microflora in cooked ham under refrigerated and temperature abuse conditions. *Food Microbiol.* 28(3): 350–355.

Jameson, J. E. 1962. A discussion of the dynamics of *Salmonella* enrichment. *J. Hyg.* 60: 193–207.

Kakagianni, M., Kalantzi, K., Beletsiotis, E., Ghikas, D., Lianou, A., and Koutsoumanis, K. P. 2018. Development and validation of predictive models for the effect of storage temperature and pH on the growth boundaries and kinetics of *Alicyclobacillus acidoterrestris* ATCC 49025 in fruit drinks. *Food Microbiol.* 74: 40–49.

345

Kelly, A. F., Martínez-Rodríguez, A. M., Bovvil, R. A., and Mackey, B. M. 2003. Description of a "Phoenix" Phenomenon in the growth of *Campylobacter jejuni* at temperature close to the minimum for growth. *App. Environ. Microbiol.* 69(8): 4975–4978.

Kim, H. W., Lee, K., Kim, S. H., and Rhee, M. S. 2018. Predictive modeling of bacterial growth in ready-to-use salted napa cabbage (*Brassica pekinensis*) at different storage temperatures. *Food Microbiol.* 70: 129–136.

Koseki, S., Takizawa, Y., Miya, S., Takahashi, H., and Kimura, B. 2011. Modeling and predicting the simultaneous growth of *Listeria monocytogenes* and natural flora in minced tuna. *J. Food Protect.* 74(2): 176–187.

Koutsoumanis, K., and Angelidis, A. S. 2007. Probabilistic modelling approach for evaluating the compliance of ready-to-eat foods with new European Union safety criteria for *Listeria monocytogenes*. *Appl. Environ. Microbiol.* 73(15): 4996–5004.

Koutsoumanis, K., Taoukis, P. S., Drosinos, E. H., and Nychas, G. J. E. 2000. Applicability of an Arrhenius model for the combined effect of temperature and CO_2 packaging on the spoilage microflora of fish. *Appl. Environ. Microbiol.* 66(8): 3528–3534.

Koutsoumanis, K., Pavlis, A., Nychas, G. J. E., and Xanthiakos, K. 2010. Probabilistic model for *Listeria monocytogenes* growth during distribution, retail storage, and domestic storage of pasteurized milk. *Appl. Environ. Microbiol.* 76(7): 2181–2191.

Koutsoumanis, K., Lianou, A., and Gougouli, M. 2016. Latest developments in foodborne pathogens modelling. *Curr. Opin. Food Sci.* 8: 89–98.

Labuza, T. P., and Riboh, D. 1982. Theory and application of Arrhenius kinetics to the prediction of nutrient losses in foods. *Food Techn.* 36: 66–74.

Lardeux, A. L., Guillier, L., Brasseur, E., Doux, C., Gautier, J., and Gnanou-Besse, N. 2015. Impact of the contamination level and the background flora on the growth of *Listeria monocytogenes* in ready-to-eat diced poultry. *Lett. Appl. Microbiol.* 60(5): 484–490.

Le Marc, Y., Huchet, V., Bourgeois, C. M., Guyonnet, J. P., Mafart, P., and Thuault, D. 2002. Modelling the growth kinetics of *Listeria* as a function of temperature, pH and organic acid concentration. *Int. J. Food Microbiol.* 73(2–3): 219–237.

Le Marc, Y., Valík, L., and Medveďová, A. 2009. Modelling the effect of the starter culture on the growth of *Staphylococcus aureus in milk*. *Int. J. Food Microbiol.* 129(3): 306–311.

Levine, L., and Boehmer, E. 1997. Dough processing systems. In: *Food Engineering Practice*, eds. K. J. Valentas, E. Rotstein, and R. P. Singh, 487–534. Boca Raton: CRC Press.

Longhi, D. A., Dalcanton, F., Aragão, G. M. F., Carciofi, B. A. M., and Laurindo, J. B. 2013. Assessing the prediction ability of different mathematical models for the growth of *Lactobacillus plantarum* under non-isothermal conditions. *J. Theor. Biol.* 335: 88–96.

Longhi, D. A., Tremarin, A., Carciofi, B. A. M., Laurindo, J. B., and Aragão, G. M. F. 2014. Modeling the growth of *Byssochlamys fulva* on solidified apple juice at different temperatures. *Braz. Arch. Biol. Technol.* 57(6): 971–978.

Longhi, D. A., Silva, N. B., Martins, W. F., Carciofi, B. A. M., Aragão, G. M. F., and Laurindo, J. B. 2018. Optimal experimental design to model spoilage bacteria growth in vacuum-packaged ham. *J. Food Eng.* 216: 20–26.

López, F. N. A., Quintana, M. C. D., and Fernández, A. G. 2007. Use of logistic regression with dummy variables for modelling the growth-no growth limits of *Saccharomyces cerevisiae* IGAL01 as a function of sodium chloride, acid type, and potassium sorbate concentration according to growth media. *J. Food Prot.* 70(2): 456–465.

Lotka, A. J. 1956. *Elements of Mathematical Biology.* New York: Dover Publications.

Mafart, P., and Leguerinel, I. 1998. Modeling combined effects of temperature and pH on heat resistance of spores by a Linear-Bigelow equation. *J. Food Sci.* 63(1): 6–8.

Malakar, P. K., Barker, G. C., Zwietering, M. H., and Van't Riet, K. 2003. Relevance of microbial interactions to predictive microbiology. *Int. J. Food Microbiol.* 84(3): 263–272.

Mandal, P. K., Biswas, A. K., Choi, K., and Pal, U. K. 2011. Methods for rapid detection of foodborne pathogens: An overview. *Am. J. Food Technol.* 6(2): 87–102.

Masana, M. O., and Baranyi, J. 2000. Growth/no growth interface of *Brochothrix thermosphacta* as a function of pH and water activity. *Food Microbiol.* 17(5): 485–493.

Mataragas, M., Drosinos, E. H., Vaidanis, A., and Metaxopoulos, I. 2006. Development of a predictive model for spoilage of cooked cured meat products and its validation under constant and dynamic temperature storage conditions. *J. Food Sci.* 71(6): M157–M167.

Maturin, L., and Peeler, J. T. 2001. Aerobic plate count. In: *Bacteriological Analytical Manual.* https://www.fda.gov/food/foodscienceresearch/laboratorymethods/ucm063346.htm.

McDonald, K., and Sun, D. W. 1999. Predictive food microbiology for the meat industry: A review. *Int. J. Food Microbiol.* 52(1–2): 1–27.

McKellar, R. C., and Lu, X. 2001. A probability of growth model for *Escherichia coli* O157:H7 as a function of temperature, pH, acetic acid, and salt. *J. Food Prot.* 64(12): 1922–1928.

McKellar, R. C., and Lu, X. 2003. *Modeling Microbial Responses in Foods.* Boca Raton: CRC Press.

McMeekin, T., Olley, J., Ratkowsky, D., Corkrey, R., and Ross, T. 2013. Predictive microbiology theory and application: Is it all about rates? *Food Control* 29(2): 290–299.

McMeekin, T. A., Olley, J. N., Ross, T., and Ratkowsky, D. A. 1993. *Predictive Microbiology: Theory and Application.* Taunton, UK: Research Studies Press

McMeekin, T. A., and Ross, T. 2002. Predictive microbiology: Providing a knowledge-based framework for change management. *Int. J. Food Microbiol.* 78(1–2): 133–153.

McMeekin, T. A., Chandler, R. E., Doe, P. E., Garland, C. D., Olley, J., Putro, S., and Ratkowsky, D. A. 1987. Model for combined effect of temperature and salt concentration/water activity on the growth rate of *Staphylococcus xylosus. J. Appl. Bacteriol.* 62(6): 543–550.

McMeekin, T. A., Presser, K., Ratkowsky, D., Ross, T., Salter, M., and Tienungoon, S. 2000. Quantifying the hurdle concept by modelling the bacterial growth/no growth interface. *Int. J. Food Microbiol.* 55(1–3): 93–98.

Mejlholm, O., and Dalgaard, P. 2007. Modeling and predicting the growth of lactic acid bacteria in lightly preserved seafood and their inhibiting effect on *Listeria monocytogenes. J. Food Prot.* 70(11): 2485–2497.

Mejlholm, O., and Dalgaard, P. 2015. Modelling and predicting the simultaneous growth of *Listeria monocytogenes* and psychrotolerant lactic acid bacteria in processed seafood and mayonnaise-based seafood salads. *Food Microbiol.* 46: 1–14.

Mejlholm, O., Gunvig, A., Borggaard, C., Blom-Hanssen, J., Mellefont, L., Ross, T., Leroi, F., Else, T., Visser, D., and Dalgaard, P. 2010. Predicting growth rates and growth boundary of *Listeria monocytogenes* — An international validation study with focus on processed and ready-to-eat meat and seafood. *Int. J. Food Microbiol.* 141(3): 137–150.

Mejlholm, O., Bøknæs, N., and Dalgaard, P. 2015. Development and validation of a stochastic model for potential growth of *Listeria monocytogenes* in naturally contaminated lightly preserved seafood. *Food Microbiol.* 45(B): 276–289.

Mellefont, L. A., McMeekin, T. A., and Ross, T. 2005. Viable count estimates of lag time responses for *Salmonella typhimurium* M48 subjected to abrupt osmotic shifts. *Int. J. Food Microbiol.* 105(3): 399–410.

Mellefont, L. A., McMeekin, T. A., and Ross, T. 2008. Effect of relative inoculum concentration on *Listeria monocytogenes* growth in co-culture. *Int. J. Food Microbiol.* 121(2): 157–168.

Membré, J. M., Ross, T., and McMeekin, T. 1999. Behaviour of *Listeria monocytogenes* under combined chilling processes. *Lett. Appl. Microbiol.* 28(3): 216–220.

Membré, J. M., Kubaczka, M., and Chene, C. 2001. Growth rate and growth–no-growth interface of *Penicillium brevicompactum* as functions of pH and preservative acids. *Food Microbiol.* 18(5): 531–538.

Membré, J. M., Leporq, B., Vialette, M., Mettler, E., Perrier, L., Thuault, D., and Zwietering, M. 2005. Temperature effect on bacterial growth rate: Quantitative microbiology approach including cardinal values and variability estimates to perform growth simulations on/in food. *Int. J. Food. Microbiol.* 100(1–3): 179–186.

Mertens, L., Van Derlinden, E., Dang, T. D. T., Cappuyns, A. M., Vermeulen, A., Debevere, J., Moldenaers, P., Devlieghere, F., Geeraerd, A. H., and Van Impe, J. F. 2011. On the critical evaluation of growth/no growth assessment of *Zygosaccharomyces bailii* with optical density measurements: Liquid versus structured media. *Food Microbiol.* 28(4): 736–745.

Messens, W., Hempena, M., and Koutsoumanis, K. 2018. Use of predictive modelling in recent work of the panel on biological hazards of the European Food Safety Authority. *Microb. Risk Anal.* 10: 37–43.

Miles, D. W., Ross, T., Olley, J., and McMeekin, T. A. 1997. Development and evaluation of a predictive model for the effect of temperature and water activity on the growth rate of *Vibrio parahaemolyticus*. *Int. J. Food Microbiol.* 16(2–3): 133–142.

Møller, C. O. A., Nauta, M. J., Christensen, B. B., Dalgaard, P., and Hansen, T. B. 2012. Modelling transfer of *Salmonella* Typhimurium DT104 during simulation of grinding of pork. *J. Appl. Microbiol.* 112(1): 90–98.

Møller, C. O. A., Ilg, Y., Aabo, S., Christensen, B. B., Dalgaard, P., and Hansen, T. B. 2013. Effect of natural microbiota on growth of *Salmonella* spp. in fresh pork – A predictive microbiology approach. *Food Microb.* 34(2): 284–295.

Møller, C. O. A., Sant'Ana, A. S., Hansen, S. K. H., Nauta, M. J., Silva, L. P., Alvarenga, V. O., Maffei, D., Pacheco, F., Lopes, J., Franco, B. D. G. M., Aabo, S., and Hasen, T. B. 2016. Robustness of a cross contamination model describing transfer of pathogens during grinding of meat. *Proc. Food Sci.* 7: 97–100.

Mounier, J., Gelsomino, R., Goerges, S., Vancanneyt, M., Vandemeulebroecke, K., Hoste, B., Scherer, S., Swings, J., Fitzgerald, G. F., and Cogan, T. M. 2005. Surface microflora of four smear-ripened cheeses. *Appl. Environ. Microbiol.* 71(11): 6489–6500.

Mytilinaios, I., Salih, M., Schofield, H. K., and Lambert, J. W. 2012. Growth curve prediction from optical density data. *Int. J. Food Microbiol.* 154(3): 169–176.

Nauta, M. 2002. Modelling bacterial growth in quantitative microbiological risk assessment: Is it possible? *Int. J. Food Microbiol.* 73(2–3): 297–304.

Nauta, M. 2008. The Modular Process Risk Model (MPRM): A structured approach to food chain exposure assessment. In: *Microbial Risk Analysis of Food*, ed. D. Schaffner, 99–136. Washington, DC: AMS Press.

Nauta, M. J. 2001. A modular process risk model structure for quantitative microbiological risk assessment and its application in an exposure assessment of *Bacillus cereus* in a REPFED. Report 149106 007, RIVM, Bilthoven. www.rivm.nl/bibliotheek/rapporten/149106007.pdf.

Nauta, Maarten J. 2005. Microbiological risk assessment models for partitioning and mixing during food handling. *Int. J. Food Microbiol.* 100(1–3): 311–322.

Nauta, M. J., Evers, E. G., Takumi, K., and Havelaar, A. H. 2001. Risk assessment of Shiga toxin producing *Escherichia coli* O157 in steak tartare in the Netherlands. Report 257851 003, National Institute for Public Health and the Environment (RIVM), Bilthoven. https://www.rivm.nl/bibliotheek/rapporten/257851003.pdf.

Nicolai, B. M., and Van Impe, J. F. 1996. Predictive food microbiology: A probabilistic approach. *Math. Comput. Simul.* 42(2–3): 287–292.

Niranjan, K. 2009. Food mixing. In: *Food Engineering*, ed. G. Barbosa-Cánovas, 414–427. Oxford: Eolss Publishers.

Oscar, T. P. 2005. Validation of lag time and growth rate models for *Salmonella* Typhimurium: Acceptable prediction zone method. *J. Food Sci.* 70(2): M129–M137.

Østergaard, N. B., Eklöw, A., and Dalgaard, P. 2014. Modelling the effect of lactic acid bacteria from starter- and aroma culture on growth of *Listeria monocytogenes* in cottage cheese. *Int. J. Food Microbiol.* 188: 15–25.

Paganini, C. C., Severo, D., Menezes, N. M. C., Carciofi, B. A. M., and Aragão, G. M. F.. Modeling the effect of the inoculum physiological state and serotype of *Salmonella enterica* in the occurrence of the Phoenix Phenomenon under osmotic stress. Submitted to *Int. J. Food Microbiol.*

Patil, S., Bourke, P., Frias, J. M., Tiwari, B. K., and Cullen, P. J. 2009. Inactivation of *Escherichia coli* in orange juice using ozone. *Innov. Food Sci. Emerg. Technol.* 10(4): 551–557.

Peleg, M. 1997. Modeling microbial populations with the original and modified versions of the continuous and discrete logistic equations. *Crit. Rev. Food Sci. Nutrit.* 37(5): 471–490.

Peleg, M. 2003. Calculation of the non-isothermal inactivation patterns of microbes having sigmoidal isothermal semi-logarithmic survival curves. *Crit. Rev. Food Sci. Nutrit.* 43(6): 645–658.

Peleg, M. 2006. *Advanced Quantitative Microbiology for Foods and Biosystems: Models for Predicting Growth and Inactivation.* Boca Raton: CRC Press.

Peleg, M., and Cole, M. B. 1998. Reinterpretation of microbial survival curves. *Crit. Rev. Food Sci. Nutrit.* 38(5): 353–380.

Pérez-Rodríguez, F. 2013. Development and application of predictive microbiology models in foods. In: *Mathematical and Statistical Methods in Food Science and Technology*, eds. D. Granato and G. Ares, 321–361. Oxford: Wiley Blackwell.

Pérez-Rodríguez, F., and Valero, A. 2013. *Predictive Microbiology in Foods.* New York: Springer.

Pérez-Rodríguez, F., Valero, A., Carrasco, E., García, R. M., and Zurera, G. 2008. Understanding and modelling bacterial transfer to foods: A review. *Trends Food Sci. Technol.* 19(3): 131–144.

Perni, S., Andrew, P. W., and Shama, G. 2005. Estimating the maximum growth rate from microbial growth curves: Definition is everything. *Food Microbiol.* 22(6): 491–495.

Plaza-Rodríguez, C., Haberbeck, L. U., Desvignes, V., Dalgaard, P., Sanaa, M., Nauta, M., Filter, M., and Guillier, L. 2017. Towards transparent and consistent exchange of knowledge for improved microbiological food safety. *Cur. Op. Food Sci.* 18: 1–9.

Pouillot, R., Albert, I., Cornu, M., and Denis, J. B. 2003. Estimation of uncertainty and variability in bacterial growth using Bayesian inference. Application to *Listeria monocytogenes*. *Int. J. Food Microbiol.* 81(2): 87–104.

Pouillot, R., Miconnet, N., Afchain, A.-L., Delignette-Muller, M. L., Beaufort, A., Rosso, L., Denis, J. B., and Cornu, M. 2007. Quantitative risk assessment of *Listeria monocytogenes* in French cold-smoked salmon: I. Quantitative Exposure Assessment. *Risk Anal.* 27(3): 683–700.

Powell, M., Schlosser, W., and Ebel, E. 2004. Considering the complexity of microbial community dynamics in food safety risk assessment. *Int. J. Food Microbiol.* 90(2): 171–179.

Presser, K. A., Ross, T., and Ratkowsky, D. A. 1998. Modelling the growth limits (growth/no growth interface) of *Escherichia coli* as a function of temperature, pH, lactic acid concentration, and water activity. *Appl. Environ. Microbiol.* 64(5): 1773–1779.

Quinto, E. J., Marín, J. M., and Schaffner, D. W. 2016. Effect of the competitive growth of *Lactobacillus sakei* MN on the growth kinetics of *Listeria monocytogenes* Scott A in model meat gravy. *Food Control* 63: 34–45.

Ratkowsky, D. A., Olley, J., McMeekin, T. A., and Ball, A. 1982. Relationship between temperature and growth rate of bacterial cultures. *J. Bacteriol.* 149(1): 1–5.

Ratkowsky, D. A., Lowry, R. K., McMeekin, T. A., Stokes, A. N., and Chandler, R. E. 1983. Model for bacterial culture growth rate throughout the entire biokinetic temperature range. *J. Bacteriol.* 154(3): 1222–1226.

Reij, M. W., and Den Aantrekker, E. D. 2004. Recontamination as a source of pathogens in processed foods. *Int. J. Food Microbiol.* 15(1): 1–11.

Rodrigues, L., Teixeira, J., Oliveira, R., and Van Der Mei, H. 2006. Response surface optimization of the medium components for the production of biosurfactants by probiotic bacteria. *Process Biochem.* 41(1): 1–10.

Ross, T. 1996. Indices for performance evaluation of predictive models in food microbiology. *J. Appl. Bacteriol.* 81(5): 501–508.

Ross, T., and Dalgaard, P. 2004. Secondary models. In: *Modeling Microbial Responses in Foods*, eds. R. C. McKellar and X. Lu, 62–122. New York: CRC Press.

Ross, T., Ratkowsky, D. A., Mellefont, L. A., and McMeekin, T. A. 2003. Modelling the effects of temperature, water activity, pH and lactic acid concentration on the growth rate of *Escherichia coli*. *Int. J. Food Microbiol.* 82(1): 33–43.

Rosso, L., Bajard, S., Flandrois, J. P., Lahellec, C., Fouraud, J., and Veit, P. 1996. Differential growth of *Listeria monocytogenes* at 4 and 8°C: Consequences for the shelf life of chilled products. *J. Food Protect.* 59(9): 944–949.

Sanaa, M., Coroller, L., and Cerf, O. 2004. Risk assessment of listeriosis linked to the consumption of two soft cheeses made from raw milk: Camembert of Normandy and brie of meaux. *Risk Anal.* 24(2): 389–399.

Sant'Ana, A. S., Franco, B. D. G. M., and Schaffner, D. W. 2012. Modeling the growth rate and lag time of different strains of *Salmonella enterica* and *Listeria monocytogenes* in ready-to-eat lettuce. *Food Microbiol.* 30(1): 267–273.

Shoemaker, S., and Pierson, D. M. 1977. Phoenix phenomenon in the growth of *Clostridium perfringens*. *Appl. Envrion. Microbiol.* 32: 803–807.

Sieuwerts, S., Bok, F. A. M., Hugenholtz, J., and Vlieg, J. E. T. H. 2008. Unraveling microbial interactions in food fermentations: from classical to genomics approaches. *Appl. Environ. Microbiol.* 74(16): 4997–5007.

Silva, A. P., Rosa, Longhi, D. A., Dalcanton, F., and Aragão, G. M. F. 2018. Modelling the growth of lactic acid bacteria at different temperatures. *Braz. Arch. Biol. Techn.* 61: 1–11.

Silva, A. R., Sant'Ana, A. S., and Massaguer, P. R. 2010. Modelling the lag time and growth rate of *Aspergillus* section *Nigri* IOC 4573 in mango nectar as a function of temperature and pH. *J. App. Microbiol.* 109(3): 1105–1116.

Silva, N. B., Longhi, D. A., Martins, W. F., Laurindo, J. B., Aragão, G. M., and Carciofi, B. A. 2017. Modeling the growth of *Lactobacillus viridescens* under non-isothermal conditions in vacuum-packed sliced ham. *Int. J. Food Microbiol.* 240: 97–101.

Skandamis, P. N., Stopforth, J. D., Kendall, P., Belk, K. E., Scanga, J., Smith, G. C., and Sofos, J. N. 2007. Modeling the effect of inoculum size and acid adaptation on growth/no growth interface of *Escherichia coli* O157:H7. *Int. J. Food Microbiol.* 120(3): 237–249.

Slongo, A. P., Rosenthal, A., Camargo, L. M. Q., Deliza, R., Mathias, S. P., and Aragão, G. M. F. 2009. Modeling the growth of lactic acid bacteria in sliced ham processed by high hydrostatic pressure. *LWT Food Sci. Technol.* 42(1): 303–306.

Smith, B. A., Fazil, A., and Lammerding, A. M. 2013. A risk assessment model for *Escherichia coli* O157:H7 in ground beef and beef cuts in Canada: Evaluating the effects of interventions. *Food Control* 29(2): 364–381.

Stamati, I., Akkermans, S., Logist, F., Noriega, E., and Van Impe, J. F. 2016. Optimal experimental design for discriminating between microbial growth models as function of suboptimal temperature: From in silico to in vivo. *Food Res. Int.* 89(1): 689–700.

Sumner, E. R., and Avery, S. V. 2002. Phenotypic heterogeneity: Differential stress resistance among individual cells of the yeast *Saccharomyces cerevisiae*. *Microbiology* 148(2): 345–351.

Tarlak, F., Ozdemir, M., and Melikoglu, M. 2018. Mathematical modelling of temperature effect on growth kinetics of *Pseudomonas* spp. on sliced mushroom (*Agaricus bisporus*). *Int. J. Food Microbiol.* 266: 274–271.

Tenenhaus-Aziza, F., and Ellouze, M. 2015. Software for predictive microbiology and risk assessment: A description and comparison of tools presented at the ICPMF8 Software Fair. *Food Microbiol.* 45(B): 290–299.

Tremarin, A., Longhi, D. A., Salomão, B. C. M., and Aragão, G. M. F. 2015. Modeling the growth of *Byssochlamys fulva* and *Neosartorya fischeri* on solidified apple juice by measuring colony diameter and ergosterol content. *Int. J. Food Microbiol.* 193: 23–28.

USA Food and Drug Administration [FDA]. 2014. *Are You Storing Food Safely? Consumer Health Information Handout.* US Food and Drug Administration. https://www.fda.gov/ForConsumers/ConsumerUpdates/ucm093704.htm.

Valenti, D., Giuffrida, A., Ziino, G., Giarratana, F., Spagnolo, B., and Panebianco, A. 2013. Modelling bacterial dynamics in food products: Role of environmental noise and interspecific competition. *J. Mod. Phys.* 4(8): 1059–1065.

Van Boekel, M. A. J. S. 2002. On the use of the Weibull model to describe thermal inactivation of microbial vegetative cells. *Int. J. Food Microbiol.* 74(1–2): 139–159.

Van Damme, I., De Zutter, L., Jacxsesn, L., and Nauta, M. J. 2017. Control of human pathogenic *Yersinia enterocolitica* in minced meat: Comparative analysis of different interventions using a risk assessment approach. *Food Microbiol.* 64: 83–95.

Van Derlinden, E., Bernaerts, K., and Van Impe, J. F. 2008. Accurate estimation of cardinal growth temperatures of *Escherichia coli* from optimal dynamic experiments. *Int. J. Food Microbiol.* 128(1): 89–100.

Van Derlinden, E., Cappunyns, A. M., Mertens, L., Van Impe, J. F., and Valdramidis, V. P. 2012. Modeling microbial responses during decontamination processes. In: *Decontamination of Fresh and Minimally Processed Produce*, ed. V. M. Gómez-López, 487–519. Oxford: Wiley -Blackwell.

Vereecken, K. M., Dens, E. J., and Van Impe, J. F. 2000. Predictive modeling of mixed microbial populations in food products: Evaluation of two-species models. *J. Theor. Biol.* 205(1): 53–72.

Vermeulen, A., Gysemans, K. P. M., Bernaerts, K., Geeraerd, A. H., Debevere, J., Devlieghere, F., and Van Impe, J. F. 2009. Modelling the influence of the inoculation level on the growth/no growth interface of *Listeria monocytogenes* as a function of pH, aw and acetic acid. *Int. J. Food Microbiol.* 135(2): 83–89.

Versyck, K. J., Bernaerts, K., Geeraerd, A. H., and Van Impe, J. F. 1999. Introducing optimal experimental design in predictive modelling: A motivating example. *Int. J. Food Microbiol.* 51(1): 39–51.

Virto, R., Sanz, D., Álvarez, I., and Condón, J. R. 2005. Inactivation kinetics of *Yersinia enterocolitica* by citric and lactic acid at different temperatures. *Int. J. Food Microbiol.* 103(3): 251–257.

Wang, J., Membré, J. M., Ha, S. D., Bahk, G. J., Chung, M. S., Chun, H. S., Hwang, I. G., and Oh, D. H. 2012. Modeling the combined effect of temperature and relative humidity on *Escherichia coli* O157:H7 on lettuce. *Food Sci. Biotechnol.* 21(3): 859–865.

Whiting, R., and Buchanan, R. 1993. A classification of models in predictive microbiology - reply. *Food Microbiol.* 10(2): 175–177.

Whiting, R. C., and Buchanan, R. L. 1997. Development of a quantitative risk assessment model for *Salmonella enteritidis* in pasteurized liquid eggs. *Int. J. Food Microbiol.* 36(2–3): 111–125.

Wijtzes, T., Rombouts, F. M., Kant-Muermans, M. L., Van 't Riet, K., and Zwietering, M. H. 2001. Development and validation of a combined temperature, water activity, ph model for bacterial growth rate of *Lactobacillus curvatus*. *Int. J. Food Microbiol.* 22: 57–64.

Xanthiakos, K., Simos, D., Angelidis, A. S., Nychas, G. J. E., and Koutsoumanis, K. 2006. Dynamic modeling of *Listeria monocytogenes* growth in pasteurized milk. *J. Appl. Microbiol.* 100(6): 1289–1298.

Ye, K., Wang, H., Jiang, Y., Xu, X., Cao, J., and Zhou, G. 2014. Development of inter-specific competition models for the growth of *Listeria monocytogenes* and *Lactobacillus* on vacuum-packaged chilled pork by quantitative real-time PCR. *Food Res. Int.* 64: 626–633.

Zhou, K., George, S. M., Metris, A., Li, P. L., and Baranyi, J. 2011. Lag phase of *Salmonella enterica* under osmotic stress conditions. *Appl. Environ. Microbiol.* 77(5): 1578–1762.

Zurera-Cosano, G., García-Gimeno, R. M., Rodríguez-Pérez, R., and Hervás-Martínez, C. 2006. Performance of response surface model for prediction of *Leuconostoc mesenteroides* growth parameters under different experimental conditions. *Food Control* 17(6): 429–438.

Zwietering, M. H., and Den Besten, H. M. W. 2011. Modelling: One word for many activities and uses. *Food Microbiol.* 28(4): 818–822.

Zwietering, M. H., Jongenburger, I., Rombouts, F. M., and Van't Riet, K. 1990. Modeling of the bacterial growth curve. *Appl. Environ. Microbiol.* 56(6): 1875–1881.

12

Modelling Cross-Contamination in Food Processing

Cleide O. de A. Møller, Maarten J. Nauta, and Tina B. Hansen

Contents

The main objective of this chapter is to give an overview of modelling approaches applied for cross-contamination in large-scale processing, such as slaughter procedures, grinding and slicing. In addition, the challenges in developing a generic mechanistic model are discussed. Lastly, it outlines future perspectives for improvement and the development of more efficient models to examine the effect of cross-contamination and to improve quantitative microbial risk assessments (QMRAs).

12.1 INTRODUCTION: TRANSFER AND CROSS-CONTAMINATION

Next to bacterial growth, bacterial transfer is an important concern in microbiological food safety that may turn uncontaminated foods into contaminated foods. The transfer can take place from, and to, a variety of contact surfaces, such as food, hands and equipment, and via air, e.g. as aerosols or splashes. Prevention of microbiological transfer is crucial for food safety and is usually tackled by the implementation of a set of food hygiene measures. However, transfer can also be an inevitable side effect of food processing and food handling, for example during cutting or grinding of meat or vegetables, or during a slaughter process. Often, the term *cross-contamination* is used for describing a combination of transfer events that leads to contamination of foods, sometimes even in the case of a single transfer event. According to Pérez-Rodríguez et al. (2008), cross-contamination is "a general term which refers to the transfer, direct or indirect, of bacteria or virus from a contaminated product to a non-contaminated product."

In microbiological risk assessment, cross-contamination is recognized as a process that may increase the prevalence of contaminated servings, thereby potentially increasing the consumer exposure and health risk (Nauta 2002). Yet, in predictive modelling in foods, cross-contamination usually gets less attention than growth and inactivation, and as a consequence, only a limited number of cross-contamination models are available for performing risk assessment.

The reasons for this are probably that the effects of cross-contamination are usually highly variable, cross-contamination can occur via very many (combinations of) routes, reproducible experiments that are representative of real-life food handling processes are more difficult to design, and plausible mechanistic models are not easily constructed. Van Asselt et al. (2008) have already pointed out the lack of modelling approaches able to describe and explain the mechanisms involved in transfer of pathogens during food processing. Yet, good models for cross-contamination processes are crucial for microbiological risk assessment.

One class of cross-contamination models used in risk assessments builds on single transfer events. These events can be studied experimentally and provide estimates of transfer rates from, for example, food product to hand, hand to food product, food product to cutting board, cutting board to food product, etc. (Chen et al. 2001; Kusumaningrum et al. 2004; Montville and Schaffner 2004). For risk assessment, models can be constructed that combine the relevant transfer events by taking into account the likelihood of each event and the effect of the event (i.e. the transfer rate). This has, for example, been done to describe the bacterial transfer during the preparation of a chicken salad and provides fit-for-purpose cross-contamination models (Christensen et al. 2005; Luber et al. 2006; Mylius, Nauta, and Havelaar 2007; van Asselt et al. 2008). These models typically describe potential scenarios for food preparation that are useful for exploring the impact of food handling on the consumer health risk. However, if food handling takes place in uncontrolled environments like the consumer kitchen, the number of potential scenarios is large, and the effect of cross-contamination in terms of net transfer of microorganisms to the food product is highly dependent on events like hand washing, change of cutting board, etc. That implies that a large variety of models can be developed by different researchers, as for example nicely illustrated by Chapman et al. (2016), who compare consumer phase models for *Campylobacter* in broiler meat.

Another class of cross-contamination models is those that describe "large-scale processes", like industrial processing of animals after slaughter, industrial-scale cutting of vegetables, or cutting and grinding of meat or meat products. These processes typically include a sequence of repeated transfer events that potentially leads to a large number of contaminated products. Models for these processes can be predominantly empirical, when contamination data before and after the processes are fitted to a statistical model (Hartnett 2001; Rosenquist et al. 2003; Smid et al. 2012). Other models include knowledge or assumptions on the effects of the different transfer and removal processes involved in a more complex cross-contamination event and combine them in compartmental and mechanistic models. In a review on the available knowledge related to microbial cross-contamination dynamics, Possas et al. (2016) concluded that there is a need for such models, which are able to explain the influence of relevant processing aspects on the overall transfer dynamics of microorganisms during food processing.

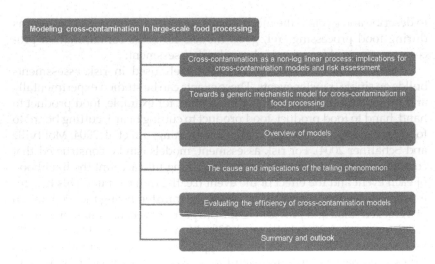

Figure 12.1 Schematic representation of the topics included in modelling of cross-contamination in large-scale food processing.

In this chapter, an overview of cross-contamination models related to large-scale food processing will be presented, and as indicated in Figure 12.1, special attention will be given to the non-linearity that characterizes cross-contamination events. The steps taken so far towards achieving a generic mechanistic cross-contamination model are summarized, and the implications of the often observed tailing phenomenon are discussed.

Applications of cross-contamination models will be given along with suggestions for how to evaluate the performance of cross-contamination models. Finally, future perspectives will be outlined for improvement and the development of more efficient models.

12.2 CROSS-CONTAMINATION AS A NON-LOG-LINEAR PROCESS: IMPLICATIONS FOR CROSS-CONTAMINATION MODELS AND RISK ASSESSMENT

From a modelling perspective, an important difference between cross-contamination models and predictive models for growth and/or inactivation is that transfer should not be modelled on a log scale. Bacterial growth is typically described as a sequence of doublings, leading to exponential growth, and bacterial inactivation is typically described by exponential decay. Such processes are linear on a log scale, which implies that growth and inactivation are additive processes on a log scale. A general description could be:

$$Log\left(N_{after}\right) = Log\left(N_{before}\right) + f(.) \tag{12.1}$$

where N_{before} and N_{after} are the concentrations before and after the growth or inactivation processes, and f(.) gives the increase or decrease in logs as described by a defined function. Usually, f(.) is a function of time, and for example, f(.) = μt in an exponential growth model, f(.) = $-t/D$ in the Bigelow inactivation model, and f(.) = $a(T - T_{min})^2t$ in a square root secondary growth model.

However, such log-linear models usually cannot be used for modelling transfer and cross-contamination. Imagine, for example, a food product with 4 log (=10,000) colony forming units (cfu) and a transfer rate of 1% to the hand. After transfer, the food product contains 3.996 log (=9900) cfu, and the hands contain 2 log (=100) cfu. If the same hand then handles a second food product with the same contamination level, the total amount on the hand increases to 2.3 log (=200) cfu. Clearly, on a log scale, these calculations are not additive; it is much more intuitive, and easier, to perform them in terms of cfu instead of log cfu.

An interesting consequence of this phenomenon is that it complicates calculations in risk assessments. If a series of processes is to be modelled, as in a modular process risk model, growth and inactivation steps are modelled on a log scale, but cross-contamination steps are not (Zwietering 2015). This means that one should be careful when applying the International Commission on Microbiological Specifications for Foods (ICMSF) equation that simplifies the risk assessment approach to evaluate whether a Food Safety Objective is met (Zwietering, Stewart, and Whiting 2010). This equation reads:

$$H_0 + \sum G - \sum R + \sum C < FSO \tag{12.2}$$

where H_0 is the initial concentration in a food product, ΣG is the sum of the effect of growth in the process steps that allow growth, ΣR is the sum of the effect of inactivation (reduction) in the process steps that lead to inactivation, ΣC is the increase due to (re-)contamination and FSO is the Food Safety Objective, the highest level still considered safe at the time of consumption. This equation is written in the same format as Equation 12.1 and implicitly assumes that all units are given on a log scale. However, this is only true for ΣG and ΣR, not for ΣC. An example may illustrate how this complicates the calculation:

For $H_0 = 1$ log cfu, $\Sigma G = 1$ log increase due to growth, $\Sigma R = 2$ log decrease due to inactivation and ΣC is 100 cfu = 2 log cfu added from an external source via transfer, $H_0 + \Sigma G - \Sigma R + \Sigma C = 1 + 1 - 2 + 2 = 2$ log cfu. The same equation with $\Sigma R = 4$ log decrease gives $H_0 + \Sigma G - \Sigma R + \Sigma C = 1 + 1 - 4 + 2 = 0$ log cfu.

If the transfer takes place *after* the growth and inactivation step, the first calculation is approximately correct, as the total contamination is $\log(1 + 100) = \log(101) = 2.004$ log cfu, but the second is not because the total contamination in the end would still be 2 log cfu. If the transfer occurred after the growth but *before* the inactivation, you would get $\log(200/100) = \log(2) = 0.3$ log cfu in the first scenario and $\log(200/10,000) = -1.7$ log cfu in the second.

These are not intuitive results, and therefore, they are relevant to consider. The ICMSF equation is obviously intended as a simplified approach to evaluate the safety of a food production process, and as such, it has proven to be very useful (Zwietering and Nauta 2007). However, when cross-contamination events are important, extra care is needed to be sure that the simplified approach is sufficient. Interestingly, the authors are not aware of any published example that illustrates the use of the ICMSF equation that includes cross-contamination.

As explained in Section 12.3, another consequence is that log-linear models are not expected to adequately describe the effects of cross-contamination. This is relevant, for example, if one considers using a regression model to describe the concentrations after a cross-contamination process, which are dependent on the concentrations before the process.

12.3 TOWARDS A GENERIC MECHANISTIC MODEL FOR CROSS-CONTAMINATION IN FOOD PROCESSING

Since cross-contamination in large-scale processing may involve diverse transfer and removal events, the dynamics of the process are not yet fully understood, and different approaches have been used to model it. Overall, it would be advantageous to have a universal model that is able to effectively describe the cross-contamination events occurring in large-scale food processing. Such a model should include the dynamics involved in the transfer of microbiological contaminants. Ideally, the model should also have a simple and user-friendly approach and still be effective enough. However, such a model is not available yet due to the complexity inherent in cross-contamination in food processing.

Here it is described how these approaches have evolved, starting with models for *Campylobacter* during broiler processing and explaining how a semi-mechanistic approach developed for this process has been used and developed further into a generic model for cross-contamination during grinding and slicing.

12.3.1 Models for Cross-Contamination during Industrial Broiler Processing

From the start of this century, several quantitative risk assessments for *Campylobacter* in broiler meat use industrial processing models that are primarily based on (scarcely available) data, assuming that the effects of the processing could be characterized by the observed increase or decrease in log concentrations as measured on the carcasses and the variation therein (Hartnett 2001; Rosenquist et al. 2003; FAO/WHO 2009; Nauta et al. 2009; Hayama et al. 2011). The advantage of this approach is that it reflects the observations and describes the dynamics in the observed slaughterhouses well. The disadvantage is that the observed changes in *Campylobacter* concentrations on the carcasses need not be similar if the concentrations at one or more of the processing stages change, for example due to interventions before or during processing. This makes it questionable to use the model for the evaluation of intervention strategies in risk assessment.

The argument here is that cross-contamination is not well described by a log-linear model, as explained in Section 12.2. An example would be the relation between *Campylobacter* concentrations on broiler carcasses before and after defeathering. If a 1 log decrease in concentration during the defeathering step is observed for carcasses with 4 log cfu/g on skin samples before defeathering, the same decrease need not occur when the concentration before defeathering is 2 log cfu/g instead. If the main reason for the change from 4 to 3 log cfu/g during defeathering is not a 1 log removal but the combined effect of several log removal due to the plucking, the water spraying applied during defeathering and the addition of 3 log cfu/g from faecal leakage from the birds' intestines, the observed 3 log cfu/g after defeathering will also be found if the concentration on the skin samples before defeathering is 2 log cfu/g. Hence, the observed 1 log reduction, which a log-linear model would typically use, is not characteristic of the defeathering process.

Being aware of this challenge, several so-called "semi-mechanistic" models for cross-contamination during broiler processing were developed

(Hartnett 2001; Nauta et al. 2005; Nauta, Jacobs-Reitsma, and Havelaar 2007). These models take some elementary process characteristics into account, without including all factors that may be relevant for the physical processes. Hartnett (2001) and FAO/WHO (2009) proposed different specific "semi-mechanistic" models for transport of broilers from the farm to the processing plant and for the defeathering and the evisceration steps.

As an alternative approach, Nauta et al. (2005) and Nauta, Jacobs-Reitsma, and Havelaar (2007) developed a generic approach for cross-contamination in a continuous processing environment and applied it to risk assessment of *Campylobacter* in broiler meat. The model is built on some hypotheses on how the dynamics involved in cross-contamination events can be explained. As shown in Figure 12.2, the model considers all processing surfaces (*pr*) as one environment/compartment. The transfer rates of the microbiological contaminant (*M*) between food matrix *i* and processing surfaces (*pr*), forwards and backwards, is then described by two parameters, *a* and *b*. Inactivation rates of the microbiological contaminant, inherent in processing (c_1) and in the food matrix *i* itself (c_2), are also incorporated in the model.

Model equations are:

$$\begin{cases} M_{i,\,after} = (1-a)(1-c_2)\,M_{i,\,before} + b\,pr_{i-1} \\ pr_i = aM_{i,\,before} + (1-b)\,(1-c_1)\,pr_{i-1} \end{cases} \tag{12.3}$$

The model is one of the first initiatives in explaining the dynamics involved in transfer of microbiological contaminants in food processing. It was able to describe the transfer and survival of the microbiological contaminant during the industrial processing of broiler carcasses and has been applied for risk assessment (Nauta, Jacobs-Reitsma, and Havelaar 2007). It was found that, due to cross-contamination, the relation between log concentrations on broiler skin samples before and after defeathering (and other carcass processing steps) is not expected to be linear but curved (Nauta et al. 2005), which confirms the inadequacy of a log-linear model in this context.

This particular modelling approach has subsequently been used in risk assessment models for processing of pig carcasses (Titus 2007; Swart et al. 2016). In these models, more detailed characteristics are included for some process steps, such as the process time and specific surface-to-surface transfer dynamics during cutting.

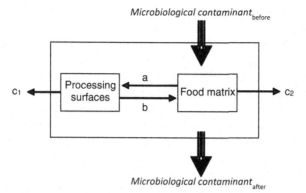

a: transfer of *Microbiological contaminant* from food matrix to processing surfaces

b: transfer of *Microbiological contaminant* from processing surfaces to food matrix

c_1: *Microbiological contaminant* inactivated on surfaces during processing

c_2: *Microbiological contaminant* inactivated on the food matrix

Figure 12.2 Schematic explanation of the parameters involved in the generic version of the cross-contamination model structure suggested by Nauta et al. (2005). In the model, consecutive food matrices (broiler carcasses) are processed in a processing environment (e.g. scalding tank, plucker, etc.). During the processing, microbiological contaminants (*Campylobacter*) are transferred between each consecutive food matrix and the processing surfaces (with rates a and b). Also, there may be inactivation or removal of microbiological contaminants during the processing (with rates c_1 and c_2).

12.3.2 Models for Cross-Contamination during Grinding and Slicing

Another category of models describing cross-contamination during food processing comprises those for slicing and grinding processes. The first approach in this classification was a binomial model developed by Ivanek et al. (2004) to describe the transfer of *Listeria monocytogenes* during slicing of smoked ready-to-eat fish. To describe the transfer of the same pathogen during slicing of a meat product, Vorst, Todd, and Ryser (2006) developed a log-linear model. Exponential models have since been developed by different authors (Aarnisalo et al. 2007; Sheen 2008; Sheen and Hwang 2010; Wang and Ryser 2016) for describing the transfer of various pathogens during slicing of different food products. Weibull models have also been successfully developed and applied to describe cross-contamination during slicing (Pérez-Rodríguez et al. 2007; Chaitiemwong et al. 2014; Possas et al. 2016). However, the first attempt to take a more mechanistic modelling approach in this category was performed by Møller et al. (2012).

Møller et al. (2012) applied the generic cross-contamination model suggested by Nauta et al. (2005) (Figure 12.2) to describe the transfer of *Salmonella* during grinding of pork inoculated with *Salmonella* followed by grinding of *Salmonella*-free pork. Figure 12.3 shows the goodness-of-fit of the Nauta et al. (2005) model (grey line) when fitted to the observed transfer (dots). The model was able to describe the transfer during grinding of the inoculated pieces (black dots) as well as the first part of the transfer during grinding of the non-inoculated pieces resulting in highly contaminated portions (grey dots). However, the model of Nauta et al. (2005) could not describe the second part of the transfer during grinding of the *Salmonella*-free pieces, represented by the tail of low-contaminated portions (open dots) in Figure 12.3. This shows that the Nauta et al. (2005) model was too simple to describe the microbial dynamics in the investigated grinding process, and an adaptation of the model was required.

Based on the observations in Figure 12.3, it was, therefore, hypothesized that cross-contamination in grinding-like processing follows two main and dissimilar dynamics. The first dynamic starts immediately after the input of the microbiological contaminant (black symbols) and is related to the transfer of microorganisms at higher levels and at a faster rate (grey symbols). In the second dynamic, the transfer of microorganisms happens at a slower rate and lower levels (open symbols).

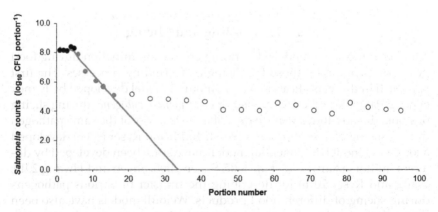

Figure 12.3 Transfer of *Salmonella* during grinding of pork, observed (dots) by Møller et al. (2012) and described (grey line) by the model of Nauta et al. (2005).

Focusing on a good description of the two cross-contamination environments, but still following a mechanistically driven approach, an alternative model was suggested by Møller et al. (2012). This alternative model can be considered an extension of the Nauta et al (2005) model, hypothesizing two processing environments instead of one (see Figure 12.4 in comparison to Figure 12.2). In this chapter, the model equation and structure are described in a more general form, not focusing only on describing the transfer of *Salmonella* in pork grinding:

$$\begin{cases} M_{i,\,\text{after}} = \left(1-a_1\right)\left(1-a_2\right)\left(1-c_2\right) M_{i,\,\text{before}} + \left(b_1\, pr_{1,i-1}\right) + \left(b_2\, pr_{2,i-1}\right) \\ pr_{1,i} = a_1 M_{i,\,\text{before}} + \left(1-b_1\right)\left(1-c_1\right) pr_{1,i-1} \\ pr_{2,i} = a_2 M_{i,\,\text{before}} + \left(1-b_2\right)\left(1-c_3\right) pr_{2,i-1} \end{cases} \tag{12.4}$$

The model has seven parameters, a_1, a_2, b_1, b_2, c_1, c_2 and c_3, which are explained in Figure 12.4. However, as contamination of the food matrix happens in the first place before processing, it is believed that the *microbiological contaminant* is already adapted to the food matrix, which in most cases, is an excellent substrate for survival of the *microbiological contaminant*. Therefore, inactivation in the food matrix itself often can be assumed not to take place, and then, c_2 can be set to zero. Similarly, inactivation of the *microbiological contaminant* in Environment 1 on the processing surfaces is assumed unlikely to occur, and c_1 is set to zero. In this case, the assumption is based on the fact that transfer in this environment happens too fast for the *microbiological contaminant* to be inactivated. With c_1 and c_2 set to zero, the model can be considered a five-parameter model (all with values between 0 and 1), which considers k unprocessed food matrix

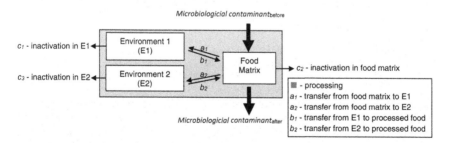

Figure 12.4 Schematic explanation of the parameters involved in the generic version of the cross-contamination model structure suggested by Møller et al. (2012).

pieces that are processed to k portions of processed food matrix ($I = 1, 2, \ldots$ k). The ith piece carries M_i *microbiological contaminant* (cfu per piece), and the resulting processed portion from piece i carries M_i *microbiological contaminant* (cfu per portion). The "contamination status" of the processing surfaces is pr_i (cfu per processing surface). The probability of transfer per *microbiological contaminant* cell from food matrix to processing surfaces, Environment 1 and to processing surfaces, Environment 2 is represented by a_1 and a_2, respectively. The backward transfer probabilities from the processing surfaces (Environments 1 and 2) to the processed food matrix are given by b_1 and b_2. The survival in Environment 2 of the processing surfaces is represented by ($1 - c_3$).

As explained by Møller et al. (2012), their model is very well able to describe the transfer of microbiological contaminants during grinding. Figure 12.5 shows the results of a model validation experiment in which contaminated and uncontaminated slices were alternately added to the grinder.

In 2010, Sheen and Hwang developed an empirical model to describe similar dynamics to those shown in Figure 12.3 in a study on cross-contamination during slicing of ham. Their model was also able to include both main transfer dynamics: 1) the transfer of microorganisms at higher levels and at a faster rate, which is commonly the focus of most available

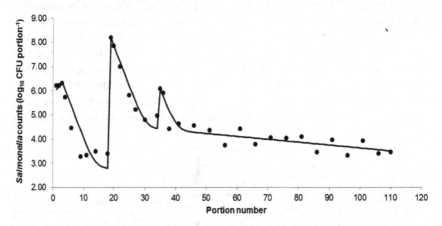

Figure 12.5 Observed transfer of *Salmonella* Typhimurium DT 104 during grinding of pork pieces (200 g). Pork pieces 1, 2, 3 and 35 were contaminated before processing with *Salmonella* at the level of $10^7 \log_{10}$ cfu g^{-1}, while piece 19 carried a level of $10^9 \log_{10}$ cfu g^{-1}. Parameter estimates obtained by Møller et al. (2012) were able to predict (continuous line) the level of transferred microbiological contaminant in all processed portions investigated.

cross-contamination models, and 2) the transfer of microorganisms happening at a slower rate and lower levels, the tailing phenomenon. Interestingly, their model is mathematically similar to that of Møller et al. (2012) (see Appendix 12.1). As the Sheen and Hwang (2010) model is purely empirical, it lacks a clear explanation for the dynamics involved in the cross-contamination event. The Møller et al. (2012) model (Figure 12.4) has the advantage that it specifically includes the pieces of food matrix, contaminated before the processing, and gives clear explanations of all parameters included in the model, generating insights for a better understanding of the dynamics related to the cross-contamination events.

The Møller et al. (2012) approach has been successfully applied to other investigations of transfer of pathogens during food processing, such as Zilelidou et al. (2015) describing transfer of *Escherichia coli* O157:H7 and *L. monocytogenes* during cutting of fresh salads. This confirms that the model may have the potential to be used as a generic cross-contamination model for large-scale food processing conditions. However, even though the model has been applied and successfully described the transfer of different microorganisms in different food matrices and for different processing conditions, there are still cross-contamination scenarios to be investigated, and consequently, the efficiency of the model needs to be evaluated in those conditions.

Another promising modelling approach, also arising from the search for a more mechanistic approach, is the work developed by Aziza et al. (2006). The model is able to describe cross-contamination during mechanical smearing of cheeses. The main advantage of this model is the inclusion of variability and uncertainty, which are inherent in the process and the experimental data. Through the application of Monte Carlo simulations, the final level of contamination is determined by means of distributions of transfer parameters. Variability and uncertainty were also analysed by Smid et al. (2013) when investigating cross-contamination rates of *Salmonella* during a pork cutting process.

12.4 OVERVIEW OF MODELS IN LARGE-SCALE FOOD PROCESSING

Table 12.1 provides an overview of published large-scale cross-contamination modelling approaches, including those introduced in the preceding sections. Many of the available models shown in Table 12.1 are purely empirical. They were developed by fitting the observed processing data to

Table 12.1 Examples of Modelling Approaches Applied to Describe the Effect of Cross-Contamination in Large-Scale, Industrial-Type Food Processing

Process	Food Matrix	Microorganism	Modelling Approach[a]	Reference
Slaughtering and carcass processing	Chicken	Campylobacter	Empirical (log-linear) and semi-mechanistic	Hartnett 2001
	Broiler	Campylobacter	Empirical (log-linear)	Rosenquist et al. 2003
	Broiler	Campylobacter	Semi-mechanistic	Nauta et al. 2005
	Pork	Salmonella, E. coli, Campylobacter	Semi-mechanistic	Titus 2007
	Broiler	Campylobacter	Empirical (log-linear) and semi-mechanistic	FAO/WHO 2009
	Poultry	Campylobacter	Data based	Hayama et al. 2011
	Pork	Salmonella	Bayesian belief network	Smid et al. 2012
	Pig	Salmonella	Semi-mechanistic	Swart et al. 2016
	Pork	Salmonella	Semi-mechanistic	Costa et al. 2017
Slicing	Smoked ready-to-eat fish	L. monocytogenes	Semi-mechanistic	Ivanek et al. 2004
	Meat product	L. monocytogenes	Empirical (log-linear)	Vorst, Todd, and Ryser 2006
	Gravad salmon	L. monocytogenes	Empiric (exponential)	Aarnisalo et al. 2007
	Cooked meat product	E. coli O157:H7 and S. aureus	Empiric (log-linear and Weibull)	Pérez-Rodríguez et al. 2007
	Salami	L. monocytogenes	Empiric (exponential)	Sheen 2008

(Continued)

Table 12.1 (Continued) Examples of Modelling Approaches Applied to Describe the Effect of Cross-Contamination in Large-Scale, Industrial-Type Food Processing

Process	Food Matrix	Microorganism	Modelling Approach[a]	Reference
	Ready-to-eat ham	E. coli O157:H7	Empiric (exponential)	Sheen and Hwang 2010
	Ready-to-eat meat and fish products	L. monocytogenes	Empirical and deterministic compartmental	Hoelzer et al. 2012
	Tomatoes	Norovirus	Empirical (non-linear regression)	Shieh et al. 2014
	Cooked ham	L. monocytogenes	Empirical (different models)	Chaitiemwong et al. 2014
	Tomatoes	Salmonella Typhimurium	Empirical (exponential)	Wang and Ryser 2016
	Onions	L. monocytogenes	Empirical (exponential)	Scollon, Wang, and Ryser 2016
	Ready-to-eat turkey product	Salmonella Enteritidis	Empirical (Weibull and modified Weibull)	Possas et al. 2016
Mechanical smearing	Cheese	Brevibacterium linens	Stochastic compartmental	Aziza et al. 2006
Processing line	Fresh-cut lettuce	E. coli O157:H7	Semi-mechanistic	Pérez Rodríguez et al. 2011
Grinding	Pork	Salmonella Typhimurium	Semi-mechanistic	Møller et al. 2012
Cutting and shredding	Fresh-cut salads	E. coli and L. monocytogenes	Semi-mechanistic	Zilelidou et al. 2015
Bowl chopping	Pork	L. monocytogenes	Deterministic compartmental	Jiang et al. 2018

[a] Modelling approaches: "Empirical" models are descriptive empirical models that use the available data without explicitly considering the cross-contamination dynamics, using a statistical model that fits well with the data; "semi-mechanistic" models explicitly define the cross-contamination dynamics and fit these to the data, either based on a sequence of transfer ratio or by including more detailed dynamics; "compartmental" models describe dynamic changes through different modules/compartments of the model.

specific statistical models (log-linear, exponential, Weibull and binomial) that described the overall effect of the process. A limited number of studies have taken semi-mechanistic approaches whereby the relevant transfer and removal effects are modelled in more detail. A few studies have developed alternative statistical approaches based on generated data, especially for describing the transfer of microbiological contaminants in complex and large-scale processing such as pig and broiler slaughtering. Hoelzer et al. (2012) have applied the same principle to data obtained from cross-contamination studies of a slicing process. These authors, as well as Aziza et al. (2006), have applied compartmental modelling approaches as alternatives to a better understanding of the dynamics of cross-contamination, and they also intend to move to a more universal approach.

It is important to keep in mind that the desired degree of complexity of a cross-contamination model depends mainly on the purpose of its use. Depending on the objective, a simple model, such as transfer ratio, can be sufficient for helping food processors and risk managers in their decision-making process.

12.5 CAUSE AND IMPLICATIONS OF THE TAILING PHENOMENON

As described above and illustrated in Figure 12.2, Møller et al. (2012) found that the Nauta et al. (2005) model, which assumes one environmental compartment with constant transfer and inactivation rates, could not explain the tailing observed in the grinding process. Apparently, the cross-contamination process during grinding of meat is too complex to be described by a relatively simple one-compartment model. Interestingly, the tailing phenomenon has been shown during cross-contamination investigations of numerous pathogens (*L. monocytogenes*, *E. coli* O157:H7, *Staphylococcus aureus*, Norovirus and *Salmonella*) in different food processes such as slicing of ready-to-eat meat products, tomatoes, fresh salads and onions (Vorst, Todd, and Ryser 2006; Aarnisalo et al. 2007; Sheen 2008; Keskinen, Todd, and Ryser 2008; Sheen and Hwang 2010; Shieh et al. 2014; Zilelidou et al. 2015; Possas et al. 2016; Wang and Ryser 2016; Scollon, Wang, and Ryser 2016). It is suggested that it may be a generic phenomenon and that the absence of tailing may be an indication that the investigation of the cross-contamination event has not been performed over a long enough period or that the amount of processed food matrix is not sufficient to observe the whole transfer dynamics (Møller et al. 2016). Low starting levels could also be an explanatory factor for the lack of the tailing phenomenon when investigating cross-contamination events (Møller et al. 2012) due to the low sensitivity of the available detection methods. Møller et al. (2012, 2016) also stress other aspects impacting the transfer of pathogens during processing, such as composition of the food product, e.g. fat content or connective tissue (Lorenzo et al. 2014).

It seems that at certain points of the transfer of microorganisms, the levels of contamination are lower than the detection limits of the available techniques. Until the development and application of more sensitive techniques for detection of microorganisms, the dynamics of transfer of microorganisms remains to be investigated at low levels, which is more closely related to the concentrations reported in most real-life processing. This will allow more accurate modelling of the transfer of microorganisms. At the moment, the only alternative to avoid the continuing transfer of microorganisms would be increasing the frequency of cleaning and sanitizing regimes. Hoelzer et al. (2012) have tested the impact of sanitation on cross-contamination of *L. monocytogenes* during events between foods and surfaces or during slicing. In addition, Lim and Harrison (2016) reported the effectiveness of UV light in reducing *Salmonella* contamination on tomatoes and food contact surfaces. Furthermore, Jiang et al. (2018) have demonstrated the impact of different cleaning methods on reducing the transfer of *L. monocytogenes* from a meat-chopping process. The hygienic design of equipment is also important as well as the application of good manufacture practices during food processing.

Before discussing the implications of tailing, it is important to sort out whether tailing could be a result of artefacts, i.e. errors in the experimental design that lead to tailing. This has been demonstrated, for example, in heat inactivation studies with open tubes heated in a water bath (Tierney and Larkin 1978). These authors showed that mixing of a culture in a tube will leave a sort of coating/film on the inside surface of the tube, and some bacteria will then be situated in the substrate and others in the coating. If the tube is not fully submerged under water, bacteria in the surface coating will not experience the same temperature as bacteria in the substrate. Consequently, tailing is observed, because it is not the whole population getting the same treatment but two sub-populations getting different treatments. The same can be said for vacuum-packaged food samples. If the bags are too big, and they are not sealed close to the food, juice from the food will spread out through the whole bag, dragging part of the bacterial population out of the food into the juice. Again, two sub-populations are unintentionally created, giving rise to two different survival dynamics in the same sample. If the sample then floats in the water surface of the water bath, or the part with juice is above the water surface, survival dynamics get even more complicated. When you remove the sample from the bag to enumerate the survivors, the sample is "recontaminated" with the bacteria sitting in the juice sticking up over the water surface. Situations like this can also be seen for open tubes or flasks where condensation is formed in the lid during heating. This can lead to recontamination of the sample through drips. As illustrated in Figure 12.6, these types of errors typically show up as tailing.

When considering processes like grinding, it can be hypothesized that something similar happens. When a contaminated sample moves through the grinder, it leaves some of the bacteria inside the grinder. These can then "drip" onto the next samples moving by. In heat inactivation experiments, this has been solved by working with small, individual samples, tightly closed and fully submerged in water, that can neither recontaminate themselves nor cross-contaminate each other. However, it is not possible to take such an approach for grinding experiments, and it could be said that grinding is the epitome of cross-contamination. The same can be argued for cutting and slicing when the same knife is applied for cutting several pieces or slices.

Often, contamination through drip results in tailing with a very flat slope, almost a plateau (Figure 12.6a), which is not always the case for grinding, etc., where the tail can have many different slopes (Møller et al.

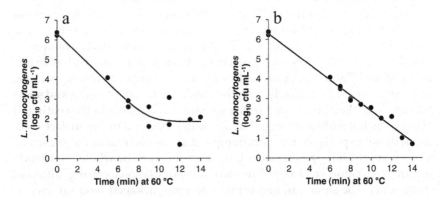

Figure 12.6 Survival dynamics of *Listeria monocytogenes* when heated in BHI broth at 60 °C in a flask with a lid creating room for drip contamination through condensate formation inside the lid (a) and a repetition without using the lid (b). Data kindly provided by Al-Khuori and Milanovic (2018).

2016). This indicates that cross-contamination is not only a result of drip contamination; other factors also play a role. For meat grinding processes, it has been shown that the bigger the piece of meat being ground, the less steep the slope of the tail. Thus, compared with a smaller piece of meat, a larger number of bacteria are transferred from the grinder to the meat, which may be explained by a larger contact surface and greater pressure between the meat piece and grinder surfaces with bigger meat pieces. Thereby, more bacteria are "torn" from the grinder surfaces into the meat being ground, suggesting cross-contamination to be a purely physical event.

A difference between types of equipment for grinding has also been demonstrated (Møller et al. 2016). A grinder with a rougher surface resulted in the tail establishing earlier in the grinding process and having a higher level of bacteria transferred than a grinder with smoother stainless steel surfaces. It can, therefore, be speculated whether food residues are more prone to stick to a grinder with a rough surface, especially when a constant flow of food is passing through. A constant flow of nutrients is for example utilized for building biofilms in flow cell systems (Weiss Nielsen et al. 2011). This is also well known for fluids in stainless steel tubes, and a surface roughness around $Ra = 0.9$ µm has been shown to enhance bacterial entrapment (Flint, Brooks, and Bremer 2000). Thus, a grinder with smooth surfaces will not entrap as many bacterial cells from the food passing through as a grinder with rough surfaces. Therefore, fewer bacterial cells will be transferred to the next pieces being ground, which will result in a tail being established at a lower bacterial concentration. A similar phenomenon can be seen if equipment is not maintained and e.g. the knife is dull (Møller et al. 2016). In this particular situation, it could also be imagined that friction plays a role. Friction can increase processing time and temperature. Depending on the temperature, the growth of bacteria in food residues in the equipment may occur, which will shift the dynamics of transfer towards drip contamination and, thereby, plateau-like tailing.

It is not only the nature of equipment surfaces that influences cross-contamination dynamics; the construction itself also appears to affect tailing. It is, for example, well known that deadlegs, cracks, crevices, grooves or pockets can result in accumulation of food residues in stagnant areas, where bacteria can grow if the process time is sufficiently long (Chmielewski and Frank 2003). The authors of the present chapter have looked further into the construction of a grinder to see whether it is possible to identify loci with loosely and tightly entrapped bacteria, respectively. If that were the case, it could explain the tailing observed during grinding as being the result of two compartments/environments inside the grinder, the one with tightly entrapped bacteria being responsible for the tail. After grinding of 5 meat pieces contaminated with *Salmonella* followed by grinding of 0, 2, 4, 7, 10, 15, 20, 30 and 40 pieces of non-contaminated meat, respectively, the nine loci shown in Figure 12.7 were swabbed to recover and enumerate *Salmonella* cells present.

As illustrated in Figure 12.7, the feed pan, part of the cylinder and the ring were pointed out as loci with high potential for tight entrapment of bacteria, meaning that *Salmonella* counts per square centimetre in those loci changed at a slower rate during grinding than in others. A closer look at those specific loci to some extent revealed signs of poor hygienic design, such as crevices in the feed pan, where the meat is fed into the grinder, a deadleg in the back of the cylinder and grooves in the ring.

Finally, biological factors may also play a role in the cross-contamination dynamics creating a tail. For heat inactivation of bacteria, it has been demonstrated that tailing can be a result of bacteria in a population not

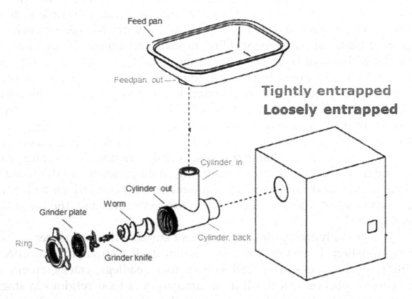

Figure 12.7 The nine loci in the meat grinder from which *Salmonella* was recovered. Loci in light grey text indicate that *Salmonella* cells were tightly entrapped; loci in dark grey text indicate that *Salmonella* cells were loosely entrapped. Data kindly provided by Hansen (2014).

all having the same tolerance to heat (Peleg and Cole 1998). Rather, heat resistance in a bacterial population follows a distribution where a smaller fraction has higher tolerance towards heat than the average bacteria in the population, which shows as a biphasic survival dynamics, resulting in a tail. As a parallel to this, tailing in cross-contamination dynamics could be interpreted as two sub-populations reacting differently to the environmental stresses encountered during processing.

Biphasic survival dynamics can also be explained biologically by natural selection or physiological adaptation. The parallel to cross-contamination would in that case revolve around the ability of bacteria to attach to and survive in the environment of the processing equipment. Take, for example, the illustration in Figure 12.4, which explains tailing as a result of two environmental compartments in which different rates of transfer and inactivation can occur, leading to different transfer possibilities from each environmental compartment. For inactivation to occur in a bacterial population, the bacteria have to be subjected to some kind of stress-factor. It is unclear what is causing the stress during grinding, cutting or slicing, but if the bacteria experienced stress, it would have to occur in an environment/compartment where bacteria are entrapped for a longer period of time. This could, for example, be in loci with tight entrapment of bacteria as experienced for the grinder mentioned earlier (Figure 12.7). If the temperature increases significantly and reaches lethal levels in such loci, e.g. due to friction from the knife or heat transfer from the motor, inactivation and natural selection can be considered. If the temperature increases to sub-lethal levels, physiological adaptation can be considered.

12.6 EVALUATION OF THE PERFORMANCE OF CROSS-CONTAMINATION MODELS

Once predictive models are developed, it is important to evaluate their performance and to check the more generic applicability of the models. This has been discussed in detail for growth and inactivation models (Ross 1996; Baranyi, Pin, and Ross 1999; Mellefont, McMeekin, and Ross 2003; Van Derlinden and Van Impe 2012) but not so much for cross-contamination models, where the number of qualified options available is limited. In addition, the approaches available are not applicable to all cases.

An acceptable prediction zone concept was developed by Oscar (2005) to validate secondary lag time and growth rate predictive models. A similar approach was successfully applied for the validation of primary predictive microbial interaction models and renamed to *acceptable simulation zone* (ASZ) (Møller et al. 2013). The ASZ approach was recently applied to evaluate cross-contamination models by Møller et al. (2016), who found a reasonable way to adapt the ASZ for evaluating the predictive value of cross-contamination models (Figure 12.8).

The ASZ is a way to evaluate whether there is, or is not, a relevant difference between the performance of models, where the models can be models with different model equations or models with the same equations but different parameter values. The ASZ is, therefore, a parallel to the goodness-of-fit used for comparing models during model development.

The impact of cross-contamination can also be evaluated by using QMRA, as shown by Pérez Rodríguez et al. (2011) and Møller et al. (2016). Here, the cross-contamination parameters obtained for different conditions are tested as alternative scenarios and compared with a baseline. The challenge in such an approach is that the cross-contamination model, containing the parameters to be tested, needs to be included in a QMRA model, which may be a demanding task. Møller et al. (2016), therefore, developed the total transfer potential (TTP), a metric defined as the proportion of bacterial cells that is transferred from a single piece of contaminated food matrix to the total processed food matrix, assuming that the processing will continue indefinitely. In their study, this proportion correlated well with the human

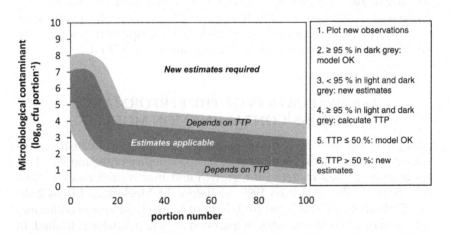

Figure 12.8 Acceptable Simulation Zone and Total Transfer Potential concepts applied to evaluate model performance. The simulated values are compared in relation to the observed transfer of *Salmonella* Typhimurium DT 104 during grinding of pork (Møller et al. 2016).

health risk associated with the cross-contamination taking place during grinding, which suggests that it may be used as a surrogate for a more complicated QMRA. It has been recognized that TTP may be a relevant alternative for evaluating cross-contamination models (Possas et al. 2016).

QMRA and/or TTP may be applied to check whether the performance of the models is similar if they are applied in risk assessment. If the tested models have a similar TTP and/or risk estimate, they perform similarly; if not, they are truly different. When comparing different models and scenarios, Møller et al (2016), found a clear relationship between risk estimate and TTP (Figure 12.8). The authors compared different approaches for model evaluation, and they observed that none of them could stand alone; they had to be used in combination as presented in Figure 12.8. When most of the approaches were in agreement, it was shown that models/scenarios could be considered equally acceptable/risky by presenting a TTP ≤50%. Different groupings of the ASZ and TTP were obtained by applying statistical analysis with the risk estimates as response variable. A TTP higher than 50% would result in underestimation of the risk if the evaluated model were applied (Møller et al. 2016).

12.7 SUMMARY AND OUTLOOK

As indicated in this chapter, microbiological cross-contamination may be a complicated combination of transfer processes. It is not an easy task to quantitatively describe the dynamics involved in transfer of microbiological contaminants, and usually, it is not possible with the obtained data from the investigated processing only. Empirical and semi-mechanistic approaches have been applied, but no fully mechanistic model that captures all relevant physical and biological details involved, has been developed yet. Investigating the transfer of microbiological contaminants in food processing is labour intensive, and results are often not generally applicable, as hard-to-measure details like the frequency of human errors, the frequency and intensity of contact between food items and the environment, and uncertainty about the number and relevance of different transfer routes vary from case to case.

Despite these challenges, various successful approaches have been taken to model cross-contamination in large-scale processes. The models obtained are useful for the specific situation for which they are developed, but it has been harder to construct more generic predictive models, like those available for growth and inactivation. Still, some generic insights have been obtained. One of these insights is that log-linear models are often not applicable for cross-contamination, which for example implies that linear regression models for slaughter and industrial processing of carcasses should be interpreted with care. Another insight is that tailing is a frequently observed phenomenon, which is possibly an intrinsic part of any sequential cross-contamination process. Even though several hypotheses exist, the reason for tailing is not fully understood, and it is not clear whether there is one generic explanation for this phenomenon.

Adequate cross-contamination models are relevant for several reasons. Cross-contamination and bacterial transfer are, for example, highly relevant in the food industry and the hygienic design of processing equipment or a processing line. As in many cases it is not feasible to completely prevent cross-contamination (as, for example, during broiler chicken processing), models can be useful to identify where the most efficient interventions against cross-contamination should be implemented. The same is basically true for the development of food preparation equipment for consumers.

One large area of application of cross-contamination models is, of course, quantitative microbiological risk assessment. Here, cross-contamination is one of the basic processes that may result in an increase in prevalence and consequently, increase the risk. This is typically the area where generic, well-studied cross-contamination models would be needed most. However, to date, these models are not widely available, as many models are process specific, and especially, the generic applicability of consumer phase models is unclear.

As mentioned by Pérez-Rodríguez and Valero (2013), many aspects related to cross-contamination in food processing remain to be investigated, such as the influence of the nature and inoculum level of a larger variety of microbiological contaminants, the type and volume of processed food matrices, aspects inherent in processing (e.g. roughness of equipment surfaces, pressure exerted on the food matrices through the machines), etc. By clarifying such aspects, it will be possible to develop models of generic character able to match the ideal universal model, which is an aspiration for both risk assessors and industry.

To further develop cross-contamination models, one of the options is to generate more data on cross-contamination and transfer by doing more experiments in specific relevant case studies. It may, however, be difficult to use the results of these case studies to obtain a general insight into the impact of cross-contamination and to develop more advanced generic models. Another approach is, therefore, to do more targeted studies into the cross-contamination dynamics so that more (semi-)mechanistic models that build on measurable parameters can be constructed.

Also, there is the possibility of more general theoretical studies in which the performance of different cross-contamination models is compared in terms of their impact on risk assessments. Examples of such studies are the comparison of consumer phase models for *Campylobacter* on chicken meat performed by Nauta and Christensen (2011) and the comparison of detailed consumer phase models with a simple surrogate effect measure by Neves, Mungai and Nauta (2018). These studies show that differences in the design and results of different models that include cross-contamination events are not always relevant for risk assessment purposes, which may simplify the performance of QMRA. On the other hand, these studies may indicate when simple approaches are not good enough and when more complex models are needed.

As explained in Sections 12.3 and 12.5, an interesting result of the study of Møller et al (2016) was that the relatively simple model of Nauta et al (2005) could not explain the tailing observed in the grinding process. Also, the tailing seems to be typical for all similar cross-contamination studies, such as, for example, the studies where exponential models are used to describe the data in slicing studies (Aarnisalo et al. 2007; Sheen 2008; Sheen and Hwang 2010; Wang and Ryser 2016). This observation may raise the question of whether the original Nauta et al (2005) model is actually appropriate to describe the effect of cross-contamination during the industrial processing of broiler chicken carcasses. This model has been implemented in a risk assessment where the parameter estimates for the model were based on expert opinion, not on observational data (Nauta, Jacobs-Reitsma, and Havelaar 2007). The predicted model dynamics have not (and could not) be tested against data, which means that there are no data showing that the model with one environmental compartment (Figure 12.2) is actually appropriate. It could well be that a two-compartmental model (Figure 12.4) would be needed to describe the process accurately, because tailing takes place that is not described by the model. This may not be an issue if the model is applied to describe the dynamics of *Campylobacter* transfer and survival during processing of a contaminated flock, but it may be relevant for the cross-contamination from a positive flock to an uncontaminated flock that is processed immediately after. Based on, among others, the Nauta, Jacobs-Reitsma, and Havelaar (2007) model, this between-flock cross-contamination has been considered irrelevant, which is an important conclusion for risk management and the potential effectivity of scheduled processing (or logistic slaughter) as a risk mitigation strategy (Nauta et al. 2009). However, if the cross-contamination model used by Nauta, Jacobs-Reitsma, and Havelaar (2007) is incorrect, and tailing does occur, this conclusion may be challenged. Observations made by Seliwiorstow et al. (2016), who observed higher levels of cross-contamination to a negative flock than expected, actually point in that direction. Clearly, these observations show that it may be necessary to analyse the importance of a model with two environmental compartments in risk assessment for *Campylobacter* in broiler chickens.

It can be concluded that in the last two decades, considerable progress has been made in the development of cross-contamination models. The variety of modelled processes and modelling approaches has shown that cross-contamination models are often process specific but also that generic approaches may be developed to capture the generic phenomena observed with cross-contamination. Further development is necessary and possible by the performance of more case studies as well as studies targeted at the development of generic approaches. In the end, this will provide us with a library of predictive models comparable to those available for growth and inactivation.

APPENDIX 12.1 COMPARISON OF THE MODELS USED BY SHEEN AND HWANG (2010), NAUTA ET AL (2005) AND MØLLER ET AL (2012)

Sheen and Hwang (2010) apply an empirical model for slicing that is formulated as an exponential model:

$$Y = C \exp(-X / D)$$

The Møller et al (2012) model for grinding is given by Equation 12.4:

$$\begin{cases} M_{i,\,after} = (1 - a_1)(1 - a_2)(1 - c_2)\, M_{i,\,before} + (b_1\, pr_{1,i-1}) + (b_2\, pr_{2,i-1}) \\ pr_{1,i} = a_1 M_{i,\,before} + (1 - b_1)(1 - c_1) pr_{1,i-1} \\ pr_{2,i} = a_2 M_{i,\,before} + (1 - b_2)\,(1 - c_3)\, pr_{2,i-1} \end{cases}$$

where $M_{i,after}$ is the portion of ground meat from piece of meat i.

If $M_{i,before} = 0$ (the uncontaminated slices) and $c_1 = 0$, it becomes

$$\begin{cases} M_{i,\,after} = b_1 pr_{1,i-1} + b_2 pr_{2,i-1} \\ pr_{1,i} = \left(1-b_1\right) pr_{1,i-1} \\ pr_{2,i} = \left(1-b_2\right)\left(1-c_3\right) pr_{2,i-1} \end{cases}$$

or

$$M_{i,after} = b_1 (1-b_1)\, pr_{1,i-2} + b_2 (1-b_2)(1-c_3)\, pr_{2,i-2}$$

For ground meat portion k:

$$M_{k,after} = b_1 (1-b_1)^{k-1} pr_{1,0} + b_2 [(1-b_2)(1-c_3)]^{k-1} pr_{2,0}$$

If

$$\alpha_1 = b_1 (1-b_1) \times pr_{1,0} \text{ and } \beta_1 = (1-b_1)$$

$$\alpha_2 = b_2/[(1-b_2)(1-c_3)] \times pr_{2,0} \text{ and } \beta_2 = (1-b_2)(1-c_3)$$

then

$$M_{k,after} = \alpha_1\beta_1{}^k + \alpha_2\beta_2{}^k$$

which equals

$$M_{k,after} = \alpha_1 e^{k\ln(\beta 1)} + \alpha_2 e^{k\ln(\beta 2)}$$

This means that the Nauta et al. (2005) model, which is a simplified version of the Møller et al. model with only one environment, can be written as the Sheen and Hwang model, because it has one term less ($\alpha_2 e^{k\ln(\beta 2)} = 0$) and with

$$C = \alpha_1 \text{ and } -1/D = \ln(\beta_1),$$
$$Y = M_{k,after} \text{ and } X = k$$

The models are identical. The Sheen and Hwang model can therefore be considered a simplified version of the Møller et al (2012) model.

REFERENCES

Aarnisalo, K., S. Sheen, L. Raaska, and M. Tamplin. 2007. "Modelling Transfer of *Listeria monocytogenes* During Slicing of 'Gravad' Salmon." *International Journal of Food Microbiology* 118(1). Elsevier: 69–78. doi:10.1016/J.IJFOOD MICRO.2007.06.017.

Al-Khuori, S., and M. Milanovic. 2018. *Microbiological Challenges of Sous Vide Treatment in a Hospital Kitchen.* (In Danish). Technical University of Denmark, Kgs. Lyngby, Denmark.

van Asselt, E.D., A.E.I. de Jong, R. De Jonge, and M.J. Nauta. 2008. "Cross-Contamination in the Kitchen: Estimation of Transfer Rates for Cutting Boards, Hands and Knives." *Journal of Applied Microbiology* 105(5): 1392–1401.

Aziza, F., E. Mettler, J.J. Daudin, and M. Sanaa. 2006. "Stochastic, Compartmental, and Dynamic Modelling of Cross-Contamination During Mechanical Smearing of Cheeses." *Risk Analysis* 26(3): 731–745. Doi:10.1111/j.1539-6924. 2006.00758.x.

Baranyi, József, Carmen Pin, and Thomas Ross. 1999. "Validating and Comparing Predictive Models." *International Journal of Food Microbiology* 48(3): 159–166. Doi:10.1016/S0168-1605(99)00035-5.

Chaitiemwong, N., W.C. Hazeleger, R.R. Beumer, and M.H. Zwietering. 2014. "Quantification of Transfer of *Listeria monocytogenes* Between Cooked Ham and Slicing Machine Surfaces." *Food Control* 44. Elsevier Ltd: 177–184. Doi:10.1016/j.foodcont.2014.03.056.

Chapman, B., A. Otten, A. Fazil, N. Ernst, and B.A. Smith. 2016. "A Review of Quantitative Microbial Risk Assessment and Consumer Process Models for *Campylobacter* in Broiler Chickens." *Microbial Risk Analysis* 2–3. Elsevier B.V.: 3–15. Doi:10.1016/j.mran.2016.07.001.

Chen, Y., K.M. Jackson, F.P. Chea, and D.W. Schaffner. 2001. "Quantification and Variability Analysis of Bacterial Cross-Contamination Rates in Common Food Service Tasks." *Journal of Food Protection* 64(1): 72–80.

Chmielewski, R.A.N., and J.F. Frank. 2003. "Biofilm Formation and Control in Food Processing Facilities." *Comprehensive Reviews in Food Science and Food Safety* 2(1): 22–32. Doi:10.1111/j.1541-4337.2003.tb00012.x.

Christensen, B.B., H. Rosenquist, H.M. Sommer, N.L. Nielsen, S. Fagt, N.L. Andersen, and B. Nørrung. 2005. "A Model of Hygiene Practices and Consumption Patterns in the Consumer Phase." *Risk Analysis* 25(1): 49–60.

Costa, E.de F., L.G. Corbellini, A.P.S.P. da Silva, and M. Nauta. 2017. "A Stochastic Model to Assess the Effect of Meat Inspection Practices on the Contamination of the Pig Carcasses." *Risk Analysis* 37(10): 1849–1864. Doi:10.1111/risa.12753.

van Derlinden, E. and J.F. Van Impe. 2012. "Modelling Growth Rates as a Function of Temperature: Model Performance Evaluation with Focus on the Suboptimal Temperature Range." *International Journal of Food Microbiology* 158(1). Elsevier B.V.: 73–78. Doi:10.1016/j.ijfoodmicro.2012.05.015.

FAO, and WHO. 2009. Risk Assessment of Campylobacter spp. In: *Broiler Chickens: Interpretative Summary.* Microbial Risk Assessment Series.

Flint, S.H., J.D. Brooks, and P.J. Bremer. 2000. "Properties of the Stainless Steel Substrate, Influencing the Adhesion of Thermo-Resistance Streptococci." *Journal of Food Engineering* 43(4): 235–242.

Hansen, S.K.H. 2014. *Location and Transfer of Pathogens in a Meat Grinder During Grinding of Pork.* Technical University of Denmark, Kgs. Lyngby, Denmark.

Hartnett, E. 2001. "Human Infection with *Campylobacter* spp. From Chicken Consumption: A Quantitative Risk Assessment." *PhD Thesis*, University of Strathclyde, Glasgow, Scotland.

Hoelzer, K., R. Pouillot, D. Gallagher, M.B. Silverman, J. Kause, and S. Dennis. 2012. "Estimation of *Listeria monocytogenes* Transfer Coefficients and Efficacy of Bacterial Removal Through Cleaning and Sanitation." *International Journal of Food Microbiology* 157(2). Elsevier B.V.: 267–277. Doi:10.1016/j.ijfood micro.2012.05.019.

Ivanek, R., Y.T. Grohn, M. Wiedmann, and M.T. Wells. 2004. "Mathematical Model of *Listeria monocytogenes* Cross-Contamination in a Fish Processing Plant." *Journal of Food Protection* 67(12): 2688–2697. http://www.jfoodprotection.org/doi/pdf/10.4315/0362-028X-67.12.2688.

Jiang, R., X. Wang, W. Wang, Y. Liu, J. Du, Y. Cui, C. Zhang, and Q. Dong. 2018. "Modelling the Cross-Contamination of *Listeria monocytogenes* in Pork During Bowl Chopping." *International Journal of Food Science and Technology* 53(3): 837–846. Doi:10.1111/ijfs.13660.

Keskinen, L.A., E.C.D. Todd, and E.T. Ryser. 2008. "Impact of Bacterial Stress and Biofilm-Forming Ability on Transfer of Surface-Dried *Listeria monocytogenes* During Slicing of Delicatessen Meats." *International Journal of Food Microbiology* 127(3): 298–304. Doi:10.1016/j.ijfoodmicro.2008.07.021.

Kusumaningrum, H.D., E.D. van Asselt, R.R. Beumer, and M.H. Zwietering. 2004. "A Quantitative Analysis of Cross-Contamination of *Salmonella* and *Campylobacter* spp. Via Domestic Kitchen Surfaces." *Journal of Food Protection* 67(9): 1892–1903.

Lim, W., and M.A. Harrison. 2016. "Effectiveness of UV Light as a Means to Reduce *Salmonella* Contamination on Tomatoes and Food Contact Surfaces." *Food Control* 66. Elsevier Ltd: 166–173. Doi:10.1016/j.foodcont.2016.01.043.

Lorenzo, J.M., M. Pateiro, M.C.G. Fontán, and J. Carballo. 2014. "Effect of Fat Content on Physical, Microbial, Lipid and Protein Changes During Chill Storage of Foal Liver Pâté." *Food Chemistry* 155: 57–63. Doi:10.1016/j.foodchem.2014.01.038.

Luber, P., S. Brynestad, D. Topsch, K. Scherer, and E. Bartelt. 2006. "Quantification of *Campylobacter* Species Cross-Contamination During Handling of Contaminated Fresh Chicken Parts in Kitchens." *Applied and Environmental Microbiology* 72(1): 66–70. Doi:10.1128/AEM.72.1.66-70.2006.

Mellefont, L.A., T.A. McMeekin, and T. Ross. 2003. "Performance Evaluation of a Model Describing the Effects of Temperature, Water Activity, pH and Lactic Acid Concentration on the Growth of *Escherichia coli*." *International Journal of Food Microbiology* 82(1): 45–58. Doi:10.1016/S0168-1605(02)00253-2.

Møller, C.O.A., M.J. Nauta, B.B. Christensen, P. Dalgaard, and T.B. Hansen. 2012. "Modelling Transfer of *Salmonella* Typhimurium DT104 During Simulation of Grinding of Pork." *Journal of Applied Microbiology* 112(1): 90–98.

Møller, C.O.A., Y. Ilg, S. Aabo, B.B. Christensen, P. Dalgaard, T.B. Hansen. 2013. "Effect of Natural Microbiota on Growth of *Salmonella* spp. In Fresh Pork: A Predictive Microbiology Approach." *Food Microbiology* 34: 284–295.

383

Møller, C.O.A., A.S. Sant'Ana, S.K.H. Hansen, M.J. Nauta, L.P. Silva, V.O. Alvarenga, D. Maffei, et al. 2016. "Evaluation of a Cross Contamination Model Describing Transfer of *Salmonella* spp. And *Listeria monocytogenes* During Grinding of Por and Beef." *International Journal of Food Microbiology* 226: 42–52.

Montville, R., and D.W. Schaffner. 2004. "Statistical Distributions Describing Microbial Quality of Surfaces and Foods in Food Service Operations." *Journal of Food Protection* 67(1): 162–167.

Mylius, S.D., M.J. Nauta, and A.H. Havelaar. 2007. "Cross-Contamination During Food Preparation: A Mechanistic Model Applied to Chicken-Borne *Campylobacter*." *Risk Analysis* 27(4): 803–813. Doi:10.1111/j.1539-6924.2006.00872.x.

Nauta, M., and B. Christensen. 2011. "The Impact of Consumer Phase Models in Microbial Risk Analysis." *Risk Analysis* 31(2): 255–265. Doi:10.1111/j.1539-6924.2010.01481.x.

Nauta, M., A. Hill, H. Rosenquist, S. Brynestad, A. Fetsch, P. van der Logt, A. Fazil, et al. 2009. "A Comparison of Risk Assessments on *Campylobacter* in Broiler Meat." *International Journal of Food Microbiology* 129(2). Elsevier B.V.: 107–123. doi:10.1016/j.ijfoodmicro.2008.12.001.

Nauta, M., I. Van Der Fels-Klerx, A. Havelaar, and W. Jacobs-Reitsma. 2005. "A Poultry Processing Model for Quantitative Microbiological Risk Assessment." *Risk Analysis* 25(1): 23–26.

Nauta, M.J. 2002. "Modelling Bacterial Growth in Quantitative Microbiological Risk Assessment: Is It Possible?" *International Journal of Food Microbiology* 73(2): 297–304.

Nauta, M.J., W.F. Jacobs-Reitsma, and A.H. Havelaar. 2007. "A Risk Assessment Model for *Campylobacter* in Broiler Meat." *Risk Analysis* 27(4): 845–861. Doi:10.1111/j.1539-6924.2006.00834.x.

Neves, M.I., S.N. Mungai, and M.J. Nauta. 2018. "Can Stochastic Consumer Phase Models in QMRA Be Simplified to a Single Factor?" *Microbial Risk Analysis* 8. Elsevier: 53–60. Doi:10.1016/j.mran.2017.09.001.

Oscar, T.P. 2005. "Validation of Lag Time and Growth Rate Models for *Salmonella* Typhimurium: Acceptable Prediction Zone Method." *Journal of Food Science* 70(2): 129–137.

Peleg, Micha, and Martin B. Cole. 1998. "Reinterpretation of Microbial Survival Curves." *Critical Reviews in Food Science and Nutrition* 38(5): 353–380. Doi:10.1080/10408699891274246.

Pérez Rodríguez, F., D. Campos, E.T. Ryser, A.L. Buchholz, G.D. Posada-Izquierdo, B.P. Marks, G. Zurera, and E. Todd. 2011. "A Mathematical Risk Model for *Escherichia coli* O157:H7 Cross-Contamination of Lettuce During Processing." *Food Microbiology* 28(4). Elsevier Ltd: 694–701. Doi:10.1016/j.fm.2010.06.008.

Pérez-Rodríguez, F., and A. Valero. 2013. *Predictive Microbiology in Foods*, edited by R.W. Hartel, J.P. Clark, J.W. Finley, D. Rodriguez-Lazaro, and D. Topping. London: Springer New York Heidelberg Dordrecht.

Pérez-Rodríguez, F., A. Valero, E. Carrasco, R.M. Garcia, and G. Zurera. 2008. "Understanding and Modelling Bacterial Transfer: A Review." *Trends in Food Science and Technology* 19(3): 131–144.

Pérez-Rodríguez, F., A. Valero, E.C.D. Todd, E. Carrasco, R.M. García-Gimeno, and G. Zurera. 2007. "Modelling Transfer of *Escherichia coli* O157:H7 and *Staphylococcus aureus* During Slicing of a Cooked Meat Product." *Meat Science* 76(4): 692–699. Doi:10.1016/j.meatsci.2007.02.011.

Possas, A.M.M., G.D. Posada-Izquierdo, F. Pérez-Rodríguez, and R.M. García-Gimeno. 2016. "Modelling the Transfer of *Salmonella* Enteritidis During Slicing of Ready-to-Eat Turkey Products Treated with Thyme Essential Oil." *Journal of Food Science* 81(11): M2770–2775. Doi:10.1111/1750-3841.13506.

Rosenquist, H., N.L. Nielsen, H.M. Sommer, B. Nørrung, and B.B. Christensen. 2003. "Quantitative Risk Assessment of Human Campylobacteriosis Associated with Thermophilic *Campylobacter* Species in Chickens." *International Journal of Food Microbiology* 83(1): 87–103. Doi:10.1016/S.

Ross, T. 1996. "Indices for Performance Evaluation of Predictive Models in Food Microbiology." *The Journal of Applied Bacteriology* 81(5): 501–508. Doi:10.1111/j.1365-2672.1996.tb03539.x.

Scollon, A.M., H. Wang, and E.T. Ryser. 2016. "Transfer of *Listeria monocytogenes* During Mechanical Slicing of Onions." *Food Control* 65. Elsevier Ltd: 160–167. Doi:10.1016/j.foodcont.2016.01.021.

Seliwiorstow, T., J. Baré, I. Van Damme, I.G. Algaba, M. Uyttendaele, and L. de Zutter. 2016. "Transfer of *Campylobacter* from a Positive Batch to Broiler Carcasses of a Subsequently Slaughtered Negative Batch: A Quantitative Approach." *Journal of Food Protection* 79(6): 896–901. Doi:10.4315/0362-028X. JFP-15-486.

Sheen, S. 2008. "Modelling Surface Transfer of *Listeria monocytogenes* on Salami During Slicing." *Journal of Food Science* 73(6): E304–E311. Doi:10.1111/j.1750-3841.2008.00833.x.

Sheen, S., and C. Hwang. 2010. "Mathematical Modelling the Cross-Contamination of *Escherichia coli* O157:H7 on the Surface of Ready-to-Eat Meat Product While Slicing." *Food Microbiology* 27(1). Elsevier Ltd: 37–43. Doi:10.1016/j.fm.2009.07.016.

Shieh, Y.C., M.L. Tortorello, G.J. Fleischman, D. Li, and D.W. Schaffner. 2014. "Tracking and Modelling Norovirus Transmission During Mechanical Slicing of Globe Tomatoes." *International Journal of Food Microbiology* 180: 13–18. Doi:10.1016/j.ijfoodmicro.2014.04.002.

Smid, J., R. De Jonge, A.H. Havelaar, and A. Pielaat. 2013. "Variability and Uncertainty Analysis of the Cross-Contamination Ratios of *Salmonella* During Pork Cutting." *Risk Analysis* 33(6): 1100–1115. Doi:10.1111/j.1539-6924.2012.01908.x.

Smid, J.H., L. Heres, A.H. Havelaar, and A. Pielaat. 2012. "A Biotracing Model of *Salmonella* in the Pork Production Chain." *Journal of Food Protection* 75(2): 270–280. Doi:10.4315/0362-028X.JFP-11-281.

385

Swart, A.N., E.G. Evers, R.L.L. Simons, and M. Swanenburg. 2016. "Modelling of *Salmonella* Contamination in the Pig Slaughterhouse." *Risk Analysis* 36(3): 498–515. Doi:10.1111/risa.12514.

Tierney, J.T., and E.P. Larkin. 1978. "Potential Sources of Error During Virus Thermal Inactivation." *Applied and Environmental Microbiology* 36(3): 432–437.

Titus, S.M. 2007. *A Novel Model Developed for Quantitative Microbial Risk Assessment in the Pork Food Chain.* Massey University, Palmerston North, New Zealand.

Vorst, K.L., E.C.D. Todd, and E.T. Ryser. 2006. "Transfer of *Listeria monocytogenes* During Slicing of Turkey Breast, Bologna, and Salami with Simulated Kitchen Knives." *Journal of Food Protection* 69(3): 2939–2946. doi:10.4315/0362-028X-69.12.2939.

Wang, H., and E.T. Ryser. 2016. "Quantitative Transfer of *Salmonella* Typhimurium LT2 During Mechanical Slicing of Tomatoes as Impacted by Multiple Processing Variables." *International Journal of Food Microbiology* 234(October). Elsevier B.V.: 76–82. doi:10.1016/j.ijfoodmicro.2016.06.035.

Weiss Nielsen, Martin, Claus Sternberg, Søren Molin, and Birgitte Regenberg. 2011. "*Pseudomonas aeruginosa*; and *Saccharomyces cerevisiae*; Biofilm in Flow Cells." *Journal of Visualized Experiments* 47: 1–5. doi:10.3791/2383.

Zilelidou, E.A., V. Tsourou, S. Poimenidou, A. Loukou, and P.N. Skandamis. 2015. "Modelling Transfer of *Escherichia coli* O157: H7 and *Listeria monocytogenes* During Preparation of Fresh-Cut Salads: Impact of Cutting and Shredding Practices." *Food Microbiology* 45(B) (PB): 254–265. doi:10.1016/j.fm.2014.06.019.

Zwietering, M.H. 2015. "Risk Assessment and Risk Management for Safe Foods: Assessment Needs Inclusion of Variability and Uncertainty, Management Needs Discrete Decisions." *International Journal of Food Microbiology* 213. Elsevier B.V.: 118–123. doi:10.1016/j.ijfoodmicro.2015.03.032.

Zwietering, M.H., and M.J. Nauta. 2007. "Predictive Models in Food Risk Assessment Cambridge, UK." In *Modelling Microorganisms in Food,* edited by S. Brul, S. Van Gerwen, and M. Zwietering, 110–28. Cambridge: Woodhead Publishing Ltd.

Zwietering, M.H., C.M. Stewart, and R.C. Whiting. 2010. "Validation of Control Measures in a Food Chain Using the FSO Concept." *Food Control* 21(12 SUPPL.): 1716–1722. doi:10.1016/j.foodcont.2010.05.019.

13

Expert Systems Applied to Microbial Food Safety

Mariem Ellouze

Contents

13.1 INTRODUCTION

Predictive microbiology has emerged as an essential element of the modern food microbiology area (McMeekin and Ross 2002), making it possible to move from a hazard-based and final product testing approach to a risk-based management approach based on process control. In the former approach, the presence of a hazard at a detectable level is used as a basis for legislation, whereas in the latter approach, supported by the Codex Alimentarius, the decision is made on the assessment of the risks (probability of observing an adverse effect) to the consumers. Exposure assessment is one of the four steps of risk assessment. It uses predictive microbiology mathematical models to predict microbial behavior over time, taking into account several environmental factors such as the temperature, a_w, pH, etc. These models can thus help improve food quality and safety, reduce costs and decrease "time to market". They are therefore appealing to several stakeholders, including academia, food business operators, food safety agencies and regulators (Halder, Dhall et al. 2011). Historically, statistical tools have been used for the development of such models to run simulations of different scenarios (Nunes, Alvarenga et al. 2015). The challenge is that quality managers, regulators and other stakeholders in the food industry are generally trained in chemistry and/or microbiology, so that it can be difficult for some of them to use these models directly. To address this limitation, there have been several initiatives to develop practical software using widely recognized, peer-reviewed and validated mathematical models.

These tools allow users without advanced mathematical or programming skills to fit their data, simulate the behavior of pathogens and/or spoilers during food manufacturing processes or during the product's shelf life, and perform exposure and risk assessments throughout the food chain.

However, transferring the knowledge of the mathematical models into user-friendly tools remains a challenge. In fact, the software output should be explicit and operational so that decision makers can understand it in order to take their decisions (Guillier 2016). Yet, users need to know the accuracy and reliability of the tools, the hypothesis supporting the simulations, and the limitations to make the best use of this information in the decision-making process.

With the vast amount of software available today, guidance is therefore welcome so that users can identify the most relevant software to meet their specific requirements or needs, according to their own level of expertise. General presentations of selections of software can be found in Plaza-Rodríguez, Thoens et al. (2015) and Tenenhaus-Aziza, Daudin et al. (2014). Regular presentations are also given at conferences and scientific events to disseminate the knowledge and use of these software among users from academia, industry, government and food safety agencies (Ellouze, Tenenhaus-Aziza et al. 2015).

13.2 SOFTWARE PRESENTATION

A list of contact points for risk assessment and predictive microbiology software was generated based on information given in the United States Department of Agriculture Agricultural Research Service (USDA-ARS)

Predictive Microbiology Information Portal, the "OpenML for Predictive Modelling in Food" (http://sourceforge.net/p/microbialmodelingexc hange/wiki/Tools/) and a literature search.

Figure 13.1 gives an overview of the software creation dates. The first, the Pathogen Modelling Program or PMP, was launched in 1980. It remained the only predictive microbiology software available for a decade, and then, two commercial software packages, Nizo Premia and Forecast, were issued in 1990 and 1995, respectively. Driven by intensive research work in the field, increasing numbers of software releases were observed from 2003, when the famous ComBase, Sym'Previus, GInaFIT and Refrigerated Index calculator were made available. A peak in software releases was reached in 2012, with the launch of nine different software packages in the same year. Another set of nine software packages was developed between 2012 and 2018.

The software can be classified into two categories: some require a level of modelling and computer programming expertise, while others would be more appropriate for beginners and intermediate users without prior expertise in mathematical modelling or programming.

The majority of software developers address their tools to a large audience, including representative from academia, industry and governments (Table 13.1). Only a few software packages are targeted to a specific audience, such as Nizo Premia (industry), Oyster Refrigeration Index and Listeria Meat Model (industry and government). Most of the packages are accessible via the internet, which eases the updating process and allows the user to work on the latest version at all times. Some of the packages are provided free of charge, others are commercial and some are internal software developed and used by service-providing companies for their assessments; these tools are then not made available to the public (except for Forecast from Campden, Nizo Premia from Nizo or Creme Microbial from Creme Global).

The main software applications include shelf-life predictions, hazard analysis and critical control point (HACCP) studies, thermal inactivation simulations, new product development, sampling plans and risk assessment. Some tools are generic and can address several needs by using dedicated modules (e.g. Baseline, ComBase, GroPin and Sym'Previus), while others are specific to precise applications, such as the Oyster Refrigeration Index (shelf-life predictions). The modules include growth and inactivation predictors, fitting tools for growth or inactivation, growth/non-growth simulators and databases (Table 13.2). The main modelling approach used by the different software is deterministic. This means that a single input

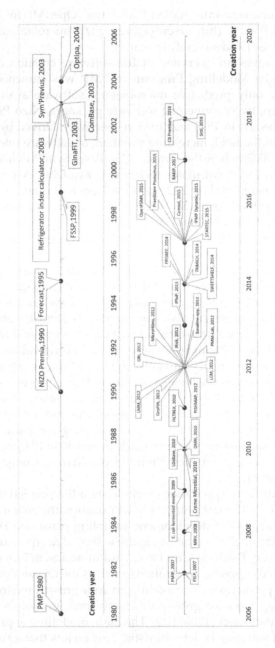

Figure 13.1 Software launch timeline.

Table 13.1 General Presentation of the Studied Software and Tools

Software	Creation Date / Last Update	Expected Users				Contact (Affiliation)	Access
		A	I	G	F		
Baseline-app	2012/2015	√	√	√	√	Antonio Valero (University of Cordoba)	www.microhibro.com
CB Premium	2018/2019	√	√	√	√	Mark Tamplin (University of Tasmania)	www.cbpremium.org/
ComBase	2003/2017	√	√	√	√	Mark Tamplin (University of Tasmania)	www.combase.cc/index.php/en/
Comsol	2015/2016	√	–	–	–	Ashim K Datta (Cornell University)	http://blogs.cornell.edu/edusim/
Creme Microbial	2010	√	√	–	–	Cian O' Mahony (Creme Global)	www.cremeglobal.com/modelling-software/creme-microbial
DMRI predictive models for meat	2016/2016	√	√	√	√	Annemarie Gunvig (Danish Meat Research Institute)	http://dmripredict.dk
FDA-iRisk	2012/2017	√	√	√	√	Yuhuan Chen (U.S. FDA)	https://irisk.foodrisk.org/
FILTREX	2010/2015	√	√	–	–	Jean-Pierre Gauchi (INRA)	http://informatique-mia.inra.fr/logiciels/node/230
FISHMAP	2011/2013	√	√	√	√	Begoña Alfaro (AZTI-Tecnalia)	www.azti.es/network/shelf-life-prediction-software/
Forecast	1995/2014	–	√	√	√	Linda Everis (Campden BRI)	Software not publicly available, used internally only
FRISBEE Tool	2014/2014	√	√	√	√	Annemie Geeraerd (KU Leuven)	http://frisbeetool.eu/FrisbeeTool/about.html

(Continued)

Table 13.1 (Continued)   General Presentation of the Studied Software and Tools

Software	Creation Date / Last Update	Expected Users				Contact (Affiliation)	Access
		A	I	G	F		
FSLP	2007/2008	√	√	√	√	Begoña Alfaro (AZTI-Tecnalia)	www.azti.es/network/shelf-life-prediction-software/
FSSP	1999/2013	√	√	√	–	Paw Dalgaard (DTU, Technical University of Denmark)	http://fssp.food.dtu.dk/
GInaFiT	2003/2012	√	√	√	√	Annemie Geeraerd (KU Leuven)	www.biw.kuleuven.be/biosyst/mebios/mebiosdownloads
GroPIN	2012/2015	√	√	√	√	Panagiotis N. Skandamis (Agricultural University of Athens)	www.aua.gr/psomas/gropin/
IPMP Dynamic Prediction	2015/2015	√	√	√	√	Lihan Huang (USDA-ARS Eastern Regional Research Center)	www.ars.usda.gov/Main/Docs.htm?docid=25312
LCM2012 Listeria Control Model 2012	2012/2015	√	√	√	√	Olav Sliekers (Corbion)	lcm.purac.com
Lemgo D- and z-value Database for food (LDzBase)	2010/2011	√	√	√	√	Knut Schwarzer (Institute for Food Technology.NRW)	www.hs-owl.de\fb4\ldzbase
Listeria Meat Model	2012	–	√	√	–	Jan Van Impe (KU Leuven)	Not possible to download or access directly
Microbial Responses Viewer, MRV	2008/2015	√	√	√	√	Shige Koseki (Hokkaido University)	http://mrviewer.info/

(Continued)

Table 13.1 (Continued) General Presentation of the Studied Software and Tools

Software	Creation Date / Last Update	Expected Users			Contact (Affiliation)	Access
		A	I	G	F		
MicroHibro	2012/2015	√	√	√	√	Fernando Perez Rodriguez (University of Cordoba)	www.microhibro.com
MLA *E. coli* inactivation in fermented meats model	2009/2009	√	√	–	–	Tom Ross (Meat Live Stock Australia)	www.foodsafetycentre.com.au/fermenter.php
MLA refrigerator index calculator	2003/2015	√	√	√	√	Tom Ross (Meat and Livestock Australia)	www.foodsafetycentre.com.au/refrigerationindex.php
NIZO Premia	1990/2015	–	√	–	–	Erik Smit (Nizo Food Research)	Write to the contact person
open Food Safety Model Repository (openFSMR)	2015/2015	√	√	√	–	Matthias Filter (Federal Institute for Risk Assessment, BfR)	https://sites.google.com/site/openfsmr
Optipa	2004/2016	√	–	–	–	Maarten Hertog (KU Leuven)	www.optipa.be
Oyster refrigeration index (ORI)	2012/2012	–	√	√	–	Mark Tamplin (Australian Seafood Cooperative Research Centre)	http://vibrio.foodsafetycentre.com.au/
PMM-Lab	2012/2016	√	√	√	√	Matthias Filter (Federal Institute for Risk Assessment, BfR)	http://sourceforge.net/projects/pmmlab/
Praedicere Possumus	2015/2015	√	√	√	√	Pierluigi Polese (University of Udine)	http://praedicere.uniud.it/index.php

(Continued)

393

Table 13.1 (Continued) General Presentation of the Studied Software and Tools

Software	Creation Date / Last Update	Expected Users				Contact (Affiliation)	Access
		A	I	G	F		
RAKIP	2017	✓	✓	✓	✓	Matthias Filter (Federal Institute for Risk Assessment, BfR)	https://foodrisklabs.bfr.bund.de/rakip-web-portal/
STARTEC	2015/2015	✓	✓	✓	✓	Taran Skjerdal (Norwegian Veterinary Institute/IRIS)	Prototype, link can be obtained on request to the developers
SWEETSHELF	2014/2014	✓	✓	–	–	An Vermeulen (Gent University)	info@cpmf2.be
Sym'Previus	2003/2018	✓	✓	✓	✓	Yvan Le Marc (ADRIA development)	symprevius.eu
TRiMiCri	2014	✓	✓	✓	✓	Maarten Nauta (DTU: Technical University of Denmark)	http://tools.food.dtu.dk/apex/f?p=158:1:0:::::
USDA IPMP 2013	2013/2014	✓	✓	✓	✓	Lihan Huang (USDA ARS Eastern Regional Research Center)	www.ars.usda.gov/Main/docs.htm?docid=23355
USDA-ARS Pathogen Modeling Program (PMP)	1980/2016	✓	✓	✓	✓	Vijay Juneja (USDA ARS-Eastern Regional Research Center)	http://ars.usda.gov/services/docs.htm?docid=6786
USDA-ARS Predictive Microbiology Information Portal (PMIP)	2015/2007	✓	✓	✓	✓	Vijay Juneja (USDA ARS Eastern Regional Research Center)	http://portal.errc.ars.usda.gov/

A: academia; F: food safety agencies; I: industry; G: government.

Table 13.2 Specific Information about the Software

Software	Accessibility	Cost	Publications	Modelling Approach	Main Application	Database	G/NG	GF	GS	TIF	TIS	NTIF	NTIS	Training?	Help?
Baseline-app	Internet	Free	NA	D	Sampling plans, shelf-life studies	–	–	+	–	+	–	–	–	+	+
CB Premium	Internet	Free	NA	D	Validation studies including shelf life	–	–	–	+	–	+	–	+	+	+
ComBase	Internet	Free	(Baranyi and Tamplin 2004)	D	Validation studies including shelf life	+	–	+	+	+	+	+	+	+	+
Comsol	Download	Commercial	NA	D	Thermal processing applications	–	–	–	–	–	+	–	–	–	–
Creme Microbial	Internet	Commercial	NA	P	New product development	+	+	+	+	+	+	+	+	+	+
DMRI predictive models for meat	Internet	Free	(Mejlholm, Gunvig et al. 2010)	D+P	Food safety	–	+	–	+	–	–	–	+	+	+
FDA-iRisk	Internet	Free	(Chen, Dennis et al. 2013)	D+P	Risk assessment	–	–	–	+	–	+	–	–	+	+
FILTREX	Download	Free	(Gauchi and Vila 2013)	P	Growth and inactivation studies	–	–	+	+	–	–	–	–	–	+
FISHMAP	Internet	Free	(Alfaro, Hernández et al. 2013)	D	Shelf-life studies	–	–	+	+	–	–	–	–	+	+
Forecast	Consultancy	Commercial	NA	D+P	Shelf-life studies	–	+	–	–	–	–	–	–	+	–
FRISBEE Tool	Download	Free	(Gwanpua, Verboven et al. 2015)	P	Shelf-life studies	–	–	–	+	–	–	–	–	+	+
FSLP	Download	Free	(Nuin, Alfaro et al. 2008)	D	Shelf-life studies	–	–	+	–	–	–	–	–	+	–

(Continued)

Table 13.2 (Continued) Specific Information about the Software

Software	Accessibility	Cost	Publications	Modelling Approach	Main Application	Database	G/NG	GF	GS	TIF	TIS	NTIF	NTIS	Training?	Help?
PSSP	Download	Free	(Mejlholm, Boknaes et al. 2015)	D	Shelf-life studies	–	+	–	+	–	–	–	–	+	–
GInaFiT	Download	Free	(Geeraerd, Valdramidis et al. 2005)	D	Thermal processing applications	–	–	–	–	+	–	–	–	+	+
GroPIN	Download	Free	(Psomas, Nychas et al. 2012)	D+P	New product development	–	+	–	+	–	+	–	+	–	+
IPMP Dynamic Prediction	Download	Free	(Huang 2015)	D+P	HACCP studies	–	–	–	+	–	–	–	–	–	+
LCM2012 Listeria Control Model 2012	Internet	Free	(Mejlholm, Gunvig et al. 2010)	D+P	New product development	–	–	–	+	–	–	–	–	+	+
Lengo D- and z-value Database for food (LDzBase)	Internet	Free	(Schwarzer, Schneider et al. 2010)	D	Thermal processing applications	+	–	–	–	–	–	–	–	–	+
Listeria Meat Model	Consultancy	Commercial	NA	D	Shelf-life studies	–	+	–	+	–	–	–	–	–	–
Microbial Responses Viewer, MRV	Internet	Free	(Koseki 2013)	D+P	New product development	–	+	–	–	–	–	–	–	–	–
MicroHibro	Internet	Free	NA	D+	Shelf-life and risk assessment	+	–	–	+	–	+	–	+	+	+
MLA E. coli inactivation in fermented meats model	Internet	Free	NA	D	HACCP studies	–	–	–	+	–	–	–	–	–	–

(Continued)

Table 13.2 (Continued) Specific Information about the Software

Software	Accessibility	Cost	Publications	Modelling Approach	Main Application	Database	G/NG	GF	GS	TIF	TIS	NTIF	NTIS	Training?	Help?
MLA refrigerator index calculator	Internet	Free	(Mellefont, McMeekin et al. 2003)	D	Shelf-life studies	-	-	-	+	-	-	-	-	-	-
NIZO Premia	Download/consultancy	Commercial	NA	D	Process optimization	+	-	-	+	+	-	+	+	+	+
open Food Safety Model Repository (openFSMR)	Internet	Free	(Plaza-Rodríguez, Thoens et al. 2015)	D	Thermal processing applications	Repository of models					-			-	-
Optipia	Internet	Free	(Philips, Haest et al. 2013)	P	Any, according to user model	Depending on the models of the user				+	-			-	+
Oyster refrigeration index	Internet	Free	NA	D	Shelf-life studies	-	-	-	+	-	-	-	-	-	-
PMM-Lab	Download	Free	(Filter, Thöns et al. 2013)	D	Thermal processing applications	+	-	+	+	+	+	+	+	-	+
Praedicere Possumus	Internet	Free	(Polese, Del Torre et al. 2016)	D	New product development	-	-	-	+	+	-	+	-	-	-
RAKIP	Internet	Free	(Plaza-Rodríguez, Haberbeck et al. 2017)	D+P	Any, according to chosen model	Repository of models and data								-	+
STARTEC	Internet	Free	(Skjerdal, Gefferth et al. 2017)	D	HACCP studies	-	-	-	+	-	+	-	-	+	-
SWEETSHELF	Download	Commercial	NA	P	New product development	-	+	-	-	-	-	-	-	-	+

(Continued)

Table 13.2 (Continued) Specific Information about the Software

Software	Accessibility	Cost	Publications	Modelling Approach	Main Application	Database	G/NG	GF	GS	TIF	TIS	NTIF	NTIS	Training?	Help?
Sym'Previus	Internet	Commercial	(Couvert, Pinon et al. 2010)	D+P	Shelf-life studies, thermal processing, new product development, HACCP studies	+	+	+	+	+	+	+	+	+	+
TRiMiCri	Download	Free	(Seliwiorstow, Uyttendaele et al. 2016)	P	Risk-based microbiological criteria	-	-	-	+	-	-	-	-	-	+
USDA IPMP 2013	Download	Free	(Huang 2014)	D	Shelf-life studies	-	+	-	+	-	-	-	-	-	+
USDA-ARS Pathogen Modeling Program (PMP)	Download	Free	NA	D	HACCP studies	-	-	-	+	-	+	+	+	-	+
USDA-ARS Predictive Microbiology Information Portal (PMIP)	Internet	Free	NA	D	HACCP studies	-	-	-	+	-	+	+	+	-	+

NA: not available; D: deterministic; D+P: some modules are deterministic, some are probabilistic; P: probabilistic.
GF: Growth fitting module; G/NG: Growth/No Growth Interface; GS: Growth Simulation module; NTIF: Non Thermal Inactivation Fitting module; NTIS: Non Thermal Inactivation Simulation module; TIF: Thermal Inactivation Fitting module; TIS: Thermal Inactivation simulation module; (+): available; (−): not available.

value is given to the software, and then, a single output value is gener-ated. This approach has the advantage of showing clear results and aids communication to a non-expert audience. Some packages also propose probabilistic modules or both deterministic and probabilistic modules. In the probabilistic approach, the inputs are associated with a probability distribution that takes into account the natural variability (and sometimes the uncertainty) around the targeted factor. Simulations are then run with a high number of iterations; for each iteration, a value is sampled for each input in its distribution, and the calculations are made to provide the esti-mate for that iteration. Results from the different iterations form the out-put distribution, giving a more precise idea of the response; however, this is more complex to explain to a non-expert audience. Specific graphical representations are then helpful to help the user understand the model outcome, as the probabilistic approach, despite its complexity over deter-ministic (or point-estimate) calculations, has become the method of choice for quantitative microbial risk assessment (QMRA).

Software providing risk assessment interfaces is less common than tools providing predictive microbiology simulations; some examples are presented here.

MicroHibro is an online free solution that allows microbial growth or inactivation to be simulated for different micro-organisms in several matrices, including food and culture media. Models can be implemented and stored in the system, possibly shared with other users, to perform QMRA studies or stand-alone simulations. When performing microbial risk assessments, the tool estimates the final concentration at the moment of consumption in a probabilistic way, applies a specific dose response and derives a probability of illness as a final risk metric. This object-ori-ented tool allows the user to design the studied food chain and implement the models based on three microbial processes: reduction, growth and transfer.

FDA-iRISK® is a web-based food safety risk analysis and ranking tool designed to guide fairly experienced users to perform microbiological and chemical risk assessments (Chen, Dennis et al. 2013). The challenge in chemical risk assessment is that the exposure to chemical hazard is not caused by a single food and that the exposure is chronic and not acute. For an appropriate chemical risk assessment, a variety of developed tools are available. It is crucial to understand the mode of action of the contaminant. For those that do not exhibit genotoxic activity, a threshold can be defined to observe the adverse effects. However, for chemicals with genotoxic activity, a safety threshold cannot be established, and a risk assessment is required (Cartus and Schrenk 2017). The FDA-iRisk software takes into account multiple sources of exposure for a single contaminant and also chronic exposure to calculate the risk metrics.

Some information needs to be provided by the user, as there is no database in the current version of the software. The output is an expression of health metrics to assess the health burden on a population level; however, it is also possible to limit the output to the exposure assessment. A strong statistical background and a clear understanding of the risk assessment methodology and the significance of the output (health metrics and exposure) are needed to carry out the analysis. There are some examples of risk scenarios that are provided and that are helpful to understand the different options proposed by the software.

Finally, there are some *modelling interfaces* in which the users can code their own equations and data. This type of software requires more programming and modelling skills than fit-for-purpose software, but it also gives more flexibility. Examples include R (Core Team, 2017), MATLAB® (The Mathworks Inc, USA), SAS (SAS Institute Inc., USA) and @RISK (Palisade Corp., Ithaca).

The *R software* has gained more and more popularity over the last years. In fact, its community of users develop packages that can be used and adapted to best fit the users' needs. The risk assessment package mc2d (Pouillot et al. 2010), for example, assists the user to perform microbial risk assessment. It gives an illustration with two simplified models to perform (1) a QMRA for *Escherichia coli* O157:H7 for children under 3 years old after consumption of frozen ground beef and (2) a quantitative risk assessment to evaluate the risk of infection with *Cryptosporidium parvum* from consumption of tap water. There are also other risk assessment and predictive microbiology packages. The Bio-inactivation package published in 2019 provides functions for modelling microbial inactivation under isothermal or dynamic conditions. For a more exhaustive list of packages, readers can access the CRAN site to identify the package that best fits their needs.

Going one step further, the *RAKIP initiative* aims to ease the sharing of data and models within the food safety community. This initiative has proposed a common description language, Predictive Modelling in Food Markup Language (PMF-ML) for predictive microbial models. This has facilitated information exchange between different software tools, such as from ComBase and GroPIN to PMM-Lab or from PMM-Lab to the software R. A new FSK-ML (Food Safety Knowledge Markup Language) has also been proposed to adapt certain specifications of the PMF-ML specifically for QMRA models (Plaza-Rodríguez, Haberbeck et al. 2017).

13.3 USER PERSPECTIVE

Assessing the user experience for each software is key but remains a relatively subjective exercise. In fact, it depends on several criteria, such as the user profile and level of expertise, the targeted applications, etc. Feedback from the predictive microbiology community was collected via an online survey conducted between January and February 2016. During that period, 130 respondents from 21 different countries answered questions about their use of predictive microbiology and risk assessment software. The group consisted of 50 researchers, 34 industrialists, 16 representatives from the government, 11 representatives from technical/research institutes, nine students, six food safety agency members and four respondents coming from individual labs or R&D facilities (Figure 13.2). The majority of respondents (34%) declared that they used the software on a monthly basis, while 13% indicated that they used these tools on a weekly basis and 13% used the tools more frequently. Less frequent users reported a quarterly (17%), biannual (11%) or annual (12%) use. Of all respondents, 65% indicated that they preferred to use free software provided by the scientific community against 14% using commercial tools and 21% (mainly researchers) using their own software, developed internally using software like SAS, MATLAB or R.

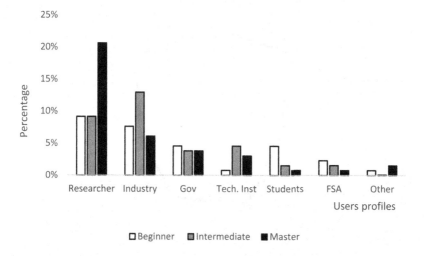

Figure 13.2   Level of expertise according to several groups: researchers, industry, government representatives (Gov), technical institutes (Tech. Inst), students, food safety agencies (FSA) and other (n = 130).

Participants were also asked to rate their level of expertise. Respondents were equally distributed among the three proposed levels: beginners represented 30%, intermediates 34% and master users 38% of the respondents (Figure 13.2). The survey participants were asked to report, through an open question, the software that they mainly used. ComBase ranked first, with 68 spontaneous citations over 130 responses, while Sym'Previus and GInaFIT ranked second and third, with 19 and 13 citations, respectively.

When asked about the main application they targeted, shelf-life studies was the most cited application (37%), followed by training and research activities (19%), thermal processing (10%), HACCP studies (9%) and new product development (9%). A few excerpts from the "other" applications include the design of experiments and risk and exposure assessments.

The survey also aimed at identifying the specific software criteria that are of interest to users. Figure 13.3 illustrates that the most important criterion is the number of environmental factors taken into account. Besides, it is important for the users that models are validated and that the software is made as user-friendly as possible. Surprisingly, less importance is given to regular updates or to specific work on model specificity.

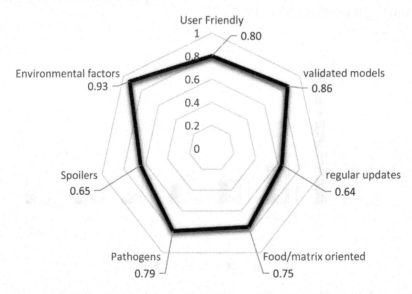

Figure 13.3 Radar chart for the software specifications that are important to users (n = 130).

13.4 CHALLENGES AND OPPORTUNITIES

Each software package has its own specificities and scope. Some focus on targeted micro-organisms in specific matrices, while others are more generic and address different objectives such as risk prioritization. Some require programming skills, while others provide a user-friendly interface for non-modelling experts.

According to their needs, the user will naturally select one package over the other, given that a comprehensive list of all the available software is provided. It can be challenging to keep such a list updated, as new modules and software are regularly released and others are not maintained. Within the RAKIP initiative, a list of software is available at https://foodrisklabs.bfr.bund.de/predictive-microbial-modelling-and-qmra-software-directory/

This initiative encourages and sets the scene for collaboration, data and model sharing between the tools. Ideally, all published or otherwise archived knowledge on microbial behavior should be available for the community and accessible by any stakeholder (McMeekin and Ross 2002).

By sharing within the food safety community, the users will get the utmost from the combination of the different software while increasing the efficiency, transparency and usability of predictive models (Plaza-Rodríguez, Thoens et al. 2015).

For risk assessors, this could close an existing gap between the developed predictive microbiology models and those used in microbial risk assessments. In fact, it is difficult for them to quickly select the appropriate predictive microbiology model among the abundant literature. Once selected, it would be much easier if the model was already implemented in a tool, so that the risk assessor could proceed with the analysis (Pouillot and Lubran 2011).

However, thinking that a person with minimal knowledge of programming and/or modelling can use a few clicks to perform a safety assessment and give sound conclusions might be simplistic. Technical expertise will still be needed to understand the hypothesis behind the models and to be able to interpret the results and the software output, which can sometimes be complex. The need for training is therefore increasing; demonstration sessions and dissemination activities in conferences are a good option to address this need. Tutorials and help sections available in some software are good, but some users really need more detailed messages; these could be delivered in the form of webinars or eLearnings. The newly created International Committee on Predictive

Modelling in Food (www.icpmg.org) will then have a role to play in further disseminating these tools beyond the microbial risk assessment and predictive microbiology community, targeting microbiologists, hygienists and quality managers. Commercial software providers can also join this global training exercise by giving access to their tools for a limited time so that users can discover their modules.

Moreover, the quality of such tools should be assessed through the prescription of minimum standards for validation so that the users can have confidence in the tools and software they use.

Some of this software has been assessed by local authorities; for example, the SSSP software that evolved into the actual FSSP software was approved by the Danish Veterinary and Food Administration as a means to predict *Listeria monocytogenes* growth in ready-to-eat (RTE) food and to document their compliance with the European regulation EC 2073/2005, and the Sym'Previus software was evaluated by the French Food Safety Agency (AFSSA 2009). The European Food Safety Authority (EFSA) BIOHAZ panel has also assessed a number of risk ranking and risk assessment tools (EFSA, 2015). This is not an easy exercise, as it requires the definition of common criteria to evaluate the tools and then to compare their performances while they provide versatile outputs.

Finally, the community should by now be convinced of the benefits of using previously developed models and tools. Efforts should be focused on the improvement of existing tools and closing current gaps rather than engaging in the production of more and more tools, which will be more confusing and less productive, especially for the end users.

REFERENCES

Alfaro, B., I. Hernández, Y. Le Marc and C. Pin (2013). "Modelling the effect of the temperature and carbon dioxide on the growth of spoilage bacteria in packed fish products." *Food Control* 29(2): 429–437.

ANSES (2009). Avis de l'Agence française de sécurité sanitaire des aliments relatif à une demande d'appui scientifique et technique de l'outil prévisionnel Sym'Previus version II. Accessed at: https://www.anses.fr/fr/system/files/MIC2009sa0033.pdf.

Baranyi, J. and M. L. Tamplin (2004). "ComBase: A common database on microbial responses to food environments." *Journal of Food Protection* 67(9): 1967–1971.

Cartus, A. and D. Schrenk (2017). "Current methods in risk assessment of genotoxic chemicals." *Food and Chemical Toxicology* 106(Pt. B): 574–582.

Chen, Y., S. B. Dennis, E. Hartnett, G. Paoli, R. Pouillot, T. Ruthman and M. Wilson (2013). "FDA-iRISK--A comparative risk assessment system for evaluating and ranking food-hazard pairs: Case studies on microbial hazards." *Journal of Food Protection* **76**(3): 376–385.

Couvert, O., A. Pinon, H. Bergis, F. Bourdichon, F. Carlin, M. Cornu, C. Denis, N. Gnanou Besse, L. Guillier, E. Jamet, E. Mettler, V. Stahl, D. Thuault, V. Zuliani and J. C. Augustin (2010). "Validation of a stochastic modelling approach for *Listeria monocytogenes* growth in refrigerated foods." *International Journal of Food Microbiology* **144**(2): 236–242.

EFSA (2015). Scientific opinion on the development of a risk ranking toolbox for the EFSA BIOHAZ Panel. *EFSA Journal* **13**(1): 3939.

Ellouze, M., F. Tenenhaus-Aziza and F. Carlin (2015). "Highlights from the 8th International Conference on Predictive Modelling in Food (ICPMF8)." *Food Microbiology* **45**(B): 160–161.

Filter, M., C. Thöns, A. Weiser, A. Falenski, B. Appel and A. Käsbohrer (2013). PMM-Lab? A community resource for integrated Predictive Microbial Modelling (PMM). BfR/FDA-Workshop Series on Tools for Food Defense and Safety, 4–6 September, Berlin.

Gauchi, J.-P. and J.-P. Vila (2013). "Nonparametric particle filtering approaches for identification and inference in nonlinear state-space dynamic systems." *Statistics and Computing* **23**(4): 523–533.

Geeraerd, A. H., V. P. Valdramidis and J. F. Van Impe (2005). "GInaFiT, a freeware tool to assess non-log-linear microbial survivor curves." *International Journal of Food Microbiology* **102**(1): 95–105.

Guillier, L. (2016). "Predictive microbiology models and operational readiness." *Procedia Food Science* **7**: 133–136.

Gwanpua, S. G., P. Verboven, D. Leducq, T. Brown, B. E. Verlinden, E. Bekele, W. Aregawi, J. Evans, A. Foster, S. Duret, H. M. Hoang, S. van der Sluis, E. Wissink, L. J. A. M. Hendriksen, P. Taoukis, E. Gogou, V. Stahl, M. El Jabri, J. F. Le Page, I. Claussen, E. Indergård, B. M. Nicolai, G. Alvarez and A. H. Geeraerd (2015). "The Frisbee tool, a software for optimising the trade-off between food quality, energy use, and global warming impact of cold chains." *Journal of Food Engineering* **148**: 2–12.

Halder, A., A. Dhall, A. K. Datta, D. G. Black, P. M. Davidson, J. Li and S. Zivanovic (2011). "A user-friendly general-purpose predictive software package for food safety." *Journal of Food Engineering* **104**(2): 173–185.

Huang, L. (2014). "IPMP 2013--A comprehensive data analysis tool for predictive microbiology." *International Journal of Food Microbiology* **171**: 100–107.

Huang, L. (2015). "Dynamic determination of kinetic parameters, computer simulation, and probabilistic analysis of growth of *Clostridium perfringens* in cooked beef during cooling." *International Journal of Food Microbiology* **195**: 20–29.

Koseki, S. (2013). "Alternative approaches to predicting microbial behaviour: A probabilistic modelling approach for microbial inactivation and a revised web-tool, the Microbial Responses Viewer." *Food Control* **29**(2): 416–421.

McMeekin, T. A. and T. Ross (2002). "Predictive microbiology: Providing a knowl-edge-based framework for change management." *International Journal of Food Microbiology* 78(1–2): 133–153.

Mejlholm, O., N. Boknaes and P. Dalgaard (2015). "Development and validation of a stochastic model for potential growth of *Listeria monocytogenes* in naturally contaminated lightly preserved seafood." *Food Microbiology* 45(B): 276–289.

Mejlholm, O., A. Gunvig, C. Borggaard, J. Blom-Hanssen, L. Mellefont, T. Ross, F. Leroi, T. Else, D. Visser and P. Dalgaard (2010). "Predicting growth rates and growth boundary of *Listeria monocytogenes* - An international valida-tion study with focus on processed and ready-to-eat meat and seafood." *International Journal of Food Microbiology* 141(3): 137–150.

Mellefont, L. A., T. A. McMeekin and T. Ross (2003). "Performance evaluation of a model describing the effects of temperature, water activity, pH and lactic acid concentration on the growth of *Escherichia coli*." *International Journal of Food Microbiology* 82(1): 45–58.

Nuin, M., B. Alfaro, Z. Cruz, N. Argarate, S. George, Y. Le Marc, J. Olley and C. Pin (2008). "Modelling spoilage of fresh turbot and evaluation of a time-temperature integrator (TTI) label under fluctuating temperature." *International Journal of Food Microbiology* 127(3): 193–199.

Nunes, C. A., V. O. Alvarenga, A. de Souza Sant'Ana, J. S. Santos and D. Granato (2015). "The use of statistical software in food science and technology: Advantages, limitations and misuses." *Food Research International* 75: 270–280.

Philips, J., P. J. Haest, D. Springael and E. Smolders (2013). "Inhibition of *Geobacter dechlorinators* at elevated trichloroethene concentrations is explained by a reduced activity rather than by an enhanced cell decay." *Environmental Science and Technology* 47(3): 1510–1517.

Plaza-Rodríguez, C., L. U. Haberbeck, V. Desvignes, P. Dalgaard, M. Sanaa, M. Nauta, M. Filter and L. Guillier (2017). "Towards transparent and consistent exchange of knowledge for improved microbiological food safety." *Current Opinion in Food Science* 18: 1–9.

Plaza-Rodríguez, C., C. Thoens, A. Falenski, A. A. Weiser, B. Appel, A. Kaesbohrer and M. Filter (2015). "A strategy to establish Food Safety Model Repositories." *International Journal of Food Microbiology* 204: 81–90.

Polese, P., M. Del Torre and M. L. Stecchini (2016). "A web-based application cus-tomized to food safety requirements of small-sized enterprises." *Procedia Food Science* 7: 149–153.

Pouillot, R., M. L. Delingnette-Muller, D. L. Kelly and J. B. Denis (2010). The mc2d package. Available from: http//riskassessment.r-forge.r-project.org/.

Pouillot, R. and M. B. Lubran (2011). "Predictive microbiology models vs. model-ing microbial growth within Listeria monocytogenes risk assessment: What parameters matter and why." *Food Microbiology* 28(4): 720–726.

Psomas, A. N., G.-J. Nychas, S. A. Haroutounian and P. Skandamis (2012). "LabBase: Development and validation of an innovative food microbial growth responses database." *Computers and Electronics in Agriculture* 85: 99–108.

Schwarzer, K., J. Schneider, U. Müller, B. Becker and P. Wilhelm (2010). "Lemgo database for D- and z-values: Useful for beverage spoilage microorganisms." *Brauwelt International* **28**(5): 252–256.

Seliwiorstow, T., M. Uyttendaele, L. De Zutter and M. Nauta (2016). "Application of TRiMiCri for the evaluation of risk based microbiological criteria for *Campylobacter* on broiler meat." *Microbial Risk Analysis* **2–3**: 78–82.

Skjerdal, T., A. Gefferth, M. Spajic, E. G. Estanga, A. de Cecare, S. Vitali, F. Pasquali, F. Bovo, G. Manfreda, R. Mancusi, M. Trevisiani, G. T. Tessema, T. Fagereng, L. H. Moen, L. Lyshaug, A. Koidis, G. Delgado-Pando, A. C. Stratakos, M. Boeri, C. From, H. Syed, M. Muccioli, R. Mulazzani and C. Halbert (2017). "The STARTEC decision support tool for better tradeoffs between food safety, quality, nutrition, and costs in production of advanced ready-to-eat foods." *BioMed Research International* **2017**: 6353510.

Tenenhaus-Aziza, F., J. J. Daudin, A. Maffre and M. Sanaa (2014). "Risk-based approach for microbiological food safety management in the dairy industry: The case of *Listeria monocytogenes* in soft cheese made from pasteurized milk." *Risk Analysis* **34**(1): 56–74.

14

Dose–Response Models for Microbial Risk Assessment

Moez Sanaa

Contents

14.1 INTRODUCTION

In microbial quantitative risk assessment (Chapter 9), the objective of a dose–response (DR) model is to establish a link between the level of biological hazard exposure (total ingested dose of microorganisms expressed, for example, in colony-forming units (cfu), focus-forming units, number of oocysts, etc.) and the probability of occurrence of an effect. We can consider different effects such as infection, illness and death. The aim of dose–response models is to estimate the proportion of individuals in a population expected to experience an adverse health effect when exposed to a specific amount of a biological hazard. For many microorganisms, dose–response studies in human volunteers and animal models are available and can be used to estimate the probability of an adverse effect for a given exposure to microorganisms. It is also possible to assess dose–response models from epidemiological studies on foodborne diseases, to the extent that the data collected on exposure are made available.

The biological foundation for dose–response models derives from major steps in the disease process: exposure, infection, illness and consequences (recovery, sequelae or death). The final outcome is the result of the interactions between the pathogen, the host and the food matrix. Traditionally, the concept of minimum infectious dose (MID) was used to measure the infectivity of an organism. MID is an expression of the lowest number of organisms required to initiate an infection in any individual under given circumstances. The term recognizes that each organism has the potential to cause an infection but builds upon the variation that exists in the actual dose required, the virulence of the agent and the individual's response. This traditional interpretation of dose–response information assumes the existence of a threshold level of pathogens that must be ingested in order for the microorganism to produce infection or disease. Attempts to define the numerical value of such thresholds in test populations have typically been unsuccessful. An alternative hypothesis, due to the potential for microorganisms to grow within the host, is that a single viable infectious pathogenic organism is able to induce infection ("single-hit concept"). Mathematically, there is always a non-zero probability of infection or illness when a host is exposed to an infectious pathogenic organism. The minimal infectious dose is accordingly one single microorganism, and the probability at this dose is in general very small. Practically, experimentally testing the existence or absence of a threshold, at both the individual and the population level, is impossible. Experiments are usually designed to assess response at relatively high doses (microbial detection limit and limited number of experimental units), and a small frequency of infection or illness cannot be precisely assessed. If one assumes the existence of a threshold, it cannot be considered as constant and will vary depending on the outcome of the interaction between the pathogen, the host and the food matrix.

The threshold model implies a definitive threshold below which no infection would occur. So, the non-threshold model is a more cautious and more appropriate approach than is the threshold model in some circumstances. Therefore, we believe that prudent public health protection requires the application of non-threshold approaches to the assessment of microbial dose–response relationships. A dose-response model gives the probability of illness according to the number of ingested pathogenic microorganisms. Among d ingested microorganisms, some might survive human host barriers and subsequently initiate infection and cause illness. Illness probability is defined as the probability of achieving this sequence of events. If each ingested microorganism has the same probability of provoking illness, r, then the number of microorganisms surviving different barriers follows a binomial distribution. If each microorganism is capable of inducing illness, then the probability of illness given d ingested microorganisms is the complement of the probability of absence of illness:

$$P(d) = 1 - (1 - r)^d \tag{14.1}$$

14.2 EXPONENTIAL MODEL

Generally, the assessment of a dose–response relationship requires measurement of responses at different dose levels. Given a fixed number of host individuals available for a dose–response study, the investigator determines the necessary number of different dose levels, the location of the dose levels within the dose range, and the proportions of host individuals for each of the dose levels. Statistical design optimization theory covers in general the choice of the dose levels and the number of measurements for each dose level in order to estimate parameters of interest as precisely as possible.

Experimental design for dose–response model assessment defines the following:

- Dose levels: $d_1, d_2, \ldots, d_i, \ldots, d_k$, where k is the number of dose levels
- Number of individuals per dose level: $n_1, n_2, \ldots, n_i, \ldots, n_k$

At the end of the experiment, the investigator counts the number of individuals showing the response under study, e.g. infection, illness or death:

- Responses per dose level: $y_1, y_2, \ldots, y_i, \ldots, y_k$

The probability of observing the adverse effect for a given host individual j exposed to the level d_i is

$$P(d_{ij}) = 1 - (1 - r_{ij})^{d_{ij}} \qquad (14.2)$$

where d_{ij} is the actual exposure dose and r_{ij} is the output of the interaction between the individual host and the pathogen.

The actual dose d_{ij} is, however, unknown in practice, and only its probability distribution can be assessed. Indeed, in experimental challenge trials, only the dose level (expected dose) is known. A certain amount of pathogen isolated from previous clinical cases for the experiment is mixed within a liquid medium, and subsequent serial 10× dilutions are then made in the same solvent using glass bottles. After each dilution, the bottle is vigorously agitated using a mechanical shaker. In these conditions and in the absence of clumping it is acceptable to assume a Poisson distribution of organisms ingested by each individual host. Thus, instead of assessing the probability of illness given a specific dose (conditional probability), we calculate an expected probability given the Poisson distribution of the doses:

$$P(d_i) = 1 - \exp(-r_{ij} \times d_i) \qquad (14.3)$$

Moreover, in the context of experimental challenge trials, it is very common to use a very homogeneous group of animals, and all the individuals are exposed to the same strain of the pathogen. Accordingly, it is expected that the parameter r_{ij} will not vary too much, and it is thus considered as a constant, to end up with the exponential model:

$$P(d_i) = 1 - \exp(-r \times d_i) \qquad (14.4)$$

Therefore, it is important to keep in mind that the exponential model does not estimate the risk at individual level but at population level, and the model assumes that the interaction between the hazard and the host is constant for every individual in the population.

14.3 BETA-POISSON MODEL

In the beta-Poisson model, we consider in addition to the variability between doses the variability between individual r. Not all hosts respond in a similar way to exposure to a similar dose. Some hosts may have a relatively more severe reaction to exposure to a known dose than the others. Assuming that r follows a beta distribution of parameters α and β, the beta-Poisson model is derived on solving the following integral:

$$P(d_i) = 1 - \int_0^1 \exp(-r \times d_i) f(r) dr = 1 - 1F1(\alpha, \beta, -d_i) \qquad (14.5)$$

where 1F1 is the confluent hypergeometric function. This function plays an important role in distribution theory in multivariate analysis, but from a practical point of view, it is difficult to approximate. Various approaches approximating this confluent hypergeometric function are implemented in the R package, such as the BAS package (Clyde, 2018).

A widely accepted, user-friendly approximate beta-Poisson model formula, however, exists for ($\alpha \ll \beta$ and $\beta \gg 1$):

$$P(d_i) = 1 - \left(1 + \frac{d_i}{\beta}\right)^{-\alpha} \qquad (14.6)$$

A more practical rule of thumb that defines the conditions to ensure that the approximate model (Equation 14.6) is in fact an accurate approximation of the exact beta-Poisson model (Equation 14.5) is $\beta > \sqrt{22\alpha}$ or $0.02 < \alpha < 2$ (Xie et al, 2017).

The beta-Poisson model captures, in addition to the Poisson variation in the ingested amount of hazard, the variation in the response to this dose among the hosts. The latter variation does not only describe the variation between individual hosts but also includes the variation between strains of the microorganism.

14.4 FITTING CHALLENGE TRIALS DATA TO EXPONENTIAL AND BETA-POISSON MODELS

The usual assumption made in the analysis of challenge trials data is that the number of responses y_i is generated from a binomial distribution with an unknown probability of response p_i assumed to be constant across individuals receiving the same dose level d_i and known sample size dose n_i, i.e.

$$y_i \sim Binomial(p_i, n_i) \qquad (14.7)$$

The probability p_i is linked with the dose level d_i using Equation 14.3 or 14.5, respectively, for the exponential and the beta-Poisson model.

The objective of the statistical analysis is to estimate the exponential model or the beta-Poisson parameters. Once the parameters have been estimated, the predicted responses $\widehat{p}_i$ can be obtained using Equation 14.3 or 14.5. To estimate the parameters, r for the exponential model and α and β for beta-Poisson model, we minimize a quantity known as *deviance*:

$$\text{Deviance} = -2 \sum_{i=1}^{k} y_i \ln \frac{y_i}{n_i \widehat{p}_i} + (n_i - y_i) \ln \frac{n_i - y_i}{n_i - n_i \widehat{p}_i} \qquad (14.8)$$

Table 14.1 Dose Levels and Response to the CJN Strain of Human Rotavirus (Ward, et al., 1986)

Dose Level	Response (Infection)	Exposed
9×10^{-3}	0	7
0.09	0	7
0.9	1	7
9	8	11
90	6	7
900	7	8
9000	5	7
9×10^3	3	3

Table 14.2 Estimates and Standard Errors for the Parameters from Both Exponential and Beta-Poisson Models

Parameters	Estimates	Standard Errors	Deviance
Exponential			125.474
$\ln(\widehat{r})$	−6.908	0.215	
Beta-Poisson			6.198
$\ln(\alpha)$	−1.374	0.324	
$\ln(\widehat{\beta})$	−0.852	1.169	

Iterative algorithms are used to find the parameter estimates $(\hat{a}, \hat{\beta})$ that minimize the deviance.

To help demonstrate the method, let us use a set of data on rotavirus challenge trials (Ward, et al., 1986) and the *"mle"* R function (Team, 2018). In this study, the CJN rotavirus strain was fed to healthy 18–45-year-old men with 0.2 g of $NaHCO_3$. The measured outcome was infection (i.e. increased antibody titer to the CJN strain), symptoms and detectable shedding of rotavirus. Eight dose levels were tested in this study (Table 14.1).

The estimates and standard errors for the parameters from both the exponential and beta-Poisson models are shown in Table 14.2.

The beta-Poisson model fits the data better than the Poisson model, because its deviance is lower than the deviance obtained using the Poisson model (Table 14.2). Thus, the beta-Poisson model is the model of choice to best describe the dataset in Table 14.1. From the parameter estimates, the median infectious dose (ID50) can be estimated:

$$\text{exponential model}: D50 = -\frac{\ln(0.5)}{r} = -\frac{\ln(0.5)}{\exp(-6.908)} = 693$$

$$\text{Beta Poisson model ID50} = \hat{\beta}(2^{\frac{1}{\hat{a}}} - 1) = \exp(-0.852)(2^{\frac{1}{\exp(-1.374)}} - 1) = 6.171$$

Once the ID50 is calculated, it is important to assess its statistical confidence. The confidence interval can be constructed using the variance/covariance matrix of the parameter estimates. However, the analytic solution is complex, and the probability distributions of the estimates cannot always be approximated by a normal distribution. One way of dealing with this problem is to use the bootstrap method for calculating the confidence interval. The bootstrap method treats the dataset as a distribution to be sampled from with replacement in order to create a new dataset. The number of infected individuals for each ith dose level is randomly generated from a binomial distribution with number of trials equal to n_i and probability equal to y_i/n_i to obtain a new set of responses:

$$y'_1, y'_2, \ldots, y'_i, \ldots, y'_k$$

The dose–response model is then fitted to the set of responses to estimate the model parameters and retrieve an ID50. This process is repeated I times (I is the number of iterations). The 95% confidence interval is

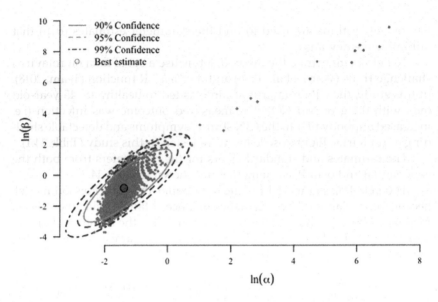

Figure 14.1 Beta-Poisson bootstrap estimates (10,000 iterations – data in Table 14.1).

Table 14.3 Quantiles of Beta-Poisson Model Parameter Estimates (Bootstrap with 10 000 Iterations – Data in Table 14.1)

Estimates	Quantiles						
	0.50%	2.50%	5%	50%	95%	97.50%	99.50%
$\hat{\alpha}$	0.128	0.152	0.163	0.26	0.518	0.635	17.153
$\hat{\beta}$	0.059	0.092	0.116	0.472	2.222	3.194	114.161
$\widehat{ID}_{50}$	1.577	2.159	2.507	6.327	18.787	24.382	42.918

easily calculated by finding the 2.5% and 97.5% quantiles of the different estimates.

Figure 14.1 shows the bootstrap estimates of the two beta-Poisson models fitted to the Table 14.1 dataset. Table 14.3 provides quantiles of the parameter and ID50 estimates. The 95% confidence interval for the ID50 is (2.159,24.382).

Figure 14.2 displays the 95% and 99% confidence bands for the dose–response curve.

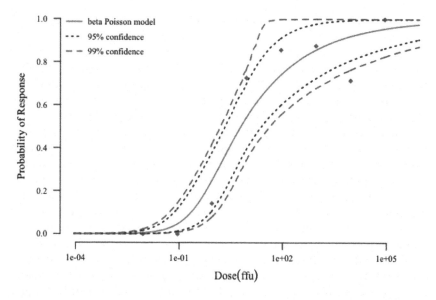

Figure 14.2 Plot of dose–response data with the best fit, bootstrap 95% and 99% confidence bands. Diamonds are observed data from Table 14.1.

As shown with the rotavirus dataset, it is relatively straightforward to derive a dose–response model from challenge trial studies. In this type of study, the number of exposed individuals at predefined dose levels and the number of responses in each group are known and can be controlled by the investigator. However, areas of limitation include the enrollment of volunteers not sufficiently representative of the target population, uncertainty linked to the choice of the dose range, and uncertainty about the variability of virulence between pathogen strains. Host individuals with underlying conditions such as pregnancy, inflammatory diseases, cancer or diabetes, elderly people, AIDS patients or individuals under immunosuppressive therapy are in principle excluded from challenge trials because of the high probability of serious or fatal responses. The choice of the dose level range is difficult to make. Excluding dose levels associated with low or high probabilities of responses may bias the dose–response model parameter estimates or increase their uncertainty. Lastly, in challenge trial studies in general, only one particular strain of pathogen or a mix of pathogen strains are used. Together, these three limitations make the extrapolation of the dose–response model to the exposed population difficult. The two limitations linked with the non-representativeness of

417

the tested host individuals and the pathogen strains can be grouped, because in the beta-Poisson model, we summarize the variability of host individual susceptibility, strain virulence and their interactions with one single probability distribution (the beta distribution). The beta distribution describes the outcome of the interaction between host individual factors and pathogen strain factors, which is the probability for one single unit of the pathogen to survive the physical and biological barriers, reach the infectious target site, and cause infection or disease. This probability is likely to be extremely variable and should be specific for each single exposure occasion (contact between a particular host individual with a particular pathogen strain carried by a particular food). In order to illustrate the absence of representativeness of the enrolled host individuals (volunteers without pre-conditions) and the use of one single pathogen strain with a moderate virulence, let's start with an example of a beta distribution with $\alpha = 3$ and $\beta = 1 \times 10^5$. The absence of representativeness can be simulated with the assumption that the experimental design includes only the lowest 80%, 50% or 20% of the possible values of the beta distribution, specifically the values below the 80th, 50th or 20th percentile (Figure 14.3).

Figure 14.4 shows that for many possible dose–response model estimates, the truncations lead to an underestimation of the probability of infection/illness. The underestimation is more obvious when the truncation is at the 20th percentile of the beta distribution compared with the situation at the 50th and 80th percentiles.

14.5 DOSE-RESPONSE MODEL FROM HUMAN OUTBREAK DATA

Data from human outbreaks can be used to assess dose–response for pathogenic microorganisms. Such data are first used to estimate the dose ingested and the attack rate. In most outbreak investigations, the actual ingested dose is not directly measurable and needs to be modeled because the number of organisms may increase or decrease in the remaining contaminated food between the time of ingestion and the collection of food samples after the onset of clinical signs. For some situations, the implicated food neither supports pathogen growth nor causes rapid decline. So, an average level of contamination can be accurately estimated and combined with the amount of food consumed. However,

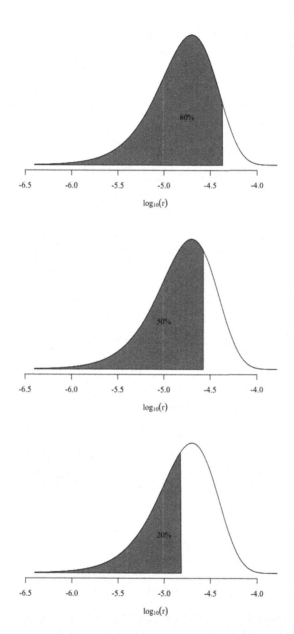

Figure 14.3 Beta distribution with $\alpha = 3$ and $\beta = 1 \times 10^5$ with different truncations. Note that the surfaces are not proportional because the x axis is at log scale.

419

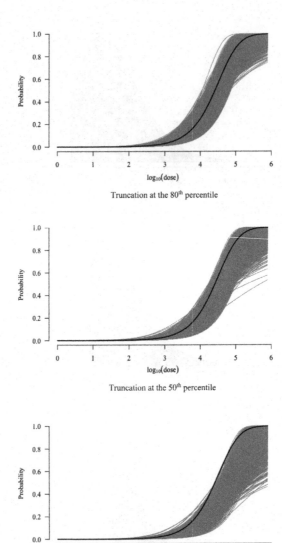

Figure 14.4 Possible dose–response model estimates with different beta distribution truncations. The beta-Poisson model is based on simulations of a challenge trial (10 000 iterations) including 30 host individuals for each of the five dose levels (1 × 10^1, 1 × 10^2, 1 × 10^3, 1 × 10^4 and 1 × 10^5).

the use of an estimated average dose may lead to bias, and further considerations are needed to assess the variability among doses consumed by individuals.

Typically, an effective single outbreak investigation provides two main pieces of information:

- Attack rate (AR): the number of cases (y) divided by the number of exposed persons (n). Case definitions are developed by the investigator to spell out which exposed persons will be included as part of the outbreak. Case definitions may include details about the features of the illness, laboratory analysis, microorganism isolation and typing, certain pathognomonic symptoms, time between exposure and disease/infection onset, etc.
- Dose estimates (d): based on samples taken from left-over foods and amount of food consumed.

In the case of homogeneous contamination of the implicated food, the number of pathogens per quantity of food consumed can be assumed to be distributed as a Poisson with a parameter $\lambda = \bar{c} \times q$, where $\bar{c}$ is the average concentration (e.g. in cfu/g) and q is the amount of consumed food in grams. The quantity q is assumed for the moment to be constant. In such a situation, only the exponential dose–response model can be estimated, because it has a single parameter (r). This parameter can be derived as follows:

$$r = \frac{-\ln(1 - AR)}{\bar{c} \times q} \tag{14.9}$$

In real-world circumstances, the food contamination cannot be assumed to be completely homogeneous as it is in challenge trials studies. The pathogen may be distributed unequally in the implicated food, and the ingested amount of food may also vary between consumers. Figure 14.5 represents as examples four situations of 1000 microorganisms within 100 food units. The distribution patterns in Figure 14.5 were randomly generated using the negative binomial (NB) distribution:

$$f(x) = \frac{\Gamma(x+s)}{\Gamma(s)x!} \left(\frac{s}{\lambda+s} \right)^s \left(\frac{\lambda}{\lambda+s} \right)^x \tag{14.10}$$

where λ and s are the mean and the shape parameter, respectively, and $\Gamma(.)$ is the gamma function. The negative binomial distribution can be expressed as a gamma mixture of Poisson distribution. It can be demonstrated that

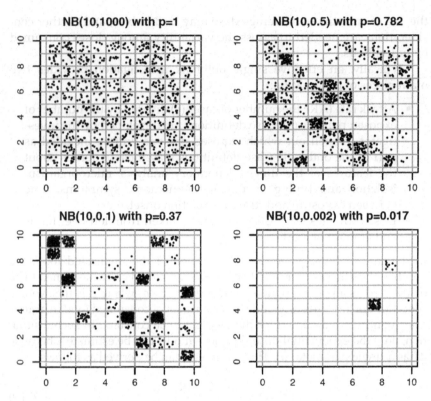

Figure 14.5 Four scenarios of spatial distribution of microorganisms. NB: Negative binomial; p: proportion of units with at least one microorganism.

$E(X) = \lambda$, and $\mathrm{Var}(X) = \lambda + \dfrac{\lambda^2}{s}$, and so $\mathrm{Var}(X) \geq E(X)$. Thus, the shape parameter s controls the amount of clustering. When s tends toward infinity, $\mathrm{Var}(X) = E(X)$, and the negative binomial model converges to a Poisson model that cannot deal with clustering.

If we consider that $f(x)$ is a negative binomial with the parameters λ and s, then the expected number of microorganisms per unit follows a gamma distribution with the same parameters λ and s (Zelterman, 2005). The overall probability of illness assuming an exponential dose–response model is then obtained by

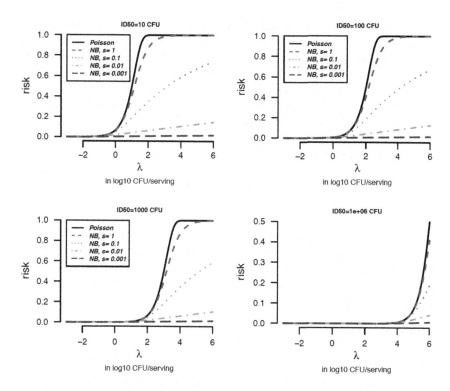

Figure 14.6 Overall risk as a function of the mean and shape of the gamma distribution. Risk: probability of illness given an exposure distributed as a negative binomial (NB[λ, s]).

$$\text{Prob}(\lambda, s) = 1 - \int_0^\infty \exp(-rx) g(x) dx \qquad (14.11)$$

where $g(x)$ is the probability density function of the gamma distribution. The integral is equivalent to the moment generating function of the gamma distribution, $E\big[\exp(tx)\big]$. (Zelterman, 2005) The probability of illness can then be derived directly by

$$\text{Prob}(\lambda, s) = 1 - \left(1 + \frac{r\lambda}{s}\right)^{-s} \qquad (14.12)$$

423

Figure 14.6 shows the impact of the parameters λ and s on the probability of illness. Given the microbial load, the same λ, the probability obtained with negative binomial distribution is in general lower than the one calculated with a Poisson distribution. The more clustered the contamination is, the lower is the overall risk. Although, the average of the microorganism concentrations per serving alone cannot satisfactory predict the probability of illness. However, the probability of illness is not impacted by the shape of the gamma distribution (s) when the average concentration is almost 100 times lower than the ID50,

$$\text{ID50} = \frac{\ln(0.50)}{r}.$$

The parameter r of the exponential model can be estimated from one single outbreak only if λ and s are known.

Heterogeneity in r (the probability of one single pathogen surviving the host barriers) is presented in Section 14.3, represented by a beta distribution with parameters α and β. The integration of the host–pathogen interaction variability in Equation 14.11 leads to:

$$\text{Prob}(\lambda, s, \alpha, \beta) = 1 - \int_{r=0}^{1}\left(1 + \frac{r\lambda}{s}\right)^{-s} h(r, \alpha, \beta)\,dr \qquad (14.13)$$

where $h(r, \alpha, \beta)$ is the beta probability density function.

Equation 14.12 integrates to obtain the following second-order hypergeometric function:

$$\text{Prob}(\lambda, s, \alpha, \beta) = 1 - 1F2\left(\alpha, s, \alpha + \beta; -\frac{\lambda}{s}\right) \qquad (14.14)$$

Naturally, information from one outbreak investigation offers a single data point on a population dose–response curve (the probability of illness given a particular dose). A collection of outbreak datasets is needed to establish more accurately the dose–response model. The combination of multiple outbreaks provides a broader range of doses and individual susceptibility and contributes to better estimates of the two main parameters α and β. Bayesian methods can be used to develop hierarchical models to combine data from different outbreaks (meta-analysis) and provide estimates of α and β by fitting Equation 14.13 simultaneously to data from various outbreaks. The hierarchical framework for estimating the parameters of the dose–response model has been applied to various pathogens (Teunis, et al., 2018; Thebault, et al., 2013; Teunis, et al., 2012; Teunis, et al., 2010; Teunis, et al., 2008).

Although the hierarchical framework for estimating dose–response relationships is very attractive, the results obtained are still difficult to use. Indeed, the large uncertainties about the parameters of the dose–response model, illustrated in the Bayesian approach by the posterior distributions, do not allow precise prediction of the probability of infection or illness. The high uncertainty may be largely attributed to the difficulty of accurately estimating the actual doses ingested by the exposed people and the fact that well-documented outbreaks concern specific groups of host individuals who are not necessary representative of the target population. In practice, the available points of observation (attack rate and dose) may correspond to groups of host individuals of very different susceptibility, which means that the exposure doses can no longer be considered independent of host susceptibility, as is the case in experimental challenge trials. In other words, the combination of data from outbreaks that occurred in different settings such as retirement institutions, primary schools and hospitals will without doubt bias the dose–response assessment. A high attack rate may be observed because of exposure to high doses or because the exposed people belong to a highly susceptible group with severe pre-conditions.

14.6 DOSE–RESPONSE MODEL COMBINING SURVEILLANCE EPIDEMIOLOGICAL DATA AND EXPOSURE DATA

Epidemiological surveillance of foodborne diseases is fundamental to any food safety program. The objectives of foodborne disease surveillance can be summarized as follows:

- Estimate the incidence of disease
- Detect trends
- Identify and monitor high-risk populations
- Detect outbreaks and attribute illnesses to specific sources, including foods, animals, food preparation practices or settings
- Prioritize interventions
- Monitor and evaluate the effectiveness of preventive programs

Table 14.4 Epidemiological Data Used to Assess the Dose–Response Model of (Pouillot, et al., 2015) Based on (Goulet, et al., 2012)

Subpopulation	Number of Individuals in France	Listeriosis Cases in France (2001–2008)	Relative Risk
Below 65 years old, no known underlying condition (i.e. "healthy adult")	48,909,403	189	1
Above 65 years old, no known underlying condition	7,038,068	377	13.9
Pregnancy	774,000	347	116.0
Nonhematological cancer	2,065,000	437	54.8
Hematological cancer	160,000	231	373.6
Renal or liver failure (dialysis, cirrhosis)	284,000	164	149.4
Solid organ transplant	25,300	16	163.7
Inflammatory diseases (rheumatoid arthritis, ulcerative colitis, giant cell arteritis, Crohn's disease)	300,674	68	58.5
HIV/AIDS	120,000	22	47.4
Diabetes (type I or type II)	2,681,000	79	7.6
Heart diseases	1,400,000	29	5.4
Total	63,757,445	1,959	

Thanks to the availability of surveillance data, it becomes possible, for example, to estimate the incidence of human listeriosis according to the underlying conditions (Table 14.4). The data show a large difference in the risk of listeriosis among subpopulations. It is therefore obvious to consider these differences by exploring the dose–response relation by subpopulation. Instead of estimating this relationship for the entire population, the stratification approach is used. It is therefore assumed that the intra-subpopulation variability is inferior to the variability between the strata.

The probability of illness/infection can be derived mathematically as follows:

$$P_g(\lambda) = 1 - \int_0^1 \exp(r\lambda) f_g(r) dr \tag{14.15}$$

where: λ is the expected dose, $P_g(\lambda)$ is the probability of illness for an individual randomly selected from the subpopulation g and exposed to a dose level λ, $f_g(r)$ is the probability density function describing the variability of the parameter r within the subpopulation g.

The probability of illness for an individual selected from the total population can be derived as follows:

$$P(\lambda) = \sum_g P_g(\lambda) h(g) \tag{14.16}$$

where $h(g)$ is the proportion of the subpopulation within the total population.

As we are using the total number of cases observed in each subpopulation, the probability of illness/infection needs to be integrated as follows:

$$P_g = \int_0^\infty P_g(\lambda) f(\lambda) d\lambda \tag{14.17}$$

where $f(\lambda)$ is the probability distribution of the expected dose levels. The latter distribution can be estimated using an exposure assessment model and should be preferably estimated for each subpopulation.

The probability within each subpopulation, P_g, can be estimated using epidemiological surveillance data: incidence in group g divided by the size of the subpopulation g. The integral in Equation 14.17 can be solved numerically using the exposure assessment model and observed cases to derive the unknown parameters of the dose–response model. In other words, we solve Equation 14.17 to assess the parameters of $f_g(r)$ in Equation 14.15.

$f_g(r)$, the distribution of the parameter r, the probability of one single pathogen surviving host barriers, as previously presented, can be a beta distribution with parameter, α_g and β_g. However, the two parameters cannot be estimated using one single observation per subpopulation (annual incidence), so we need to inform the model with data about r variability. One option is to use a lognormal distribution instead of a beta distribution. The advantage of a lognormal distribution is that it allows separation between strain virulence and host susceptibility variabilities. This approach was used to assess a *Listeria monocytogenes* dose–response model (Pouillot, et al., 2015), where $\log(r) \sim Normal(\mu,\sigma)$. The mean ($\mu$) is specific to each of the considered subpopulations (i.e. as in Table 14.4), and the standard deviation (σ) is assumed to be the same for all the population segments. It summarizes

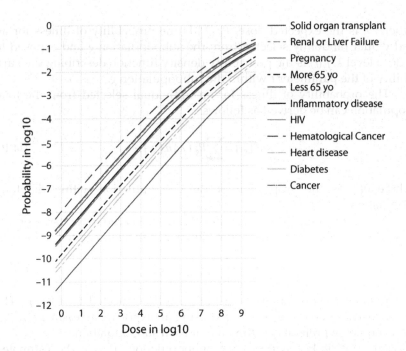

Figure 14.7 Dose–response models (probability of severe listeriosis cases conditional to the exposed dose) for each of the 11 population segments considered in (Pouillot, et al., 2016).

the variability between *L. monocytogenes* strains (σ_{sv}) and host individual susceptibilities (σ_{host}). Of the overall variability of r (σ^2), 12% is attributed to strain variability; the rest is for host individuals' variability within each population segment. The standard deviation was estimated using animal experimental trials and data from the Food and Drug Administration/Food Safety and Inspection Service (FDA/FSIS) risk assessment (www.fda.gov/Food/FoodScienceResearch/RiskSafetyAssessment/ucm183966.htm).

Figure 14.7 shows the marginal dose–response models for each of the 11 subpopulations considered (Pouillot, et al., 2015).

The same approach was used to assess a *L. monocytogenes* dose–response model considering 14 subpopulations defined by age and gender (EFSA, 2018).

14.7 CONCLUSIONS

From a biological viewpoint, the exact beta-Poisson model is perhaps the most realistic dose–response function. However, its functional form uses a confluent hypergeometric function, which is not only difficult to work with but can become numerically intractable for large values of β. Where direct calculation of the exact beta-Poisson function is not possible, one can use a numerical solution. The use of lognormal distribution can be used as an alternative to beta distribution. For the integration of a dose–response model in quantitative risk assessment, we need an accurate and non-biased estimate of the distribution that describes the variability of the parameter r: the probability of one single pathogen surviving host barriers. An unbiased and accurate estimate of the dose–response parameters requires the collection of data covering the whole diversity of host susceptibility and virulence of strains of pathogens and also a wide range of exposure doses.

A good assessment of the dose–response relationship can no longer be based on a single category of data, i.e. experimental, outbreak or epidemiological, but on the integration of data of various natures.

REFERENCES

Clyde, M., 2018. *BAS: Bayesian Variable Selection and Model Averaging Using Bayesian Adaptive Sampling.* s.l.: S.N.

EFSA Panel on Biological Hazards (BIOHAZ), Ricci, A., Allende, A., Bolton, D., Chemaly, M., Davies, R., Fernández Escámez, P.S., Girones, R., Herman, L., Koutsoumanis, K. & Nørrung, B., 2018. *Listeria monocytogenes* contamination of ready-to-eat foods and the risk for human health in the EU. *EFSA Journal*, 16(1), p. e05134.

Goulet, V., Leclercq, A., Laurent, E., King, L.A., Chenal-Francisque, L.J., Vaillant, V., Letort, M.J., Lecuit, M., & de Valk, H. 2012. Surveillance de la listériose humaine en France, 1999–2011. *Bull Epidémiol Hebdom.*

Haas, C. N., 2002. Conditional dose–response relationships for microorganisms: Development and application. *Risk Analysis*, 22(3), pp. 455–463.

Pouillot, R., Hoelzer, K., Chen, Y. & Dennis, S. B., 2015. Listeria monocytogenes dose response revisited—Incorporating adjustments for variability in strain virulence and host susceptibility. *Risk Analysis*, 35(1), pp. 90–108.

Pouillot, R., Klontz, K.C., Chen, Y., Burall, L.S., Macarisin, D., Doyle, M., Bally, K.M., Strain, E., Datta, A.R., Hammack, T.S. & Van Doren, J.M., 2016. Infectious dose of Listeria monocytogenes in outbreak linked to ice cream, United States, 2015. *Emerging Infectious Diseases*, 22(12), p. 2113.

Team, R. C., 2018. *R: A Language and Environment for Statistical Computing.* Vienna: s.n.

Teunis, P. F., Kasuga, F., Fazil, A., Ogden, I.D., Rotariu, O. & Strachan, N.J., 2010. Dose–response modeling of *Salmonella* using outbreak data. *International Journal of Food Microbiology,* 144(2), pp. 243–249.

Teunis, P. F. M., Koningstein, M., Takumi, K. & Van Der Giessen, J. W. B., 2012. Human beings are highly susceptible to low doses of *Trichinella* spp. *Epidemiology & Infection,* 140(2), pp. 210–218.

Teunis, P. F., Marinović, A.B., Tribble, D.R., Porter, C.K. & Swart, A., 2018. Acute illness from *Campylobacter jejuni* may require high doses while infection occurs at low doses. *Epidemics,* 24, pp. 1–20.

Teunis, P. F. M., Ogden, I. D. & Strachan, N. J. C., 2008. Hierarchical dose response of *E. coli* O157: H7 from human outbreaks incorporating heterogeneity in exposure. *Epidemiology & Infection,* 136(6), pp. 761–770.

Thebault, A., et al., 2013. Infectivity of GI and GII noroviruses established from oyster related outbreaks. *Epidemics,* 5(2), pp. 98–110.

Ward, R. L., et al., 1986. Human rotavirus studies in volunteers: Determination of infectious dose and serological response to infection. *The Journal of Infectious Diseases,* 154(5), pp. 871–880.

Xie, G., et al., 2017. Guidelines for use of the approximate beta-poisson dose–response model. *Risk Analysis,* 37(7), pp. 1388–1402.

Zelterman, D., 2005. *Discrete Distributions: Applications in the Health Sciences.* Chichester: John Wiley & Sons.

Section III

Chemical Risk Assessment

15

Quantitative Chemical Risk Assessment Methods

Amélie Crépet

Contents

15.1 INTRODUCTION

Risk assessment has changed considerably since its first steps in the 1980s. There have been significant efforts in terms of funding and human resources to produce representative and suitable data to refine exposure assessments (EFSA/FAO/WHO 2011). Mathematical modelling and computational capacities have also helped to introduce probabilistic methods to account for population behaviour and chemical level variability instead of using mean or maximum point estimates (Voet H and Slob 2007). Tiered exposure and risk assessment strategies have been developed to guide risk assessors in identifying priorities for risk management, taking into account available information and uncertainty (Meek et al. 2011). New data, such as biomonitoring data from human cohorts, require the use of toxicokinetic models to be linked with external exposure assessment and to be interpreted in risk management. Moreover, regulation pressures such as the REACH regulation and risk assessment recommendations have prompted risk assessors to develop high-throughput methods to assess exposure to a wide range of chemicals while accounting for all sources and routes of exposure.

This chapter aims to describe the different steps that are part of a chemical risk assessment as well as the different data and methods used to perform it. New developments and future challenges are also addressed.

15.2 CHEMICAL RISK ASSESSMENT

Chemical risk assessment is a scientifically based process that aims to quantify the risk of an adverse health effect resulting from exposure of humans to chemicals present in food or the environment over a specified period. It is generally associated with the identification of risk factors and the measurement of their impact on the risk in order to propose management options. Risk assessment is initiated by risk managers, who will decide whether to apply the proposed management options. As a result, risk assessment is preceded by a first step describing the purpose of the risk assessment and the safety questions to be answered, setting time schedules and providing the resources necessary to carry out the work. The risk assessment is divided into four main steps defined by the Codex Alimentarius (Commission 2003) and the United States National Research Council (National Research Council 1983).

15.2.1 Hazard Identification

This first step is the identification of chemical hazards that could have a negative effect on health. Depending on the various definitions, the hazard could be a chemical with the potential to cause an adverse health effect (FAO/WHO 2004) or a property associated with a chemical rather than the chemical itself (WHO/IPCS 2004). Hazard identification describes the nature and the type of effects that could be caused by the hazard and identifies the affected target organs or target tissues as well as the circumstances under which the effects may be expressed. It must be based on accurate data to evaluate the weight of evidence for adverse health effects (EFSA 2017a). Hazard identification can be done in a prospective way, using *in vivo* and *in vitro* toxicological tests, or in a retrospective way by reporting incidents or by examining epidemiological surveys.

15.2.2 Hazard Characterization

Hazard characterization consists of quantitatively defining, when possible, or alternatively qualitatively defining the relationship between the dose and the probability of the occurrence of an effect and/or its seriousness. Combined with the previous one, this step forms the hazard assessment part of the risk assessment.

There are two types of effects: (1) effects with a threshold under which there is no observed effect and (2) effects without a threshold (mutagenic, genotoxic and carcinogenic effects). For the latter, the effect is observed continuously with a greater or lesser probability or severity depending on the dose. When the toxic effect is assumed to have a threshold, the purpose of hazard characterization is to establish health-based guidance values, when the available data make this possible. For example, this threshold value is called an acceptable daily intake (ADI) for additives and pesticide residues or a tolerable intake (TI) for contaminants. These threshold values are established by international organizations (WHO, EFSA, JECFA, JMPR, etc.) or national agencies (US-EPA, RIVM, Health Canada, ANSES, etc.). Exposure to chemicals below such health-based guidance values is generally considered to be safe, and

435

conversely, exposure above them is generally thought of as unsafe. These values are derived from points of departure such as the no observed adverse effect level (NOEAL), the lowest observed adverse effect level (LOAEL) or the benchmark dose (BMD), applying uncertainty factors for possible inter-species and/or inter-individual variability. The BMD is used when a percentage effect or a probability of effect in a given population is considered. This approach is based on modelling of experimental data using a dose–response curve (EFSA 2009). Generally, it is the lower bound (BMDL) of the confidence interval of the BMD that is used to set up the health-based guidance value. Most of the time, these points of departure for a given compound are estimated from toxicological studies conducted in animals. When available, retrospective analyses of human data from epidemiological surveys can be used, integrating the evaluation of inter-individual variability. In this case, no inter-species variability factor is needed. With modern toxicology and efforts to reduce *in vivo* studies, researchers began to use *in vitro* testing to estimate points of departure, applying *in vitro* to *in vivo* extrapolation methods (IVIVE). However, these methods are adversely affected by the current lack of defined validation protocols and by the complexity of the *in vitro* to *in vivo* extrapolation process (Zgheib et al. 2017). Other uncertainty factors than those used for inter-species and/or inter-individual variability can also be applied: an uncertainty factor for the use of the LOAEL, for the extrapolation from subchronic to chronic exposure and for low data quality and/or quantity. Health-based guidance values are specific to a substance on the basis of the administered doses, a duration, a route of exposure (inhalation, ingestion and cutaneous contact), a population and a type of effect. To account for all exposure routes, it is possible to set internal health-based guidance values using absorption factors and/or PBTK models (see Sections 15.4.2 and 15.4.3).

15.2.3 Exposure Assessment

Exposure assessment consists of estimating the dose of a chemical to which an organism or a population may be exposed. The organism or the population can be exposed to a chemical from different exposure sources/media (diet, air, dust, consumer products, veterinary usages, etc.) and via different routes (ingestion, inhalation and dermal contact). Exposure, also called intake, is generally estimated by combining different types of data depending on the sources of exposure. In the case of food chemicals, dietary exposure assessment takes into consideration the consumed quantities of foods that may contain the chemical, and the amount and frequency of the chemical in those foods. Usually, a range of intake or exposure estimates are provided (e.g. for average consumers and for high consumers), and estimates may be broken down by subgroup of the population (e.g. infants, children or adults). Considering the chemical and the studied effect, exposure can be "acute" and thus calculated over a short period of time using a meal or a day, or "chronic" with an estimate over the entire lifespan. For chronic exposure, the mean values of consumed quantities and of concentrations are used, whereas for acute exposure, the consumed quantity over a meal or a day is combined with the highest observed concentration in the media.

Generally, different scenarios are studied considering different subpopulations, times of exposure, and assumptions regarding censored data and calculations.

15.2.4 Risk Characterization

The risk is defined as the probability of occurrence of known and potential adverse health effects in a given organism and/or population. Risk characterization consists in combining the hazard characterization step with the exposure assessment step in order to propose suitable advice for decision-making in risk management. For example, a comparison of the relative risks obtained with different risk management options is useful for risk managers. In practice, quantitative risk assessment consists in comparing the health-based guidance values with the estimated levels of exposure under different exposure scenarios. In this way, the risk in a population can be estimated by determining the number of individuals with a higher exposure level than the health-based guidance value divided by the total number of individuals in the population. It is possible to focus on consumers only, divided by the total number of consumers or users. Other indicators such as the hazard index (exposure divided by the health-based guidance value) or the margin of exposure (MOE, exposure divided by the NOEAL) have been developed and are commonly used to characterize the risk.

The risk assessment should include all key assumptions, the different sources and types of uncertainty and should describe the nature, relevance and magnitude of any risk to human health. It should also include, where relevant, information on susceptible sub-populations, including those with greater potential exposure or those with specific predisposing physiological conditions or genetic factors.

15.3 RISK ASSESSMENT METHODS

15.3.1 Conceptual Model of Risk Assessment

A conceptual model is a written description and visual representation of predicted relationships between the organism or the population and the hazard to which they may be exposed. It aims to represent the sequence of events that can lead to risk and to evaluate this risk qualitatively or quantitatively (Figure 15.1). A risk assessment model is composed of input variables and output variables.

The input variables of a risk model for chemical exposure are those related to:

- The medium/media (e.g. dust, soil, diet) where the chemical(s) can be found, expressed as the frequency and the level of the chemical(s) in these media.
- The behaviours of the organism or the population, for example food consumption quantity when dietary exposure is considered,

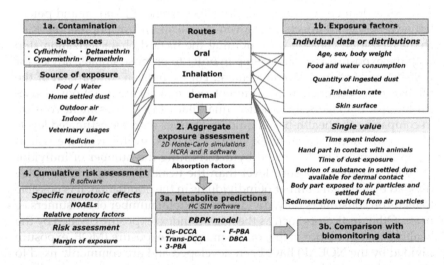

Figure 15.1 Conceptual model of general population exposure to four pyrethroids (cyfluthrin, deltamethrin, cypermethrin and permethrin). The aggregate exposure to each pesticide was estimated from different sources and cumulated using relative potency factors to calculate the margin of exposure with a BMDL value. Also, internal exposure was estimated using a PBTK model of five metabolites and compared with biomonitoring data. (From the article by Vanacker, M. et al., 2020.)

or the applied quantity of cream for cosmetic exposure assessment, or time spent outdoors and indoors when considering air exposure.

- The physiological characteristics of the individuals in the population, which are needed to estimate the exposure, for example body weight, individual size or respiratory volume.
- The health-based guidance value, if it is available, or a chosen point of departure and its corresponding uncertainty factors. Dose–response curve modelling can also be used in some cases, such as in food allergies.

The output variables are external and/or internal exposure and the associated risk. The model can become more complex, depending on the information available, by adding different sources and routes of exposure, using toxicokinetic models, and other input variables that can influence the risk. The conceptual model can be reproduced considering different exposure scenarios and sub-populations.

15.3.2 Deterministic Approach

A deterministic risk assessment consists in setting point values for the input variables of the risk assessment model. The output variables of the model are then also expressed as a point value. The values used can be the average or the median, but most often they are extreme values (the 95th, 97.5th or 99th quantiles or the maximum) and are chosen in order to propose "worst case" scenarios. The use of extreme scenarios is called the conservative approach, and the result is considered a maximized risk. This approach requires few data and technical means. As a result, it is used in data-poor situations or as a first-line approach to study whether in the "worst case", the risk can be ruled out or not. In the case where the risk cannot be ruled out, a so-called probabilistic approach is generally proposed in order to produce more precise estimates of the risk. If, on the other hand, the risk is zero with this conservative approach, it is often considered unnecessary to perform more calculations. Before concluding, however, special attention must be paid to the fact that the conservative scenario incorporates all possible uncertainties. For example, if the uncertainty related to the variability of consumption is high, then the extreme value used for this variable can be undervalued and thus skew the associated risk. This is why it is important to adopt an uncertainty analysis approach, as explained in Chapters 7 and 16.

15.3.3 Probabilistic Approach

The probabilistic approach involves the use of the set of possible values that the input variables of the risk assessment model can take. The variables are then described by probability distributions determined from observed data or from expert reports if the data are lacking (elicitation of experts' statements).

There are different approaches to determining and using the distribution of a variable from observed data. The most intuitive approach is to use the empirical distribution of data, i.e. the distribution is defined only by the observed values. In this case, it is not necessary to define beforehand the form of the distribution. In contrast, the parametric approach consists in defining a form of the distribution by the use of known probability distributions, such as the Gaussian, exponential, lognormal or Weibull distribution, etc. In this case, it is necessary to determine the values of the parameters of the distribution from the observed data. Two approaches are then possible: the frequentist approach, based solely on the maximization of the likelihood from the observed data, and Bayesian inference, which takes into account both the likelihood and also *a priori* information on the parameters, which may come from expert reports or results of previous studies. In the case of Bayesian inference, the parameters also take the form of a distribution. Once the probability distributions are defined, they are integrated into the model using Monte Carlo simulation methods. These methods involve randomly drawing multiple values in the different probability distributions of the variables. Thus, the variability and uncertainty are propagated throughout the model and make it possible to produce output variables that also take the form of probability distributions. Variability and uncertainty can be propagated separately using two-dimensional (2-D) Monte Carlo simulations. The combination of 2-D Monte Carlo with Bayesian inference as proposed in Rimbaud et al. (2010) for food allergies easily allows this separation.

Intermediate methods between the deterministic and probabilistic approaches have been developed. For example, intervals around a quantitative value or quantitative uncertainty tables are ways of integrating variability and uncertainty without going as far as a probabilistic approach. It should also be noted that using a probabilistic approach for all input variables is often not possible. In this case, a deterministic approach for some variables and a probabilistic approach for others are often combined (as proposed in Figure 15.1).

15.3.4 Tiered Approach

The tiered approach is a widely used approach in risk assessment. Its objective is to organize scientific knowledge in a situation of uncertainty in order to help the decision-maker in the implementation of management actions. This approach is of an iterative nature and can be defined as the setting up of a series of questions whose answers require the mobilization of more and more precise knowledge in terms of both hazard and exposure. It therefore provides a graduated level of response that will depend on: basic knowledge (how far is it possible to go?), the aspects that the manager needs in order to act (do we know enough to act?), the constraints of implementation in terms of means (human, financial, etc.) and a scientific approach proportionate to the needs of the question asked.

In risk assessment, the tiered approach allows gradual recognition of uncertainty. WHO/IPCS (2008), EFSA (2017b) and Meek et al. (2011) recommended a tiered approach at three levels (qualitative, deterministic and probabilistic). At the first level, simple and fast methods based on conservative hypotheses (overestimation of risk) are favoured. The purpose of these methods is to help the decision-maker to identify priority issues that require either specific regulation, data collection or specific research work. In other words, this first level draws attention to worrying situations. In this case, higher levels are implemented. These levels rely on the use of mathematical models, which are increasingly complex and/or more and more probabilistic.

15.3.5 Uncertainty Analysis

It is increasingly recommended that any risk assessment should be accompanied by an uncertainty analysis (NRC 2013, WHO/IPCS 2008, 2014, EFSA 2017b); see Chapters 7 and 16 for more details. The purpose of the uncertainty analysis is to identify and quantify the different sources of uncertainty associated with the risk being studied. The risk is then expressed as a function of its associated degree of uncertainty, thereby increasing the reliability and robustness of the result. An uncertainty analysis is broken down into several stages:

- Identify and describe the uncertainties. These two steps consist in listing all the uncertainties about the risk being studied. The sources of uncertainty are related to the incompleteness of knowledge, the representativeness and quality of data concerning the population studied and the question asked, the tools and methods of data collection, the level of complexity, the adaptability of the chosen model, the studied scenarios, etc. Tools such as uncertainty matrices make it possible to classify the various sources of uncertainty according to their dimensions and characteristics such as their origin (lack of knowledge, variability, etc.), and their location (data, model, etc.). It is not possible to address all identified uncertainties in the following steps. In general, the most important uncertainties in relation to the question asked will be considered in the assessment steps.

- Evaluate uncertainties and their impact on risk. The step for evaluating the uncertainties consists in estimating the range of the possible values of the various uncertainties relating to the problem posed. In the best case, the evaluation is carried out using data, which can come from the literature or other studies, or by referring to experts who will be able to describe the amplitude of the uncertainties. Probability distributions, whose parameters have been determined on the basis of expert data or statements, or the empirical distribution of the observed values, are then used to quantify the uncertainties. In the case where there are no data or experts who can be consulted, qualitative tools such as verbal adjectives or ordinal scales can be used to characterize their amplitude. Once the uncertainties have been assessed, they are generally propagated using Monte Carlo simulations throughout the various stages of the model in order to evaluate their impact on risk. It is important to simultaneously vary all the uncertainties to take into account their dependence on the final result. It is also beneficial to vary only one source of uncertainty at a time and to set others so as to assess the individual impact of the various uncertainties. Indeed, this step makes it possible to identify the most important sources of uncertainty and thus, to decide to put in place measures to reduce them.
- Represent and communicate uncertainties. This step consists in communicating the results of the uncertainty analysis to the managers in the form of graphs or tables in order to take into account the uncertainties relating to the risk assessment when making the decision.

15.4 CURRENT AND FUTURE CHALLENGES IN CHEMICAL RISK ASSESSMENT

15.4.1 Mixture Risk Assessment

Through their environment and diet, populations are not exposed to a single chemical but to multiple chemicals, which can interact and potentially cause diseases. Due to the complexity of chemical mixtures, the associated risk is difficult to characterize. Over the past few decades, many efforts have been made to propose concepts, methods, guidance and applications for the risk assessment of mixtures (Boobis et al. 2008, EFSA 2007, 2008, WHO/IPCS 2009, Fox, Brewer, and Martin 2017). Some of these methods are summarized here.

15.4.1.1 How to Group Substances in a Mixture

Regarding the multitude of possible combinations, the question of which substances should be assessed together remains a major challenge. One solution is to perform risk assessments for chemicals belonging to the same chemical family or considering the same target organ or type of effect. In this way, EFSA proposed a hazard-wise method based on a "common adverse outcome" to group pesticides into "cumulative assessment groups" (CAGs) (Nielsen et al. 2012, EFSA 2014). Four levels of criteria for grouping were defined, with each higher level being more refined: target organs (level 1), specific phenotypic effects (level 2), mode of action (level 3) and mechanism of action (level 4). Currently, level 1 and 2 CAGs, restricted to pesticides, have been identified in the nervous system and the thyroid (EFSA 2019a,b). Comprehensive preliminary work has been done on effects on the liver, the adrenal glands and the eyes, and on the developmental and reproductive systems (EFSA 2012, RIVM, ICPS, and ANSES 2016). However, grouping substances into a certain CAG may be based on a small number of observations, thereby introducing uncertainties regarding CAG membership and relative potency in comparison to other substances in a CAG. The mode and mechanism of action are also unknown for many pesticides, and thus, there is a strong need to study them. Moreover, as a certain CAG can contain a high number of components, it is necessary to prioritize the substances to be assessed in mixture experiments. Finally, the main disadvantage of this approach is that it does not account for the reality of the mixtures. In recent years, developments have been proposed to extract from combined exposures the most relevant mixtures to be studied as a priority for toxic effects (Crépet et al. 2013).

These methods used statistical tools related to dimension reduction and classification. Crépet and Tressou (2011) used a Bayesian non-parametric model to determine the major mixtures, classifying the population regarding their exposure profiles, and then studied correlations between pesticides. More recently, Béchaux et al. (2013) and Traoré et al. (2016) demonstrated the ability of the combination of non-negative matrix factorization (NMF, Lee and Seung (2001) with a hierarchical clustering to identify principal mixtures connected with specific diets. The NMF and clustering methods have also been used to define dietary patterns and clusters of individual diets by Zetlaoui et al. (2011), Sy et al. (2013) and Gazan et al. (2016). A modified version of the NMF method, called sparse non-negative matrix under-approximation (SNMU) (Gillis and Plemmons 2013), was applied with success by Traoré et al. (2018) to exposures of pregnant women.

Another approach is to combine exposure and hazard information to identify the most relevant mixtures of chemicals belonging to any CAG to which populations are chronically and acutely exposed. It starts from the list of substances in a defined CAG and reduces this list by using risk-based identification of co-occurring substances in the diet for a given time frame. Here, the exposures are expressed in similar toxicity terms using relative potency factors calculated from the NOEALs of the studied effect (NOEAL of each substance divided by the NOEAL of a reference substance). The method was developed in the Euromix project (No. 633172, H2020-SFS-2014-2), which aims to develop a strategy for mixture risk assessment. An example of the CAG for liver steatosis using exposure data to pesticides from several European countries is presented in Crépet et al. (2019).

15.4.1.2 How to Assess the Risk of a Chemical Mixture

Two approaches are currently proposed (EFSA 2019c): the whole mixture approach and the component-based approach. In the whole mixture approach, the mixture is treated as a single entity, similar to a single chemical, and so requires dose–response information for the mixture of concern or a sufficiently similar substance using read across methods. The whole mixture approach is particularly required with mixtures whose composition is unknown or difficult to characterize. It is also useful to apply this approach when the mixture is well known and stable over time, such as commercial formulations. The main limitation is that the results are specific to the mixture or a sufficiently similar one and cannot be used for other mixtures.

The component-based approach can be applied when the mixture components are detailed. This is defined as an approach in which the risk of a mixture is assessed based on the exposure and effect data of its individual components (EFSA 2013a). Application of the component-based approach therefore requires exposure and effect data on the individual mixture components.

Three types of mixture effect are distinguished:

- Dose or concentration addition for substances with similar mode of action: the mixture effect can be predicted by the sum of each single substance exposure present in the mixture, weighted by their toxic potency.
- Response addition for substances with independent modes of action. In this case, the statistical concept developed by Bliss (1939) can be used to predict the effect.
- Interaction, which is observed when the mixture effect is higher (synergism or potentiation) or lower (antagonism) than the sum of each single substance effect.

However, data on the mode and mechanism of action of chemical substances are sparse. The analyses reported in the literature comparing the combined effect under independent action and similar action concluded that the dose addition approach correctly estimates the combined effects of substances with independent action and even tends to overestimate them (EFSA 2013b, Kortenkamp, Backhaus, and Faust 2009). Moreover, regarding exposure dose levels, EFSA considers that the probability of interaction is low. Therefore, in practice, dose addition is the default hypothesis recommended by EFSA to assess the risk of mixtures, in accordance with WHO/IPCS (2009).

Several risk indicators are available to estimate the risk of a mixture under the dose addition assumption (EFSA 2018). Some of these are described in the following.

- The hazard index (HI) is defined as the sum of the hazard quotients of the individual components of an assessment group, in which each of the hazard quotients is calculated as the ratio between exposure to a chemical and the respective health-based guidance values (i.e. ADI, TI).

- The target organ toxicity dose (TTD) is a refined hazard index approach focusing on only one specific effect/target organ. Thus, for this specific effect/target organ, each endpoint of the mixture component associated with the uncertainty factor is used to calculate the hazard quotients.
- The margin of exposure (MOE) is the ratio between the point of departure of a reference substance and the cumulative exposure, weighted by the relative potency factors or toxic equivalent factors (WHO/IPCS 1998, US-EPA 1989). The MOE is then compared with uncertainty factors.

15.4.2 Aggregate Exposure and Biomonitoring Data

The concept of aggregate exposure is commonly used to define exposure to a chemical through the different sources where it may be found (food, water, air and consumer products) and the different routes by which it might enter the organism (ingestion, inhalation and dermal contact). This concept makes it possible to determine a complete estimate of the chemical dose that can enter an organism.

Two approaches are commonly used to quantify aggregate exposure and exposure more generally:

- An estimate of the quantities of substances that enter the body via the combination of measurements for the contamination from the different exposure media considered and exposure models, and integrating, for different scenarios, exposure parameters relating to the behaviour of individuals (space-time budget, food consumption, anthropomorphic data, etc.)
- An estimate of internal doses through the search and quantification of biomarkers of exposure in biological media (blood, urine, hair, etc.)

Measuring exposure via the use of biomarkers has the advantages of directly integrating the different exposure routes and knowing the concentration of the substance in the body. However, due to the cost and difficulty of the process, biomonitoring data and internal health-based guidance values are scarce. Moreover, biomonitoring data do not allow us to evaluate the contribution of exposure sources. In fact, to determine management options to mitigate exposure and the associated risk, the identification of exposure factors that contribute the most to total exposure is crucial. However, the modelling of aggregate exposure has many complexities. Aside from specific cases (Cao et al. 2016), it is rare that surveys collect all exposure sources for the same individual. Therefore, performing aggregate exposure often requires the mathematical modelling of data from different databases with different populations and methodologies. Aggregate exposure models combine data on the concentrations of chemicals in various media (e.g. soil, water or food) and data on individual behaviour (e.g. food consumption and time spent in an area). Depending on the chemical, different pathways of exposure can be considered. For adult humans, the most common exposure pathways are inhalation of outdoor or indoor air, or of house dust, and ingestion of food and drinking water. Exposure resulting from skin contact can also be important in some cases. For children, ingestion of dust and soil due to hand-to-mouth activities may also contribute to exposure to certain chemicals. The general principle of the mathematical models used for aggregate exposure is to create a new population from the individuals of the different surveys with Monte Carlo simulations, which are often used to combine data in risk assessment (Kennedy, van der Voet, et al. 2015, Safford et al. 2015, Paustenbach 2000, Kennedy, Butler Ellis, and Miller 2012, Kennedy, Glass, et al. 2015, Zartarian et al. 2017, Vanacker, Tressou, et al. 2020). Monte Carlo simulations make it possible to draw random samples from distributions of datasets in order to reconstruct the different sources of exposure of each individual. These methods, accounting for the variability between individuals and the uncertainty, provide a more realistic estimate of aggregate exposure to individuals across a population (Paustenbach 2000). General principles for performing aggregate exposure from heterogeneous surveys are presented in Vanacker, Tressou, et al. (2020). A number of tools have been developed in recent years to evaluate exposure via different routes and sources. The SHEDS-high-throughput tool combines direct dermal, inhalation and accidental ingestion with the ingestion of food and drinking water (Isaacs et al. 2014). In Europe, Kennedy, Butler Ellis, and Miller (2012) developed the Bystander and Residential Exposure Assessment Model (BREAM) to evaluate non-dietary exposure to pesticides for workers, bystanders and residents living near agricultural areas (Kennedy, van der Voet, et al. 2015, Kennedy, Glass, et al. 2015). In the context of the Euromix project, they proposed options to perform aggregate exposure models combining BREAM with the MCRA (Monte Carlo Risk Assessment) platform (van der Voet et al. 2015), which evaluate dietary exposure. Other European models assessing exposures from cosmetics (Dudzina et al. 2015), air and dust for the general population were linked to dietary exposure through the MCRA platform.

15.4.3 Integrating Toxicokinetic Models
in Chemical Risk Assessment

The toxicokinetic (TK) models can be defined as "mathematical descriptions simulating the relationship between external exposure level and chemical concentration in biological matrices over time" (WHO/IPCS 2010). TK models describe the ADME process, which is the absorption, distribution in the body, metabolism and excretion of chemicals and their metabolites. They vary in level of complexity and can be classified into two main categories: (1) data based (non-compartmental or compartmental) models and (2) predictive physiologically based toxicokinetic (PBTK) models.

TK models can be used in chemical risk assessment for:

- *In vitro* to *in vivo* extrapolation (IVIVE): IVIVE refers to the qualitative or quantitative transposition of experimental results or observations made *in vitro* to predict phenomena in biological organisms. The use of PBTK modelling makes it possible to improve the quantitative assessment by introducing the ADME processes into *in vitro* experiments (Quignot N 2014, Hamon J 2015).
- Reverse dosimetry to estimate external environmental exposure from measured tissue concentrations (biomarkers of internal exposure) (Lyons MA 2008, Bernillon P 2000). In recent years, advances in analytical methods have enabled the measurement of more chemicals, at lower concentrations and using smaller samples of blood or urine. As a result, biomonitoring has become more widely used in public health research and risk assessment (Choi et al. 2015). As highlighted in Albertini et al. (2006), biomonitoring data has many advantages for exposure assessment. First, it integrates exposure from all sources and routes, and provides important information for risk assessment in the case of aggregated exposure. Second, it takes into account the accumulation of the chemical in the body due to successive or long-term external exposure. Third, the collection of serial biomonitoring samples over an extended period of time can provide information regarding variability and trends in exposure. Fourth, it represents direct measurements of the dose of the chemical substance that is actually taken up from the environment, i.e. the internal dose, and therefore integrates bioavailability (the fraction of exposure dose that really enters the body) (Angerer, Ewers, and Wilhelm 2007).

- Integrating biomonitoring data to inform about past exposure. For example, (Béchaux 2014) used biomonitoring data to correct the exposure estimated with actual data over time. The European project HBM4EU (H2020 No. 733032) aims to develop methods and tools to make it possible to use biomonitoring data in risk assessment. In particular, it proposes tools based on toxicokinetic modelling to link external exposure estimated from consumer behaviour and contamination to internal doses.
- Dynamic exposure modelling to account for accumulation of chemicals over time. As previously, for chemicals with a long half-life, the burden accumulated in the body can be completely different from the dose exposure estimated from dietary intake at a fixed time or considering mean intake over the lifespan. In this case, it is necessary to integrate the TK of the chemical in the dietary exposure assessment (Verger, Tressou, and Clémençon 2007).
- Aggregate exposure from different sources and routes (cf. Section 15.4.2): in modelling the absorption process and the different organs, the PBTK models make it possible to estimate exposure from different routes and to compare it with biomonitoring results when these are available (Vanacker, Quindroit, et al. 2020).

15.5 EPIDEMIOLOGY AND CHEMICAL RISK ASSESSMENT

Samet, Schnatter, and Gibb (1998) in their paper on "Epidemiology and risk assessment" describe how epidemiological data may play a prominent role in the different steps of risk assessment. Epidemiology can be used in hazard identification and characterization. Indeed, with toxicological data, epidemiological findings can be part of the relevant lines of evidence integrated to identify hazards. Epidemiological findings can also be used to quantitatively establish the dose–response relationship and to estimate exposure. Their strength compared with toxicological data is that they come from humans, and they are in most cases directly in the exposure range of interest. Until today, toxicological surveys were often preferred in risk assessment due to the cost of epidemiological surveys, the difficulty of accurately estimating real exposure in the context of combined exposures and the fact that only a sample of the targeted population is surveyed. However, epidemiological surveys are useful to rate mathematical models and the relevance of the dataset used to estimate the risk in the risk characterization process. These epidemiological findings can be used to guide toxicological surveys to investigate the mode and the mechanism of action of a hazard. Conversely, quantitative risk assessment can help to plan epidemiological surveys to verify the adverse effects in real life.

15.6 CONCLUSION

The quantitative risk assessment of chemicals requires an integrated approach that combines hazard characterization and exposure assessment. The reliability of the results from the risk assessment is based on the representativeness and robustness of the data as well as the relevance of the model used. Risk analysis therefore depends on the state of current knowledge. This is why it is necessary to continue efforts to collect and organize data for efficient risk assessment. In addition, accompanying any risk assessment with an uncertainty analysis can identify important sources of uncertainty for which studies and tools are needed. It is also important to continue to develop methodological tools and a risk assessment approach at the national and international levels, especially to account for chemical mixtures and aggregate exposure. Multidisciplinary work between physicians, nutritionists, analysts, biostatisticians, risk assessors and economists in the framework of national and European integrative projects is the key to providing relevant assessments and management tools.

ACRONYMS

ADI	acceptable daily intake
ADME	absorption, distribution, metabolism and excretion
ANSES	French Agency for Food, Environmental and Occupational Health and Safety
BMD	benchmark dose
BREAM	Bystander and Residential Exposure Assessment Model
CAG	cumulative assessment group
EFSA	European Food Safety Authority
FAO	Food and Agriculture Organization of the United Nations
HI	hazard index
IPCS	International Programme on Chemical Safety
IVIVE	*in vitro* to *in vivo* extrapolation

JECFA	Joint Expert Committee on Food Additives
JMPR	Joint Meeting of Pesticide Residues
LOAEL	lowest observed adverse effect level
MCRA	Monte Carlo risk assessment
MOE	margin of exposure
NMF	non-negative matrix factorization
NOAEL	no observed adverse effect level
PBTK	physiologically based toxicokinetics
REACH	Registration, Evaluation, Authorisation and Restriction of Chemicals
RIVM	National Institute for Public Health and the Environment
SNMU	sparse non-negative matrix under-approximation
TI	tolerable intake
TK	toxicokinetics
TTD	target organ toxicity dose
US-EPA	United States Environmental Protection Agency
WHO	World Health Organization

REFERENCES

Albertini, R., M. Bird, N. Doerrer, L. Needham, S. Robison, L. Sheldon, and H. Zenick. 2006. "The use of biomonitoring data in exposure and human health risk assessments." *Environmental Health Perspectives* 114(11):1755–1762.

Angerer, J., U. Ewers, and M. Wilhelm. 2007. "Human biomonitoring: State of the art." *International Journal of Hygiene and Environmental Health* 210(3–4): (201–228).

Béchaux, C., A. Crépet, and S. Clémençon. 2014. "Improving dietary exposure models by imputing biomonitoring data through ABC methods." *The International Journal of Biostatistics* 10(2):277–287.

Béchaux, C., M. Zetlaoui, J. Tressou, J.C. Leblanc, F. Héraud, and A. Crépet. 2013. "Identification of pesticide mixtures and connection between combined exposure and diet." *Food and Chemical Toxicology* 59:191–198. doi: 10.1016/j.fct.2013.06.006.

Bernillon, P., and F.Y. Bois. 2000. "Statistical issues in toxicokinetic modeling: A Bayesian perspective." *Environmental Health Perspectives* 108 Supplement 5:883–893.

Bliss, C.I. 1939. "The toxicity of poisons applied jointly." *Annals of Applied Biology* 26(3):585–615.

Boobis, A.R., B.C. Ossendorp, U. Banasiak, Paul Y. Hamey, I. Sebestyen, and A. Moretto. 2008. "Cumulative risk assessment of pesticide residues in food." *Toxicology Letters* 180(2):137–150. doi: 10.1016/j.toxlet.2008.06.004.

Cao, S., X. Duan, X. Zhao, Y. Chen, B. Wang, C. Sun, B. Zheng, and F. Wei. 2016. "Health risks of children's cumulative and aggregative exposure to metals and metalloids in a typical urban environment in China." *Chemosphere* 147:404–411. doi: 10.1016/j.chemosphere.2015.12.134.

Choi, J., T.A. Mørck, A. Polcher, L.E. Knudsen, and E. Joas. 2015. "Review of the state of the art of human biomonitoring for chemical substances and its application to human exposure assessment for food." *EFSA Journal* support publication EN-724.

Commission, Codex Alimentarius. 2003. *Principles and Guidelines for the Conduct of Microbiological Risk Assessment*. Roma: FAO.

Crépet, A., and J. Tressou. 2011. "Bayesian nonparametric model for clustering individual co-exposure to pesticides found in the French diet." *Bayesian Analysis* 6(1):127–144. doi: 10.1214/11-BA604.

Crépet, A., M. Vanacker, C. Sprong, W. de Boer, U. Blaznik, M. Kennedy, C. Anagnostopoulos, D.L. Christodoulou, J. Ruprich, I. Rehurkova, J.L. Domingo, B.H. Jensen, F. Metruccio, A. Moretto, L. Jacxsens, P. Spanoghe, D. Senaeve, H. van der Voet, and J. van Klaveren. 2019. "Selecting mixtures on the basis of dietary exposure and hazard data: Application to pesticide exposure in the European population in relation to steatosis effect." *International Journal of Hygiene and Environmental Health* 222(2):291.

Crépet, A., Jessica Tressou, V. Graillot, Camille Béchaux, S. Pierlot, Fanny Héraud, and Jean-Charles Leblanc. 2013. "Identification of the main pesticide residue mixtures to which the French population is exposed." *Environmental Research* 126:125–133. doi: 10.1016/j.envres.2013.03.008.

Dudzina, Tatsiana, Christiaan J.E. Delmaar, Jacqueline W.H. Biesterbos, Martine I. Bakker, Bas G.H. Bokkers, Paul T.J. Scheepers, Jacqueline G.M. van Engelen, Konrad Hungerbuehler, and Natalie von Goetz. 2015. "The probabilistic aggregate consumer exposure model (PACEM): Validation and comparison to a lower-tier assessment for the cyclic siloxane D5." *Environment International* 79:8–16. doi: 10.1016/j.envint.2015.03.006.

EFSA. 2007. "Scientific colloquium: Cumulative risk assessment of pesticides to human health: The way forward." In *Summary Report*. 27–28 November 2006. Parma, Italy.

EFSA. 2008. "Scientific opinion of the panel on plant protection products and their residues (PPR Panel) on a request from the EFSA evaluate the suitability of existing methodologies and, if appropriate, the identification of new approaches to assess cumulative and synergistic risks from pesticides to human health with a view to set MRLs for those pesticides in the frame of Regulation (EC) 396/2005." *EFSA Journal* 704:84.

EFSA. 2009. "Use of the benchmark dose approach in risk assessment." *EFSA Journal* 15(1):4658.

EFSA. 2012. "Guidance on the use of probabilistic methodology for modelling dietary exposure to pesticide residues." *EFSA Journal* 10(2839):95.

EFSA. 2013a. "Scientifc opinion on the relevance of dissimilar mode of action and its appropriate application for cumulative risk assessment of pesticides residues in food." *EFSA Journal* 11(3472):40.

EFSA. 2013b. "Scientific opinion of the panel on plant protection products and their residues (PPR) on the identification of pesticides to be included in cumulative assessment groups on the basis of their toxicological profile." *EFSA Journal* 11(3293):131. doi: 10.2903/j.efsa.2013.3293.

EFSA. 2014. "Scientific opinion on the identification of pesticides to be included in cumulative assessment groups on the basis of their toxicological profile." *EFSA Journal* 11(3293):131.

EFSA. 2017a. "Guidance on the use of the weight of evidence approach in scientific assessments." *EFSA Journal* 15(8):4971. doi: 10.2903/j.efsa.2017.4971.

EFSA. 2017b. "Guidance on uncertainty in EFSA scientific assessment." *EFSA Journal* 16:5123.

EFSA. 2018. "Draft guidance on harmonised methodologies for human 1 health, animal health and ecological risk assessment of combined exposure to multiple chemicals." *EFSA Journal*.

EFSA. 2019a. "Establishment of cumulative assessment groups of pesticides for their effects on the thyroid." *EFSA Journal* 17(9):5801.

EFSA. 2019b. "Establishment of cumulative assessment groups of pesticides for their effects on the nervous system." *EFSA Journal* 17(9):5800.

EFSA. 2019c. "Guidance on harmonised metholgies for human health, animal health and ecological risk assessment of combined exposure to multiple chemicals." *EFSA Journal* 17(13):5634.

EFSA/FAO/WHO. 2011. "Towards a harmonised total diet study approach: A guidance document." *EFSA Journal* 9(11):2450.

FAO/WHO. 2004. *Codex Alimentarius Commission Procedural Manual*, 14th ed., edited by Food and Agriculture Organization of the United Nations. Rome: Codex Alimentarius Commission.

Fox, M.A., L.E. Brewer, and L. Martin. 2017. "An overview of literature topics related to current concepts, methods, tools and applications for cumulative risk assessment (2007–2016)." *International Journal of Environmental Research and Public Health* 14(4):389. doi: 10.3390/ijerph14040389.

Gazan, R., C. Béchaux, A. Crépet, V. Sirot, P. Drouillet-Pinard, C. Dubuisson, and S. Havard. 2016. "Dietary patterns in the French adult population: A study from the second French national cross-sectional dietary survey (INCA2) (2006–2007)." *British Journal of Nutrition* 116(2):300–315. doi: 10.1017/S0007114516001549.

Gillis, Nicolas, and Robert J. Plemmons. 2013. "Sparse nonnegative matrix underapproximation and its application to hyperspectral image analysis." *Linear Algebra and its Applications* 438(10):3991–4007. doi: 10.1016/j.laa.2012.04.033.

Hamon, J., M. Renner, M. Jamei, A. Lukas, A. Kopp-Schneider, and F.Y. Bois. 2015. "Quantitative in vitro to in vivo extrapolation of tissues toxicity." *Toxicology in Vitro* 30(1):203–216.

Isaacs, K.K., G.W. Glen, P. Egeghy, M.-R. Goldsmith, L. Smith, D. Vallero, R. Brooks, C.M. Grulke, and Halûk Özkaynak. 2014. "SHEDS-HT: An integrated probabilistic exposure model for prioritizing exposures to chemicals with near-field and dietary sources." *Environmental Science and Technology* 48(21):12750–12759. doi: 10.1021/es502513w.

Kennedy, Marc C., M. Clare Butler Ellis, and Paul C.H. Miller. 2012. "Bream: A probabilistic bystander and resident exposure assessment model of spray drift from an agricultural boom sprayer." *Computers and Electronics in Agriculture* 88:63–71. doi: 10.1016/j.compag.2012.07.004.

Kennedy, Marc C., Richard C. Glass, Bas Bokkers, Andy D.M. Hart, Paul Y. Hamey, Johannes W. Kruisselbrink, Waldo J. De Boer, Hilko van der Voet, David G. Garthwaithe, and Jacob D. van Klaveren. 2015. "A European model and case studies for aggregate exposure assessment of pesticides." *Food and Chemical Toxicology* 79:32–44. doi: 10.1016/j.fct.2014.09.009.

Kennedy, Marc C., Hilko van der Voet, Victoria Roelofs, Willem Roelofs, Richard C. Glass, Waldo J. De Boer, Johannes W. Kruisselbrink, and Andy D.M. Hart. 2015. "New approaches to uncertainty analysis for use in aggregate and cumulative risk assessment of pesticides." *Food and Chemical Toxicology* 79:54–64. doi: 10.1016/j.fct.2015.02.008.

Kortenkamp, A., T. Backhaus, and M. Faust. 2009. *State of the Art Report on Mixture Toxicity: Final Report of a Project on Mixture Toxicology and Ecotoxicology Commissioned by the European Commission*. Brussels: European Commission.

Lee, D.D., and H.S. Seung. 2001. "Algorithms for non-negative matrix factorization." In M. Press (Ed.), *Proceedings of the 13th International Conference on Neural Information Processing Systems* (Vol. 13, pp. 535–541). Denver, CO.

Lyons, M.A., R.S.H. Yang, A.N. Mayeno, and B. Reisfeld. 2008. "Computational toxicology of chloroform: Reverse dosimetry using Bayesian inference, Markov chain Monte Carlo simulation, and human biomonitoring data." *Environmental Health Perspectives* 116(8):1040–1046.

Meek, M.E., A.R. Boobis, K.M. Crofton, G. Heinemeyer, M. Van Raaij, and C. Vickers. 2011. "Risk assessment of combined exposure to multiple chemicals: A WHO/IPCS framework." *Regulatory Toxicology and Pharmacology* 60(2, Supplement):S1–S14. doi: 10.1016/j.yrtph.2011.03.010.

National Research Council. 1983. "Committee in the institutional means for assessment of risks for public health." *Risk Assessment in the Federal Government: Managing the Process*. Washington, DC: National Academy Press.

Nielsen, Elsa, Pia Norhede, Julie Boberg, Louise Krag Isling, Stine Kroghsbo, Niels Hadrup, Lea Bredsdorff, Alicja Mortensen, and John Christian Larsen. 2012. "Identification of cumulative assessment groups of pesticides." *EFSA Supporting Publications* 9(4):303.

NRC, US. 2013. "Guidance on the treatment of uncertainties associated with PRAs in risk-informed decision making." *Federal Register* 78(72):22349.

Paustenbach, D.J. 2000. "The practice of exposure assessment: A state-of-the-art review." *Journal of Toxicology and Environmental Health, Part B Critical Reviews* 3(3):179–291. doi: 10.1080/10937400050045264.

Quignot, N., J. Hamon, and F.Y. Bois. 2014. "Extrapolating in vitro results to predict human toxicity." In: *In Vitro Toxicology Systems. Methods in Pharmacology and Toxicology,* edited by Jennings, P., and Bal-Price, A., 531–550. New York: Springer Science.

Rimbaud, L., F. Héraud, S. Laveille, J.Ch. Leblanc, and A. Crépet. 2010. "Quantitative risk assessment relating to adventitious presence of allergens in food: A probabilistic model applied to peanut in chocolate." *Risk Analysis* 30(1):19.

RIVM, ICPS, and ANSES. 2016. "Toxicological data collection and analysis to support grouping of pesticide active substances for cumulative risk assessment of effects on the nervous system, liver, adrenal, eye, reproduction and development and thyroid system (GP/EFSA/PRAS/2013/02)." *EFSA Supporting Publications* 13(2):184. doi: 10.2903/sp.efsa.2016.EN-999.

Safford, B., A.M. Api, C. Barratt, D. Comiskey, E.J. Daly, G. Ellis, C. McNamara, C. O'Mahony, S. Robison, B. Smith, R. Thomas, and S. Tozer. 2015. "Use of an aggregate exposure model to estimate consumer exposure to fragrance ingredients in personal care and cosmetic products." *Regulatory Toxicology and Pharmacology : RTP* 72(3):673–682. doi: 10.1016/j.yrtph.2015.05.017.

Samet, J.M., R. Schnatter, and H. Gibb. 1998. "Epidemiology and risk assessment." *American Journal of Epidemiology* 148(10):929–936.

Sy, M.M., M. Feinberg, P. Verger, T. Barré, S. Clémençon, and A. Crépet. 2013. "New approach for the assessment of cluster diets." *Food and Chemical Toxicology* 52:180–187. doi: 10.1016/j.fct.2012.11.005.

Traoré, T., C. Béchaux, V. Sirot, and A. Crépet. 2016. "To which chemical mixtures is the French population exposed? Mixture identification from the second French Total Diet Study." *Food and Chemical Toxicology* 98(B):179–188. doi: 10.1016/j.fct.2016.10.028.

Traoré, Thiéma, Anne Forhan, Véronique Sirot, Manik Kadawathagedara, Barbara Heude, Marion Hulin, Blandine de Lauzon-Guillain, Jérémie Botton, Marie Aline Charles, and Amélie Crépet. 2018. "To which mixtures are French pregnant women mainly exposed? A combination of the second French total diet study with the EDEN and ELFE cohort studies." *Food and Chemical Toxicology* 111 (Supplement C):310–328. doi: 10.1016/j.fct.2017.11.016.

US-EPA. 1989. *Interim Procedures for Estimating Risks Associated with Exposures to Mixtures of Chlorinated Dibenzo-p-Dioxins and Dibenzofurans (CDDs and CDFs) and 1989 Update.* Washington, DC: U.S. Environmental Protection Agency, Risk Assessment Forum.

van der Voet, Hilko, Waldo J. de Boer, Johannes W. Kruisselbrink, Paul W. Goedhart, Gerie W.A.M. van der Heijden, Marc C. Kennedy, Polly E. Boon, and Jacob D. van Klaveren. 2015. "The MCRA model for probabilistic single-compound and cumulative risk assessment of pesticides." *Food and Chemical Toxicology* 79:5–12. doi: 10.1016/j.fct.2014.10.014.

Vanacker, M., P. Quindroit, C. Brochot, C. Mandin, P. Glorennec, and A. Crépet. 2020. "Cumulative risk assessment to pyrethroids for neurotoxic effect through aggregate exposure and comparison of metabolite concentrations for the French adult population." *Food and Chemical Toxicology* 143:111519.

Vanacker, M., J. Tressou, G. Perouel, P. Glorennec, and A. Crépet. 2020. "Combining data from heterogeneous surveys for aggregate exposure: Application to children's exposure to lead in France." *Environmental Research* 182:109069.

Verger, P., J. Tressou, and S. Clémençon. 2007. "Integration of time as a description parameter in risk characterisation: Application to methyl mercury." *Regulatory Toxicology and Pharmacology* 49(1):25–30.

Voet, H., van der, and W. Slob. 2007. "Integration of probabilistic exposure assessment and probabilistic hazard characterization." *Risk Analysis* 27(2):351–371.

WHO/IPCS. 1998. "Selected non-heterocyclic polycyclic aromatic hydrocarbons." Geneva: World Health Organization.

WHO/IPCS. 2004. *Risk Assessment Terminology.* Geneva: World Health Organization.

WHO/IPCS. 2008. *Uncertainty and Data Quality in Exposure Assessment.* Geneva: World Health Organization.

WHO/IPCS. 2009. *Assessment of Combined Exposures to Multiple Chemicals: Report of a WHO/IPCS International Workshop on Aggregate/Cumulative Risk Assessment.* Geneva: World Health Organization.

WHO/IPCS. 2010. *Characterization and Application of Physiologically Based Pharmacokinetic Models in Risk Assessment.* Geneva: World Health Organization.

WHO/IPCS. 2014. *Guidance Document on Evaluating and Expressing Uncertainty in Hazard Characterization.* Geneva: World Health Organization.

Zartarian, Valerie, Jianping Xue, Rogelio Tornero-Velez, and James Brown. 2017. "Children's lead exposure: A multimedia modeling analysis to guide public health decision-making." *Environmental Health Perspectives* 125(9):097009. doi: 10.1289/ehp1605.

Zetlaoui, M., M. Feinberg, P. Verger, and S. Clémençon. 2011. "Extraction of food consumption systems by nonnegative matrix factorization (NMF) for the assessment of food choices." *Biometrics* 67(4):1647–1658. doi: 10.1111/j.1541-0420.2011.01588.x.

Zgheib, E., C. Béchaux, A. Crépet, E. Mombelli, and F.Y. Bois. 2017. "High-throughput methods for toxicology and health risk assessment." *Environment Risque Santé* 16(1):44–58.

16

Uncertainty Analysis in Chemical Risk Assessment

Sandrine Fraize-Frontier, Chris Roth

Contents

16.1 INTRODUCTION

Uncertainty is an intrinsic property of risk assessment and concerns each step of the assessment process. In principle, the presence of uncertainties precludes neither risk assessment nor risk management decision-making, but it does affect the result of the assessment. Moreover, nowadays, risk assessments are subject to thorough scrutiny by scientists, decision-makers, stakeholders and the public in general (Abt et al. 2010). This increased attention reflects the growing public concern for health safety issues. It is driven not only by increased media coverage but also because inappropriate measures – insufficiently preventive, late or disproportionate – can lead, and have led, to national or even global health crises, like the asbestos disaster, the diethylstilbestrol story and the "mad cow disease" crisis, to name but a few (EEA 2001, 2013). Lessons learned from past crises have led to improved procedures for more accurate and robust risk assessments that meet the needs of decision-makers. These improvements aim to ensure that risk assessments make the best use of the best available scientific data and address uncertainty in a comprehensive and systematic way (Aven 2016). As such, uncertainty analysis is not an aim in itself but a means and an integral part of well-conducted risk assessments (NRC 2009).

Uncertainty analysis is carried out throughout all stages of the risk assessment process to identify and describe the different sources of uncertainty in the process and hence, to better understand how they impact the overall risk estimation (Van der Sluijs et al. 2003b). The goal of uncertainty analysis, by specifying all uncertainties, is to increase the transparency of risk assessments and provide the most complete information for the decision-making process. For example, only partially identifying or assessing the uncertainty is equivalent to underestimating the real uncertainty and hence overestimating the reliability of the assessment results. Uncertainty analysis is thus necessary to assess the level of confidence in the results (ANSES 2016d). Moreover, through the identification of uncertainty sources, initial methodological choices may be reconsidered and revised to improve the accuracy or robustness of the final assessment results. In such circumstances, uncertainty analysis is used as a productive means to refine the risk assessment (Verdonck et al. 2007). Uncertainty analysis can also be used as a means for further risk characterization refinement, as the main contributors to the overall uncertainty indicate which additional data should be collected or which type of research should be carried out to most refine results (Verdonck et al. 2007, ANSES 2016d). Finally, to allow a correct appreciation of risk assessment results, it is essential that the risk estimate is reported together with the results of the uncertainty analysis, both being expressed in a transparent and understandable way (EFSA 2016, ANSES 2016c). In particular, expressing the results of the uncertainty analysis in a way that is not clear or useful for decision-making may lead to inadequate risk management measures (Abt et al. 2010, ANSES 2016d).

Thus, in a risk assessment context, the aim of any uncertainty analysis is to identify, describe, then quantify or qualify and finally, communicate all the uncertainties associated with the results of the risk estimation (ANSES 2016c).

16.2 DEFINITION OF UNCERTAINTY

To estimate a health risk, risk assessors rely on and use the available scientific knowledge. This knowledge – in the form of qualitative or quantitative data or information – may relate to the health effects, the contents in various products or matrices, the concentrations emitted, the doses of exposure, etc. (NRC 2009).

Whatever the amount of available data or information and their quality, scientific knowledge has, by definition, limits and shortcomings. The data or information may, for example, be inaccurate, have measurement errors, not be completely representative with respect to the population or the exposure of interest, etc. (Walker et al. 2003). In this context, uncertainty refers to limits or shortcomings in the scientific knowledge that is available to risk assessors during the time and given the resources – in terms of methods, tools and skills – allocated to the risk assessment (ANSES 2016c).

16.3 CLASSIFICATION OF UNCERTAINTY SOURCES

In the risk assessment process, uncertainty may occur for different reasons; for example, a lack or a limit of knowledge about the exposure pathways of interest, the target population under consideration, the exposure scenario(s) on which the assessment is based, the models used and/or the data used, etc. To provide the best estimate of the risk, the full range of uncertainty has to be captured. This necessarily involves a comprehensive inventory of the different uncertainty sources (Spiegelhalter and Riesch 2011). The inventory process must be both transparent and reproducible.

For this, it is essential to use a standardized classification of uncertainty sources that corresponds to a generic checklist of the types of uncertainty sources. The use of such a checklist facilitates identifying all sources of uncertainty encountered during a risk assessment, and their individual differences, since they may need to be treated in different ways (Petersen et al. 2013). For example, probability theory is the best-known and the most widely used formalism for quantifying uncertainty. However while it may be an appropriate means of expressing some kinds of uncertainty, it is not for some others (Aven et al. 2014).

In the last 20 years, there have been several attempts to provide a generic classification for sources of uncertainty in risk assessment. The earliest developments mainly focused on uncertainty in quantitative data (emissions, dose–response, concentration, etc.) that feed risk assessment models or are used to estimate model parameters (Cullen and Frey 1999). Examples of such uncertainty include the analytical limits encountered when determining substance concentrations in food, water, etc. or the limits of pathogen amplification methods for microbiological hazard, which are usually based on models calibrated on historical data with no guarantee that the relationship still holds today. To produce relevant results in such cases, a risk assessor has to extrapolate and make further untestable assumptions. These assumptions– for example, those relating to "worst-case" situations – are typically sources of uncertainty that must be identified and discussed (Verdonck et al. 2007). Beyond the uncertainty in quantitative data, other types of uncertainty may arise from selection processes themselves (choice of scenario, model or data, etc.). Finally, more recently, the need to also define sources of uncertainty related to the assessment context, like the decision-making context, has been emphasized (U.S. EPA 2014a, Morgan et al. 2009, EFSA 2015).

Different classifications have been proposed (ANSES 2016c, IPCS 2008, van der Sluijs et al. 2003a, IOM 2013, Hayes 2011). Some of them have been developed for a specific field (e.g. climate change, chemical risks in health-environment, etc.), while others focus on a specific stage of the risk assessment process. All the proposed classifications are, however, consistent with the overall logic of the risk assessment process and include the three following broad classes of uncertainties: *Context, Method* and *Communication*.

- The *Context* class covers sources of uncertainty that relate to the scope of the assessment. Scoping an assessment implies making choices that are themselves sources of uncertainty. Some of these choices – such as the question asked, the decision-making context or the resources allocated, including the time for instruction – are made when framing the risk assessment and can be attributed to the petitioner. For example, the question asked by the petitioner may be equivocal or too broad to be fully dealt with in the allotted time or, on the contrary, too focused on one aspect to allow a proper understanding of the problem that arises. Some other contextual sources of uncertainty can be attributed to the risk assessor; for example, when reformulating the petitioner's question restricts or more generally redefines the target population or the exposure to be assessed. Current methodological guidance documents recommend distinguishing between these two subclasses (framing and reformulation), as they require different types of action to reduce uncertainties (U.S. EPA 2014a, EFSA 2015).

- The *Method* class encompasses all technical sources of uncertainty, which generally arise from three principal origins, the data, scenarios and models used for the assessment, and in particular, their representativeness with regard to the question asked; that is, whether or not the data, scenarios and models used allow the exposure and the risk of the target population to be estimated in a robust and precise manner for adequate decision-making. Thus, uncertainty analysis that explicitly characterizes and separately deals with these different aspects of the overall methodological uncertainty can lead to a more open and rational decision process. Hence, this class is generally divided into three subclasses: *data, scenarios* and *models*.
  - For *data*, the data search and selection processes, when not correctly structured to identify and retain the most appropriate data for the question being addressed, are potential sources of uncertainty. Then, the sources of uncertainty related to the methodological limitations of the data used for the evaluation (limitations in study design, in sampling or analytical methods, in questionnaires, etc.) need to be cited. Finally, the way the data is used, especially how the inherent variability of the phenomenon studied is taken into account, may also be a source of uncertainty.
  - For the *scenarios*, the possible sources of uncertainty can be revealed by the PECOTS statement, which specifies the target Population of the risk assessment (P), the Exposure to be assessed (E), the Comparator, that is, the control population or reference exposure (C), the Outcome of interest (O), the Timing of the exposure (T) and the Setting of interest (S)

461

(see Box 16.1) specific to the risk assessment question asked. Defining a PECOTS statement is strongly recommended when planning a risk assessment (see Section 16.3.1). Naturally, only those parts of the PECOTS statement that are relevant to the question asked need to be specified. An inaccurate or erroneous statement definition would be a source of uncertainty.

- For the *models*, the most often cited sources of uncertainty are those related to the mathematical equations used: do they take into account the main factors that determine the phenomenon studied (for example, exposure)? Are the possible interactions and correlations between these factors considered? Finally, as for the *data* sub-class, the model search and selection processes, if not correctly structured, can also represent a potential source of uncertainty.

- The *Communication* class focuses on uncertainties caused by the way in which the risk assessment is reported. Risk assessors have both a professional and an ethical responsibility not only to present the risk assessment results but also to clearly state the framework in which the assessment was made and limitations of their work. Indeed, an erroneous or non-exhaustive description of the assessment process, the assumptions made, the data used, and the results and conclusions may be a source of uncertainty. This may lead to a misunderstanding and hence, misuse of the results and conclusions and consequently, to inappropriate risk management decisions. When communicating a risk assessment or analyzing the uncertainties of published risk assessments (reports, articles, etc.), it is important to address the following questions. Is the *approach* or assessment process fully documented? Are the *assumptions and data* on which the assessment based precisely explained? Are the *results* and the uncertainty around them well expressed? Do the *conclusions* of the assessment take into account the uncertainty of the results?

The complete proposed generic classification of types of uncertainty sources is schematically presented in Figure 16.1.

16.4 UNCERTAINTY ANALYSIS APPROACH

A considerable variety of methodological approaches to performing uncertainty analysis in risk assessment have been developed and discussed in the literature. Most were developed and/or recommended by

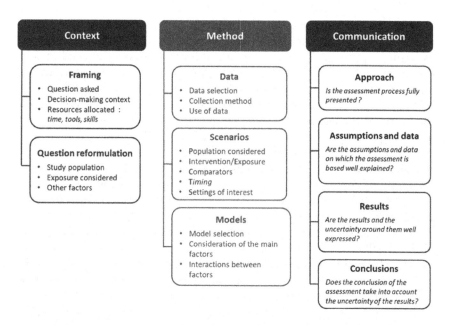

Figure 16.1 Generic classification of types of uncertainty sources in risk assessment. (Adapted from Anses, *Évaluation du poids des preuves à l'Anses : Revue critique de la littérature et recommandations à l'étape d'identification des dangers*, ANSES, Maisons-Alfort, 2016.)

regulators, health and safety agencies, and other institutions involved in risk assessment at the international level (ANSES 2016d, c, ECHA 2012, EFSA 2016, NRC 2009, IPCS 2008, 2014).

The proposed approaches are quite diverse; they may be qualitative and/or quantitative, iterative, or more or less directive. This diversity results mainly from the context and objectives behind their development. For example, the French Agency for Food, Environmental and Occupational Health & Safety (ANSES), the European Chemicals Agency (ECHA) and the European Food Safety Authority (EFSA) developed methodological approaches adapted to their own respective areas of expertise, along with the regulatory or the decision-making context of their work and their working methods (technical standards that serve as a framework for their activities, code of ethics, regulatory requirements, national and international guidelines with which they have to comply, etc.).

The different approaches all present a number of common features. On a higher level, all opt for an iterative approach, where the uncertainty analysis guides the design and refinement of the risk assessment by fixing the most appropriate level of precision needed for each step of the risk assessment process. Then, on a practical level, all stress the need to start the uncertainty analysis with a preliminary planning phase. Since uncertainty analysis is an integral part of the risk assessment itself, it has to be planned as part of the assessment. The purpose of this phase is to define the scope of the assessment and the overall process to be implemented, including that of the uncertainty analysis (U.S. EPA 2014a).

Second, all methodological approaches agree on the importance of a flexible working framework that enables the risk assessor to select the methods that most appropriately fit the purpose of their (individual) assessment (EFSA 2016). A broad range of methods and tools are currently available to carry out uncertainty analysis, with large differences in the conceptual approach, the computational effort required and the power of the results. When conducting a risk assessment, it may be difficult to decide which of them are most relevant for the problem at hand. Clearly, no one is always the best; the choice depends on different factors, such as the nature of the problem and the resources available (time, skills, tools, etc.) (EFSA 2007). Moreover, not all uncertainties can be assessed using the same methods. As a rule, uncertainties related to choices or assumptions (choice of scenarios, models, extrapolations, etc.) are assessed using qualitative methods, whereas quantitative methods are preferred for assessing the uncertainties surrounding the data. Finally, even when the nature of the data is suitable, the data available at the time of the assessment does not always allow the use of elaborate quantitative methods (ANSES 2016d). To conduct an uncertainty analysis that fits a purpose, the majority of the published guidance documents recommend proceeding in a pragmatic iterative way, refining if necessary initial analysis choices and progressively moving towards more sophisticated methods according to the specific needs of the risk assessment (ANSES 2016c).

Lastly, all current methodological approaches recommend a step-wise method for comprehensive uncertainty analysis – namely, to identify, describe, quantify and communicate the uncertainty surrounding the results of the risk assessment. While the number of steps and their focus may differ from one approach to another, the following five steps are generally always included: Identify and describe the sources of uncertainty (Step 1), Assess individual uncertainties (Step 2), Assess combined uncertainty (Step 3), Prioritize or rank sources of uncertainty (Step 4) and Communicate uncertainty analysis results (Step 5) (ANSES 2016c, EFSA 2016) – see Figure 16.2.

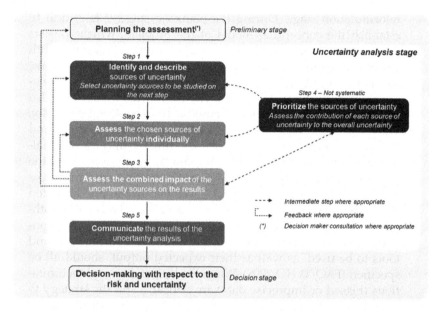

Figure 16.2 A five-step iterative approach for uncertainty analysis in risk assessment. (Adapted from Anses, *Évaluation du poids des preuves à l'Anses : Revue critique de la littérature et recommandations à l'étape d'identification des dangers*, ANSES, Maisons-Alfort, 2016.)

16.4.1 Preliminary Step: Planning the Uncertainty Analysis

Planning a risk assessment, including the embedded uncertainty analysis, is an iterative process consisting of three interdependent stages: *Framing, Question reformulation* and *Choice of the assessment method*:

- *Framing* is necessary to make explicit the context of the assessment and in particular, to define what the assessment results will be used for, the time and resources allocated, and all other elements that may influence the choice of the assessment method. When necessary, the elements in the *framing* stage can be adjusted iteratively according to the choices made in the two following stages (IPCS 2014).
- *Question reformulation* is used to clarify the question to be addressed. Any possible ambiguities are removed and, if necessary, the question rephrased. It is recommended to define the PECOTS parameters (see Box 16.1) as part of the question

465

reformulation stage. During the planning stage, it is critical to establish the conceptual model of the problem, which consists of identifying and describing the main factors relating to health effects and exposure (pathogen, sources, pathways, populations, etc.) to be considered during the assessment, as well as any relationships between these factors. The conceptual model will also help to define the assessment method to be used (U.S. EPA 2014b).

- The *Choice of the assessment method* stage defines the methods and tools used for the risk assessment, including the uncertainty analysis. The methods and tools selected should be consistent with the elements of the *framing* stage, especially the time and resources available. The conceptual model developed in the previous step should be amended to explicitly include every choice or hypothesis on which the assessment is based (target populations, exposure scenario, etc.). The data, information, methods, models and tools to be used, as well as their expected output, should all be specified (FAO/WHO 2009). When the lack of data or data limitations (biased or imprecise data) are critical, a specific strategy to collect new data via specific research work may be considered.

BOX 16.1   PECOTS PARAMETERS

To address questions about health effects of exposure in the context of a health risk assessment, the following elements have to be considered: the target population (Population), their exposure (Exposure), the reference exposure or control population (Comparator), the health effect of interest (Outcome), the time scale of the exposure (Timings) and the settings of the exposure (Settings of interest) (Counsell 1997, Rooney et al. 2014).

The acronym PECOTS represents these different elements that should be defined when planning and scoping the assessment process (ANSES 2016a). The key questions to be answered are the following (Counsell 1997, Whitlock et al. 2010, EFSA 2010, NTP 2015):

- **Population** – Which population does the assessment focus on? Is it the general population? Is it a subpopulation (e.g. workers, pregnant women, infants/children)? Are there any population characteristics that may influence the considered health effect occurrence (e.g. age, sex)?

- **Exposure** – What is the pathogen/agent of concern? What are the sources and media, the routes and the pathways of exposure that contribute most to an individual's exposure?
- **Comparator** – Which population/exposure is defined as control population/reference exposure? Is this a population unexposed to the pathogen/agent of concern? Is this a population exposed to a particular dose of the pathogen/agent?
- **Outcome** – What indicator is chosen to characterize the health effect of interest? Is it the occurrence of a disease, for example, lung cancer? Is it mortality, for example by cancer? Is it the decrease of a quantity, such as the number of red blood cells? Is it an increase in quantity, such as an increase in liver weight?
- **Timings** – What is the time scale of the exposure? Is it an acute, chronic or periodic exposure? What time interval should be considered? What is the timing of exposure (frequency and duration)?
- **Settings of interest** – Where are the population and/or exposure set (e.g. France, factory, transport, home)?

For a given risk assessment, only those key questions of the PECOTS statement that are relevant to the question asked need to be answered. The others will be left blank.

16.4.2 Step 1: Identification and Description of Uncertainty Sources

This step involves identifying and describing all sources of uncertainty so as to determine those that should be assessed in depth, in the next step.

To identify all sources of uncertainty that arise in a given risk assessment, it is essential to use a generic classification of types of uncertainty sources, such as that presented in Section 16.2. Used as a checklist, such a classification allows a structured, reproducible and exhaustive identification of uncertainty sources (Petersen et al. 2013). Each identified source is described narratively and then characterized according to a list of descriptors. The list of descriptors used varies from one risk assessment to another according to the specific needs of each assessment. Some descriptors, such as the *Magnitude* (i.e. the degree of uncertainty the source induces) or the *Direction* (according to whether an uncertainty source will lead to "overestimated" or "underestimated" results of the risk assessment), are often used to select the sources of uncertainty to be further analyzed (IPCS 2008). The *Nature* of the sources of uncertainty (for example, an "inadequate consideration of variability" or a simple "limit of knowledge") (Walker et al. 2003) and the *Qualification of basic knowledge* (for example, a problem of "precision or inaccuracy", "lack of observation or measures" or "ignorance") (Craye, Funtowicz, and van der Sluijs 2005) are also among the most commonly used descriptors.

For the sake of convenience, this information is collected and reported in a table. The identified sources of uncertainty are listed as the rows of the table according to the structure of the classification used, the description and characteristics (descriptors) being reported in the columns of the table (Knol et al. 2009). This table format is particularly convenient for presenting information on uncertainties in a simple, concise and transparent way (see Table 16.1) (ANSES 2016c). Other columns are added to the table to specify those sources of uncertainty that are retained for the next step of the process and the justification for this choice. Most often, the choice is made based on the magnitude and/or direction; for example, sources of uncertainty with a high magnitude and that lead to an underestimation of the results of the risk assessment.

16.4.3 Step 2: Individual Assessment of Uncertainties

The present step focuses on those sources of uncertainty identified at the end of the previous step for thorough assessment. The aim here is to evaluate the range of possible values of each uncertainty source. However, it is important, first and foremost, to distinguish uncertainties about quantities (concentration measurements, quantity of food consumed, frequency or duration of use of a product, etc.) from uncertainties about methodological choices or assumptions (choice of scenarios or models, extrapolations, etc.), since they are not usually treated in the same way.

Several options are available to characterize uncertainty about a quantity: a numerical description of possible values, the use of an ordinal scale or at least a descriptive text. A numerical solution is recommended wherever possible, as it is the most informative and facilitates the comparability of the results. When there is a large amount of relevant data, the traditional frequentist statistical approach is usually applied; the uncertainty (i.e. the set of possible values) is characterized using confidence intervals or a probability distribution. When limits in data preclude the use of conventional statistical techniques, subjective probabilistic estimates can be obtained using elicitation of expert knowledge (O'Hagan et al. 2006).

Table 16.1 Table of Uncertainties of the Health Risk Assessment Related to the Use of Plastic Toys for Children under 3 Years Old

Sources of Uncertainty	Description	Treatment during the Assessment	Impact on the Risk
Context – Question reformulation			
Substances to be assessed (ATBC, DEHTP, DINCH, TXIB, DOIP)	Selection based on analysis of literature and measure data	Not applicable	Not applicable
Method – Data, Scenarios, Models			
Hazard Identification			
DOIP – No toxicological reference value for DOIP	No data	Risk assessment of DOIP impossible	Not applicable
TXIB – No toxicological reference value for TXIB	Few toxicological data available	Margin of exposure approach. Selection of a low critical dose based on an adaptive effect	Overestimation of risk
ATBC, DEHTP, DINCH – Applicability of the toxicological reference value for children under 3 years	Assumption based on previous work of Anses	Reprotoxicity or multi-generation studies were considered when building the toxicological reference value	Overestimation of risk

(Continued)

Table 16.1 (Continued)   Table of Uncertainties of the Health Risk Assessment Related to the Use of Plastic Toys for Children under 3 Years Old

Sources of Uncertainty	Description	Treatment during the Assessment	Impact on the Risk
Exposure assessment			
Composition and migration tests: Choice of toys to be assessed	Composition and migration tests were carried out on a limited number of articles with limited representativeness of the French market	Extrapolation of results obtained on the studied toys to all the toys on the market	Impact on the risk unknown
Migration Testing: Study protocol	Use of saliva simulant partially reproducing human saliva characteristics; Suction dynamics not taken into account	Simulant of saliva: use of artificial saliva whose composition is recommended by the European Joint Research CentreSuction: Overlooked	Impact on the risk unknown
Evolution of migration over time	Gaps of data on the evolution of chemical migration in toys	A probabilistic approach was applied for this parameter using the migration values from the composition and migration tests as input data	Overestimation of risk

(Continued)

Table 16.1 (Continued) Table of Uncertainties of the Health Risk Assessment Related to the Use of Plastic Toys for Children under 3 Years Old

Sources of Uncertainty	Description	Treatment during the Assessment	Impact on the Risk
Altered toys (aging, deterioration of the material) that may lead to a change in migration phenomena	Gaps of data and methodology to evaluate the phenomenon	Overlooked	Impact on the risk unknown
Choice of the population: Children from 0 to 5 months were included in the exposure scenario based on spontaneous mouthing of toys behavior	Studies on children's psychomotility estimate that the behavior of voluntary mouthing appears after the age of 5 months. However, an "accidental" mouthing is also possible.	Upper bound approach	Overestimation of exposure (and risk) of children aged from 0 to 5 months
Choice of the exposure mechanism: Chemical migration by diffusion in saliva	Direct ingestion of part of the toy was not considered	Overlooked	Negligible underestimation of exposure/risk
Duration of mouthing	Experimental difficulties in accurately measuring the mouthing duration	Self-mouthing: Selection of the most powerful study that documents the parameter.Mouthing by a third party: Overlooked	Self-mouthing: Overestimation of exposure/risk. Mouthing by a third party: Impact on the risk unknown

(Continued)

471

Table 16.1 (Continued) Table of Uncertainties of the Health Risk Assessment Related to the Use of Plastic Toys for Children under 3 Years Old

Sources of Uncertainty	Description	Treatment during the Assessment	Impact on the Risk
Surface area of the toy put in mouth	Gaps of data	Use of a default value to estimate the parameter regardless of age	Impact on the risk low
Analytical limit	Low number of censored data	Imputation of the censored data by a default value and sensitivity analysis	Impact on the risk negligible
Risk Characterization			
Comparison of a whole-life toxicological reference value with exposures obtained over a shorter period (1 year)	Assumption	Upper bound approach	Strong overestimation of the risk
Mixture effects or cumulative effects	Potential cumulative or mixture effects of the studied substances not taken into account because of gaps of data	Assumption based on the absence of a common adverse effect of the substances studied	Impact on the risk unknown

Adapted from Anses, *Jouets et équipements pour enfants en matière plastique destinés aux enfants de moins de 3 ans*, ANSES, Maisons-Alfort, 2016.

Formal expert elicitation refers to a structured approach of consulting experts on an issue. Diverse expert elicitation protocols and methods exist. None of them are universally applicable, but some appear flexible enough to deal with a wide range of uncertainties that can be encountered in risk assessment (Knol et al. 2010). When considering uncertain quantity, the Sheffield method is one of the most popular. It includes four stages: selection of experts, training, individual elicitation and collective elicitation. The result of the elicitation process is a distribution to represent the aggregate judgment of the expert group (EFSA 2014). Framework and tools supporting the formal Sheffield elicitation process have been developed by T. O'Hagan and J. Oakley at the University of Sheffield, United Kingdom. Within the process, the MATCH Uncertainty Elicitation Tool freeware, for example, could be used to individually assess the uncertainty of model parameters as a distribution (Morris, Oakley, and Crowe 2014).*

When addressing uncertainty about methodological choices or assumptions, it is also customary to ask experts for their best professional judgment, but other elicitation methods, such as those developed by A.B. Knol and collaborators, may be more relevant (Knol et al. 2010). Graphical representations, like a distribution function or simply a histogram or boxplot, are valuable tools for visualizing the numerically evaluated uncertainties (Morgan 2014). If a numerical evaluation of individual uncertainties is not possible, then the use of an ordinal scale is the next preferred option, taking care to use the same scale for different uncertainty sources during an overall analysis for comparability. Whatever method is used to characterize uncertainty, it must always be accompanied by a narrative description that allows a clear and unambiguous understanding of the results.

16.4.4 Step 3: Assessing the Impact of the Combined Uncertainties

This step aims at determining the total uncertainty induced in the risk assessment results by the various sources of uncertainty identified. Generally, it starts after the prioritization of sources to be addressed, according to the results of the individual assessments of Step 2 and their contribution to the total uncertainty. Details about prioritization of uncertainty sources are provided in the next section.

---

* http://optics.eee.nottingham.ac.uk/match/uncertainty.php

Assessment of the combined impact of diverse sources of uncertainty uses the results of their individual assessment. The difficulty is that the nature or format of these results may vary from one source to another. Some results may be quantitative, whereas others may be qualitative. To overcome this, normally, the quantitatively assessed uncertainties are treated and combined separately from the others. As far as possible, the combined impact of quantified uncertainties should be assessed using a mathematical model (Rojas, Feyen, and Dassargues 2008). The treatment boils down to assessing the combined impact of the input factors of the model by implementing uncertainty propagation methods. In general, these numerical methods are based on simulation techniques (e.g. Monte Carlo or Latin hypercube) (Bogen et al. 2009, Makowski 2013). For assessing the combined impact of unquantified uncertainties, it is first necessary to ensure that the same ordinal scale has been used for the individual uncertainty assessments of Step 2. If not, then they must be harmonized. The combined impact of unquantified uncertainties is then assessed based on expert judgment, ideally through a formal elicitation process.

Determining the total uncertainty induced in the risk assessment results amounts to integrating the combined impact of quantified uncertainties with that of the unquantified uncertainties (van der Sluijs et al. 2005). This is again done through expert judgment, again ideally using a formal elicitation process. Whatever the methods involved in assessing the combined impact, it is essential that any correlations and relationships between different sources of uncertainty are taken into account so that only realistic situations and scenarios are considered.

At the end of the assessment, the total uncertainty may be too large to allow reasonable decision-making. In such situations, complementary data and/or information may be sought to reduce uncertainty in those aspects having the most impact on the total uncertainty. Otherwise, it is also possible to simplify or even remove some of the more intricate steps during the treatment of risk assessment that are highly uncertain but add very little to the assessment result itself. This should be done keeping in mind that the analysis must fit the purpose; that is, it should be as simple, clear and understandable as possible and consistent with the questions of concern (ANSES 2016c, Morgan and Henrion 1990). Depending on the situation and the choices made, as shown in Figure 16.2, it may be necessary to re-plan the whole uncertainty analysis and/or re-select the uncertainty sources to be treated in detail (Step 1) and/or more accurately characterize individual uncertain sources (Step 2) and/or refine the impact assessment of combined sources of uncertainty (current step).

16.4.5 Step 4: Prioritization of Sources of Uncertainty

The purpose of this step is to determine which uncertainty sources significantly affect the results of risk assessment. Setting a hierarchy of uncertainty sources is usually necessary at the end of Step 1 to select those sources of uncertainty to be individually assessed in Step 2, and it can also be useful at the beginning of Step 3 to select those sources of uncertainty to be considered for assessing the combined impact of sources of uncertainty.

Generally, when dealing with the uncertainty of a model's input factors, it is recommended to perform sensitivity analysis, which aims to quantify the effect of changes over a range of values of input factors on the model output (Oakley and O'Hagan 2004). It quantitatively measures the contribution of the input factors of the model to the variations of its outputs (Morgan and Henrion 1990). Hence. the uncertainty ranges and/or distributions set at Step 2 ("Individual assessment of uncertainty") for each uncertain factor are used as input for the sensitivity analysis. Thus, sensitivity analysis helps to prioritize the input factors according to how much their uncertainty contributes to the total uncertainty of the model output.

Sensitivity analysis methods can be categorized into two main branches according to the problem at hand: local and global sensitivity analyses. Local sensitivity analysis, based on derivatives of the model function evaluated at a given baseline input values (e.g. central estimates), may be used if the analysis is motivated by the need to understand the effects of slight disturbances around the baseline of the inputs of interest on the model output. Otherwise, when considering more substantial changes in inputs, global sensitivity analysis is more relevant.

To perform a global sensitivity analysis, a wide range of methods and techniques exists. For a comprehensive and accurate characterization of the uncertainty of model inputs, avoiding the pitfall of output values in an unrealistic range, the use of probability distributions is necessary. Such an approach is known as probabilistic sensitivity analysis. Although defining probability distributions for each uncertain model input may be regarded as an aspect of probabilistic sensitivity analysis, it corresponds to individual assessment of uncertainty of these factors (Step 2). This can be taken from the literature or derived from data by fitting an empirical distribution or based on an expert's opinion. It may be required to consider the correlation structure between input factors. The next step is to generate the input sample depending on the method chosen for the sensitivity analysis and then to evaluate the model used on the generated sample.

Some widely used methods of sensitivity analysis, including variance-based methods, can be seen in terms of a decomposition of the model function into the main effects of input factors and interaction effects between two or more input factors. Plotting the main effects and, at least, first-order interaction effects (i.e. interaction between two inputs) is a powerful visual tool for examining how the model output behaves in the face of each individual input and how these inputs interact in their influence on the output. For a broad overview of the subject, including selection of the method, framing of the analysis, and interpretation and presentation of the results, one may refer, for example, to the work of Saltelli and collaborators (Homma and Saltelli 1996, Saltelli et al. 2008, Saltelli et al. 2012, Saltelli and Tarantola 2002, Saltelli et al. 2004).

In other situations, especially when prioritizing unquantified uncertainties, Risk ranking methods or analytic hierarchy processes (AHP) may provide a solution. AHP is a measurement method used to derive ratio scales from paired comparisons. These pairwise comparisons reflect the relative strength of preferences and feeling, and in this way, they identify uncertainties whose impact is non-negligible. Multi-criteria decision-making is one of the widest applications of the AHP method (Saaty 1987).

16.4.6 Step 5: Communication of the Results of the Uncertainty Analysis

Analyzing uncertainty is of limited use if the results of these efforts are not communicated in a clear and understandable way. The way uncertainty is communicated determines the usefulness of the risk assessment itself as a decision-making support tool (IOM 2014, EFSA 2016).

The use of a figure or an image is no doubt the most efficient means for such communication, especially when addressing a quantified uncertainty. Nonetheless, there are pitfalls to avoid. For further reading, one can refer to Morgan and Henrion (1990), who explored the communication of uncertainty to both a scientific and a non-scientific audience.

More generally, the most recent methodological guidance documents stress the need to respect the following principles:

- *Transparency*: a presentation of the uncertainty analysis must be included with the risk assessment results as a necessary condition for an adequate interpretation of the results and their limits.

- *Understandability*: the results of the uncertainty analysis should be comprehensively reported in a specific section or chapter. All identified and described sources of uncertainty should be presented together with those that were assessed in depth and the justification for those that were not.
- *Accessibility*: the uncertainty analysis results should be communicated in a way that is accessible to all stakeholders of the risk assessment, namely risk managers and assessors and more broadly, all interested parties.

16.5 CONCLUSION

Analyzing uncertainty in risk assessment is essential, as uncertainties may strongly affect the risk management decision made. However, currently, risk assessment is too frequently conducted without the corresponding uncertainty analysis. Failure to systematically analyze uncertainty during a risk assessment leaves both the risk assessor and the risk managers unable to judge the adequacy of results and the conclusions reached.

A large range of methods and techniques for uncertainty analysis is advocated in the literature. While most of them are quantitative methods based on probability theory, other methods – including possibility theory and qualitative or hybrid approaches – exist and allow the capture of different types of information and knowledge that are equally important for decision-making. The purpose of the present chapter is not to recommend one approach over another, as they are all valuable, but rather, to suggest the use of several of them in a complementary and coherent way to lead to more comprehensive uncertainty analyses. The aim of the chapter is to go beyond the diversity of methods in looking for a unified approach.

Thus, after laying down the basic definitions and concepts, the chapter provides a general framework for analyzing uncertainties in risk assessment. The variety of decision-making situations calls for a flexible framework that enables the risk assessor to choose the most appropriate methods for the setting of the problem and the assessment context (decisional context, time and resource constraints, etc.). Hence, the framework presented includes a preliminary stage of planning the uncertainty analysis in order to explain the context of the assessment (*Framing*), clarify the question to be addressed (*Question reformulation*) and define the assessment method (*Choice of the assessment method*). Uncertainty analysis per se is a five-step approach, including first, an inventory of all uncertainty sources; second, an individual assessment of the identified uncertainties; third, an assessment of the impact of the combined uncertainties; fourth, if necessary, a prioritization of the different sources of uncertainty; and lastly, the communication of the results of the uncertainty analysis. If the combined uncertainties are too great, the risk assessment results may not be informative enough for decision-making purposes. In a pragmatic way, the framework thus proposes an iterative approach allowing the refinement, if necessary, of the initial uncertainty analysis according to the specific needs of the risk assessment.

Each step of the proposed framework is described in this chapter so as to provide the reader with the key information necessary to carry out uncertainty analyses.

REFERENCES

Abt, E., J. V. Rodricks, J. I. Levy, L. Zeise, and T. A. Burke. 2010. "Science and decisions: Advancing risk assessment." *Risk Analysis* 30(7):1029–1036.

ANSES. 2016a. *Évaluation du poids des preuves à l'Anses : Revue critique de la littérature et recommandations à l'étape d'identification des dangers.* Maisons-Alfort: ANSES.

ANSES. 2016b. *Jouets et équipements pour enfants en matière plastique destinés aux enfants de moins de 3 ans.* Maisons-Alfort: ANSES.

ANSES. 2016c. *Prise en compte de l'incertitude en évaluation des risques : Revue de la littérature et recommandations pour l'Anses.* Maisons-Alfort: ANSES.

ANSES. 2016d. *Traitement de l'incertitude dans le processus d'évaluation des risques sanitaires des substances chimiques - Réflexion sur la caractérisation et la prise en compte de l'incertitude en évaluation des risques sanitaires au sein de l'Anses.* Maisons-Alfort: ANSES.

Aven, T. 2016. "Risk assessment and risk management: Review of recent advances on their foundation." *European Journal of Operational Research* 253(1):1–13. doi: 10.1016/j.ejor.2015.12.023.

Aven, T., P. Baraldi, R. Flage, and E. Zio. 2014. *Uncertainty in Risk Assessment: The Representation and Treatment of Uncertainties by Probabilistic and Non-Probabilistic Methods, Uncertainty in Risk Assessment: The Representation and Treatment of Uncertainties by Probabilistic and Non-Probabilistic Methods.* Chichester: Wiley Blackwell.

Bogen, K. T., A. C. Cullen, H. C. Frey, and P. S. Price. 2009. "Probabilistic exposure analysis for chemical risk characterization." *Toxicological Sciences* 109(1):4–17. doi: 10.1093/toxsci/kfp036.

Counsell, C. 1997. "Formulating questions and locating primary studies for inclusion in systematic reviews." *Annals of Internal Medicine* 127(5):380–387.

Craye, M., S. O. Funtowicz, and J. P. van der Sluijs. 2005. "A reflexive approach to dealing with uncertainties in environmental health risk science and policy." *International Journal of Risk Assessment and Management* 5(2/3/4):216–236. doi: 10.1504/ijram.2005.007169.

Cullen, A. C., and H. C. Frey. 1999. *Probabilistic Techniques in Exposure Assessment. A Handbook for Dealing with Variability and Uncertainty in Models and Inputs.* New York: Plenum Press.

ECHA. 2012. "Uncertainty analysis (ECHA-12-G-25-EN)." *Guidance on Information Requirements and Chemical Safety Assessment.* Helsinki: ECHA.

EEA. 2001. "Late lessons from early warnings: The precautionary principle 1896–2000." *Environmental Issue Report.* Copenhagen: European Environment Agency (EEA).

EEA. 2013. "Late lessons from early warnings: Science, precaution, innovation." *Late Lessons from Early Warnings.* Copenhagen: European Environment Agency (EEA).

EFSA. 2007. "Guidance of the Scientific Committee on a request from EFSA related to Uncertainties in Dietary Exposure Assessment (EFSA-Q-2004-019)." *EFSA Journal* 5(1):438, 54 pp. doi: 10.2903/j.efsa.2007.438.

EFSA. 2010. "Application of systematic review methodology to food and feed safety assessments to support decision making (EFSA-Q-2008-717)." *EFSA Journal* 6(1637), 90 pp. doi: 10.2903/j.efsa.2010.1637.

EFSA. 2014. *Guidance on Expert Knowledge Elicitation in Food and Feed Safety Risk Assessment.* Parma: European Food Safety Authority (EFSA).

EFSA. 2015. "Draft report – Guidance on uncertainty in EFSA Scientific Assessment." *EFSA Journal.* Parma, Italy: European Food Safety Authority (EFSA).

EFSA. 2016. "Revised draft for internal testing – Guidance on uncertainty in EFSA Scientific Assessment." *EFSA Journal.* Parma, Italy: European Food Safety Authority (EFSA).

FAO/WHO. 2009. "Principles and methods for the risk assessment of chemicals in food." *Environmental Health Criteria* (240). Rome: FAO/WHO.

Hayes, K. R. 2011. *Uncertainty and Uncertainty Analysis Methods.* Hobart, Australia: CSIRO.

Homma, T., and A. Saltelli. 1996. "Importance measures in global sensitivity analysis of nonlinear models." *Reliability Engineering and System Safety* 52(1):1–17. doi: 10.1016/0951-8320(96)00002-6.

IOM. 2013. *Environmental Decisions in the Face of Uncertainty.* Washington, DC: The National Academies Press.

IOM. 2014. *Characterizing and Communicating Uncertainty in the Assessment of Benefits and Risks of Pharmaceutical Products: Workshop Summary.* Washington, DC: The National Academies Press.

IPCS. 2008. "Uncertainty and data quality in exposure assessment. Harmonization Project Document No. 6." *IPCS Harmonization Project Document.* Geneva: WHO Press.

Knol, A. B., A. C. Petersen, J. P. van Der Sluijs, and E. Lebret. 2009. "Dealing with uncertainties in environmental burden of disease assessment." *Environmental Health: A Global Access Science Source* 8(1):1–13.

Knol, A. B., P. Slottje, J. P. van Der Sluijs, and E. Lebret. 2010. "The use of expert elicitation in environmental health impact assessment: A seven step procedure." *Environmental Health: A Global Access Science Source* 9(1):1–16. doi: 10.1186/1476-069x-9-19.

Makowski, D. 2013. "Uncertainty and sensitivity analysis in quantitative pest risk assessments; practical rules for risk assessors." *NeoBiota* 18:157–171.

Morgan, M. G. 2014. "Use (and abuse) of expert elicitation in support of decision making for public policy." *Proceedings of the National Academy of Sciences of the United States of America* 111(20):7176–7184. doi: 10.1073/pnas. 1319946111.

Morgan, M. G., H. Dowlatabadi, M. Henrion, D. Keith, R. Lempert, S. McBride, M. Small, and T. Wilbanks. 2009. *Best Practice Approaches for Characterizing, Communicating, and Incorporating Scientific Uncertainty in Climate Decision Making*. Washington, DC: U.S. Climate Change Science Program (CCSP).

Morgan, M. G., and M. Henrion. 1990. *Uncertainty: A Guide to Dealing with Uncertainty in Quantitative Risk and Policy Analysis*. New York: Cambridge University Press.

Morris, D. E., J. E. Oakley, and J. A. Crowe. 2014. "A web-based tool for eliciting probability distributions from experts." *Environmental Modelling and Software* 52:1–4. doi: 10.1016/j.envsoft.2013.10.010.

NRC. 2009. *Science and Decision: Advancing Risk Assessment*. Washington, DC: The National Academies Press.

NTP. 2015. *Handbook for Conducting a Literature-Based Health Assessment Using OHAT Approach for Systematic Review and Evidence Integration Research*. Triangle Park, NC: NIEHS.

O'Hagan, A., C. E. Buck, A. Daneshkhah, J. R. Eiser, P. H. Garthwaite, D. J. Jenkinson, J. E. Oakley, and T. Rakow. 2006. *Uncertain Judgements: Eliciting Experts' Probabilities*. John Wiley & Sons, Ltd.

Oakley, J. E., and A. O'Hagan. 2004. "Probabilistic sensitivity analysis of complex models: A Bayesian approach." *Journal of the Royal Statistical Society: Series B: Statistical Methodology* 66(3):751–769. doi: 10.1111/j.1467-9868.2004.05304.x.

Petersen, A. C., P. H. M. Janssen, J. P. van der Sluijs, J. S. Risbey, J. R. Ravetz, J. A. Wardekker, and H. Martinson Hughes. 2013. *Guidance for Uncertainty Assessment and Communication*. 2nd ed. The Hague: PBL.

Rojas, R., L. Feyen, and A. Dassargues. 2008. "Conceptual model uncertainty in groundwater modeling: Combining generalized likelihood uncertainty estimation and Bayesian model averaging." *Water Resources Research* 44(12):1–16. doi: 10.1029/2008WR006908.

Rooney, A. A., A. L. Boyles, M. S. Wolfe, J. R. Bucher, and K. A. Thayer. 2014. "Systematic review and evidence integration for literature-based environmental health science assessments." *Environmental Health Perspectives* 122(7):711–718. doi: 10.1289/ehp.1307972.

Saaty, R. W. 1987. "The analytic hierarchy process-what it is and how it is used." *Mathematical Modelling* 9(3–5):161–176.

Saltelli, A., M. Ratto, T. Andres, F. Campolongo, J. Cariboni, D. Gatelli, M. Saisana, and S. Tarantola. 2008. *Global Sensitivity Analysis: The Primer.* Chichester, UK: John Wiley & Sons, Ltd.

Saltelli, A., M. Ratto, S. Tarantola, and F. Campolongo. 2012. "Update 1 of: Sensitivity analysis for chemical models." *Chemical Reviews* 112(5):PR1–PR21. doi: 10.1021/cr200301u.

Saltelli, A., and S. Tarantola. 2002. "On the relative importance of input factors in mathematical models: Safety assessment for nuclear waste disposal." *Journal of the American Statistical Association* 97(459):702–709. doi: 10.1198/016214502388618447.

Saltelli, A., S. Tarantola, F. Campolongo, and M. Ratto. 2004. *Sensitivity Analysis in Practice: A Guide to Assessing Scientific Models.* Chichester, UK: John Wiley & Sons, Ltd.

Spiegelhalter, D. J., and H. Riesch. 2011 (1956). "Don't know, can't know: Embracing deeper uncertainties when analysing risks." *Philosophical Transactions of the Royal Society, Series A: Mathematical, Physical and Engineering Sciences* 369(1956):4730–4750. doi: 10.1098/rsta.2011.0163.

U.S. EPA. 2014a. *Framework for Human Health Risk Assessment to Inform Decision Making.* Washington, DC: Office of the Science Advisor. U.S. Environmental Protection Agency.

U.S. EPA. 2014b. *Risk Assessment Forum White Paper: Probabilistic Risk Assessment Methods and Case Studies.* Washington, DC: Office of the Science Advisor, U.S. Environmental Protection Agency.

van der Sluijs, J. P., M. Craye, S. Funtowicz, P. Kloprogge, J. Ravetz, and J. Risbey. 2005. "Combining quantitative and qualitative measures of uncertainty in model-based environmental assessment: The NUSAP system." *Risk Analysis* 25(2):481–492.

van der Sluijs, J. P., J. S. Risbey, P. Kloprogge, J. R. Ravetz, S. O. Funtowicz, S. Corral Quintana, A. Guimarães Pereira, B. De Marchi, A. C. Petersen, P. H. M. Janssen, R. Hoppe, and S. W. F. Huijs. 2003a. "RIVM/MNP guidance for uncertainty assessment and communication: Detailed guidance." *RIVM/ MNP Guidance for Uncertainty Assessment and Communication Series.* Utrecht: Utrecht University.

Van der Sluijs, Jeroen P., James S. Risbey, Penny Kloprogge, Jerome R. Ravetz, Silvio O. Funtowicz, Serafin Corral Quintana, Angela Guimarães Pereira, Bruna De Marchi, Arthur C. Petersen, Peter H. M. Janssen, Rob Hoppe, and Simône W. F. Huijs. 2003b. *RIVM/MNP Guidance for Uncertainty Assessment and Communication: Detailed Guidance (RIVM/MNP Guidance for Uncertainty Assessment and Communication Series, Volume 3).* Utrecht: Copernicus.

Verdonck, F. A., A. Souren, M. B. van Asselt, P. A. van Sprang, and P. A. Vanrolleghem. 2007. "Improving uncertainty analysis in European Union risk assessment of chemicals." *Integrated Environmental Assessment and Management* 3(3):333–343.

481

Walker, W. E., P. Harremoës, J. Rotmans, J. P. van der Sluijs, M. B. A. van Asselt, P. Janssen, and M. P. Krayer von Krauss. 2003. "Defining uncertainty: A conceptual basis for uncertainty management in model-based decision support." *Integrated Assessment* 4(1):5–17. doi: 10.1076/iaij.4.1.5.16466.

Whitlock, E. P., S. A. Lopez, S. Chang, M. Helfand, M. Eder, and N. Floyd. 2010. "AHRQ Series Paper 3: Identifying, selecting, and refining topics for comparative effectiveness systematic reviews: AHRQ and the Effective Health-Care Program." *Journal of Clinical Epidemiology* 63(5):491–501. doi: 10.1016/j.jclinepi.2009.03.008.

WHO/IPCS. 2018. "Guidance document on evaluating and expressing uncertainty in hazard characterization." 2nd ed. Geneva: World Health Organization.

17

Examples of Quantitative Mycotoxin Risk Assessments
Use and Application in Risk Management

Sonia Marín, German Cano-Sancho

Contents

17.1 INTRODUCTION

Mycotoxins are widespread chemical hazards in crops and plant products, which pose a risk to human and animal health. Although considered as chemical hazards, they have a biological origin, as they are produced by filamentous fungi. Their presence in our food may be of concern, and there is an urgent need for control strategies based on both qualitative and quantitative risk assessment.

A number of risk assessments, commonly based on dietary exposure point estimations, have been carried out for several mycotoxins, including aflatoxins, ochratoxin A and *Fusarium* mycotoxins, in different countries. The estimates have resulted, in general, in safe risk characterization situations despite the acknowledged limitations to accounting for related uncertainties and variabilities. In this regard, more refined approaches using stochastic exposure models have revealed that the exposure among several population groups, such as babies, infants or high consumers of certain food products, may be of concern. Several examples of the application of stochastic exposure and hazard assessment models will be discussed, including the advantages and drawbacks as well as the key methodological issues.

Quantitative risk assessment (QRA), based on food chain models instead of analytical data from food samples, although hardly explored for mycotoxins, allows the estimation of different simulated scenarios of exposure throughout the food chain. This includes cropping, postharvest and food processing. In this way, it is possible to test different risk management alternatives, run the simulations and choose the alternative that results in a lower level of exposure for the population. As mycotoxins have a biological origin, the models used for the exposure assessment may include predictive mycology models based on fungal growth and mycotoxin production.

17.2 MYCOTOXINS

Mycotoxins are secondary metabolites produced by several fungal species. The worldwide contamination of foods and feeds with mycotoxins is a significant problem. From an economic and public health standpoint, the foodborne mycotoxins that are considered as being relevant are aflatoxins, fumonisins, certain trichothecene mycotoxins (including deoxynivalenol [DON] and T-2 and HT-2 toxins), ochratoxin A (OTA), patulin and zearalenone (ZEA).

Mycotoxins can cause a variety of adverse health effects in humans. Aflatoxins, including aflatoxin B1 (AFB1), are the most toxic and have been shown to be genotoxic; i.e., they can damage DNA and cause cancer in animal species. There is also evidence that they can cause liver cancer in humans. Other mycotoxins have a range of other health effects, including kidney damage, gastrointestinal disturbances, reproductive disorders and suppression of the immune system. For most mycotoxins, a tolerable daily intake (TDI) has been established, which estimates the quantity of mycotoxin to which someone can be exposed daily over a lifetime without it posing a significant risk to health.

17.3 RISK ASSESSMENT

Risk assessment is commonly structured as a conceptual framework that in the context of food chemical safety, provides a mechanism for the systematic compilation, integration and critical review of information relevant to estimating the probability of adverse health effects derived from exposure to harmful chemicals present in food. For operational reasons, the most widely implemented risk assessment frameworks for mycotoxins divide the process into four major components: hazard identification, hazard characterization (or dose–response assessment), exposure assessment and risk characterization.

The extent of precision of the analysis relies on the quality of primary data on the toxicity of the compound (hazard characterization) and concerning the population exposure (exposure assessment); hence, the overall process can include a key component in which the probability of harm is estimated. In general, the risk assessment frameworks follow a tiered approach: qualitative assessments (Tier 1), deterministic assessments (Tier 2) and finally, probabilistic assessments (Tier 3) (EFSA 2008). In this sense, the probabilistic assessments, also known as *stochastic*, refer to the quantitative analysis of variability and uncertainties related to the estimated health risks of chemicals, depicting the QRA approaches (Figure 17.1).

The QRA frameworks have been commonly developed under regulatory settings pursuing the achievement of health protection goals extending or complementing the deterministic or point estimate risk assessment

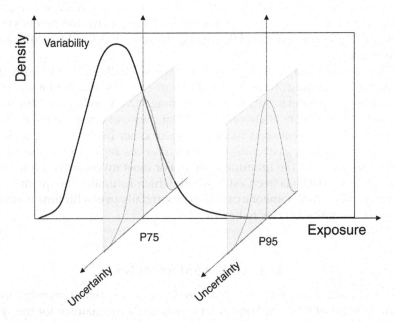

Figure 17.1 Representation of the variability of exposure estimates and the uncertainty distributions for given percentiles (e.g., 75th and 95th) estimated through Monte-Carlo Simulations. Adapted from Council et al. (2005).

frameworks. Overall, public policies developed for managing food chemical risks are generally precautionary rather than being focused on producing accurate predictions. Hence, the default approaches are mainly deterministic, based on "worst-case" scenarios, using conservative uncertainty factors, resulting in margins of safety that do not require more refined (probabilistic) evaluations. In consequence, QRA of food chemicals and specifically of mycotoxins has been mainly conducted within minor academic or scientific contexts rather than being promoted by food protection agencies. Actually, there is an acknowledged lack of standardized guidelines or frameworks for the QRA of mycotoxins delivered by scientific supporting agencies like the European Food Safety Authority (EFSA). Only one Scientific Opinion from EFSA provides insights into the application of probabilistic methods for modelling dietary exposure to pesticide residues, which may partially apply to mycotoxins (EFSA, 2012).

From a theoretical point of view, the QRA framework can be structured on the basis of the deterministic workflow, whereby a probabilistic assessment may be performed on the exposure distribution parameters and also extended to the dose–response assessment (hazard characterization), and both may be integrated into a full probabilistic risk assessment setting (Figure 17.2). Since no definite settings have been established, QRA of mycotoxins has been mainly focused on the probabilistic exposure assessment and in minor cases, on probabilistic hazard characterization.

17.3.1 Dietary Exposure Assessment of Mycotoxins

The absence of fully validated individual biomarkers for mycotoxins constrains the exposure assessments to dietary exposure modelling by integrating consumption with occurrence data.

In general, dietary exposure assessment describes the pathways through which mycotoxins are introduced into the food chain and challenged throughout the food processing and production systems, their distribution within the food commodities and finally, their consumption by the end-users. However, different objectives may motivate an exposure assessment, and consequently, the approach and methodology may vary. The aims may be, for example, to be combined with a hazard characterization as part of a risk assessment to estimate the risk associated with a mycotoxin+commodity combination, to identify foods in the diet likely to make a major contribution to human exposure to hazards, to identify where interventions or control options are likely to be most effective in reducing the level of exposure to a hazard in a given product, to compare

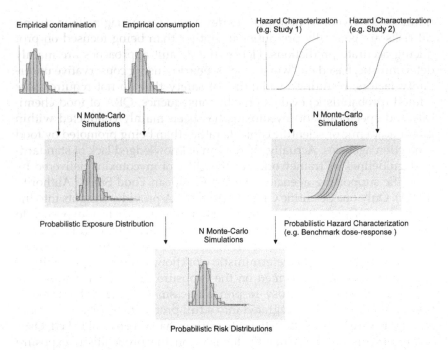

Figure 17.2 Schematic representation of a full probabilistic risk assessment framework for mycotoxins.

the efficiency of mitigation measures in reducing the exposure to a given hazard or to compare the levels of exposure resulting from different processes and food products (WHO, 2008).

Thus, depending on the scope, exposure assessment can begin with the evaluation of mycotoxin contamination at field level or in raw materials (e.g., a "farm-to-fork" risk assessment), or it can begin with the description of the mycotoxin contamination distributions at subsequent steps (e.g., as input to a food processing step) or just at the point of consumption. In any case, the main goal of risk assessment is to estimate the likelihood of mycotoxins being ingested by the consumer. By completing the pathway to the consumer and the potential adverse health effects, we incorporate the important information that may be critical for depicting the food management strategies along the food chain (i.e., modification of storage conditions to impair the fungal growth or mycotoxin production).

As introduced earlier, most of the existing dietary exposure assessments of mycotoxins have consisted of combining food consumption distributions deterministically or probabilistically with the occurrence of such mycotoxins in those target food categories; thus, they assessed the exposure of the population under the current risk management situation. It must be highlighted that exposure assessment can be conducted more accurately when mycotoxin concentrations are measured in the consumed product instead of predicted by a model. On the other hand, little research has considered the whole food chain or particular processing steps to link exposure assessment to risk management, due possibly to the lack of suitable data both on models for the prediction of mycotoxin production and on the impact of processes in mycotoxins to build risk assessment models.

17.4 QUANTITATIVE EXPOSURE ASSESSMENT OF MYCOTOXINS THROUGH DIETARY EXPOSURE MODELLING

Conceptually, the exposure to mycotoxins may be better described as a range of values rather than a single value (i.e., a deterministic approach), because each individual in the population is expected to be exposed to different concentrations of mycotoxins over time, displaying a group of random variables that may depict the classical definition of a stochastic process.

The general exposure model used to assess intakes from the diet can be written as in Equation 17.1:

$$E_i = \frac{\sum\sum Q_{i,t,g} \times C_{i,t,g}}{n_i \times bw_i} \tag{17.1}$$

where i is an index for an individual ($i = 1,...n$); E_i is the usual intake of individual i (ng/kg body weight/day); t is an index for the time window to assess food consumption and contamination (weeks [W], days [D] or single occasions [O] of consumption; g is an index for food group ($g = 1,...n$); $Q_{i,t,g}$ is the consumption of food group g on occasion t by individual i (kg); $C_{i,t,g}$ is the contamination of food group g encountered on occasion t by individual i (μg/kg); n_i is the number of days of food records available for individual i; and bw_i is the body weight of individual i (kg).

In the probabilistic exposure model, consumption ($Q_{i,t,g}$) and contamination levels ($C_{i,t,g}$) in Equation 17.1 are drawn randomly from underlying cumulative distribution functions (CDFs) in a one-dimensional Monte-Carlo simulation through two main approaches:

- Non-parametric method (NP). In the NP approach, no probability density function (*pdf*) hypothesis is made either on the consumption or on the contamination data. Each normalized consumption profile of the survey is taken into account, and each type of consumed food is assigned a value of contamination randomly drawn from the available contamination data.
- Parametric (P) and mixed parametric methods (MP). The P approaches attempt to better fit the *pdf* to the contamination and consumption datasets. In the MP method, a mixed *pdf* is fitted to each food consumption and a parametric *pdf* to each food contamination dataset.

Common approaches to evaluate the best fit of candidate distributions include the maximum likelihood method, quantile-quantile (Q-Q) plots, the Anderson–Darling statistic and the agreement between empirical and estimated quantiles.

17.4.1 Acute vs. Chronic Dietary Exposure Assessments

Some refinements may be required to the dietary exposure assessment depending on the kind of health endpoint to be compared in the risk characterization step.

17.4.1.1 Chronic Exposure Assessment Models

In many cases, the toxicological endpoint and related dose–response are developed on the basis of a chronic exposure (long-term). In those cases where long-term effects are expected, the mean chemical concentration is typically used, because is acknowledged that this value better represents the long-term average of actually encountered concentrations. For that reason, a full probabilistic exposure model including the probabilistic contamination term is often avoided, because it may lead to unrealistic scenarios such as high consumption patterns combined with high contamination distributions for long time periods.

17.4.1.2 Acute Exposure Assessment Models

Some mycotoxins (e.g., DON, T2 and HT2) have been shown to be toxic after a few days of exposure, emphasizing the application of acute exposure assessment models. In those cases, the worst-case scenario is preferred; the full probabilistic exposure modelling may help to reveal the extreme events on the tails of exposures.

17.4.2 Case Studies

17.4.2.1 Quantitative Assessment of Risk Derived from Dietary Intake of OTA

Quantitative risk analysis of mycotoxins has been more extensively applied to OTA compared with the rest of mycotoxins. The high world-wide exposure of OTA, combined with the long list of health effects of this fungal toxin (including carcinogenicity), may have motivated these regulatory actions. We have identified some major studies performing probabilistic assessment of health risks derived from dietary intake of OTA, including in Spain, China, France and Canada (Table 17.1).

A pioneering study on quantitative exposure assessment of mycotoxins was led by Jean-Pierre Gauchi and Jean-Charles Leblanc using French consumption and contamination data of OTA. The study went far beyond a simple probabilistic exposure assessment; they explored different methodological approaches and compared the performance and applicability in the OTA datasets. The report was published in 2002, followed by a complementary study extending the work with a sensitivity analysis of high percentiles, consolidating a reference work in the field. The authors compared an NP-NP method with a MP-P approach, examining three different methodologies to build the confidence intervals (CIs): pseudo-parametric bootstrap CIs, parametric (lognormal and gamma) bootstrap CIs and distribution-free CIs. In their concluding remarks, the authors proposed a) calculation of estimations of the mean and the variance, b) calculation of mean, standard deviation, median, skewness, kurtosis, and the high quantiles required from the outputs of the NP-NP and MP-P methods in order to validate them with respect to each other if the results are close and c) calculation of pseudo-parametric bootstrap CIs (Type 1) for the parameters corresponding to the preceding statistics. Beyond the methodological contributions of this work, the authors highlighted the potential high exposure estimates for children when compared with safety doses of 1.8 or 14 ng/kg bw/day derived from nephrotoxic and carcinogenic effects, respectively. The authors acknowledged that the approach did not address chronic or long-term exposures as expected for those toxicological endpoints, and a more dynamic model should be applied. Given the relevance of the high quantiles in risk assessment and the methodological limitations on their accurate estimation, the study

491

Table 17.1 Overview of Characteristics, Databases and Probabilistic Models Used in a Selection of Studies Performing Probabilistic Exposure Assessments of OTA

Reference	Country	Occurrence Data	Consumption Data	Probabilistic Exposure Assessment Models Details
Coronel et al. 2012	Spain	TDS ND = 0.5LOD	FFQ-TDS All ages/both sexes	P method (gamma) SAS software N = 10,000 iterations
Han et al. 2013	China	Occurrence Survey N = 400 2011–2012 ND = 0.5LOD	FFQ-Adults 2011–2012	P method MP method TDIs = 17, 14 and 5 ng/kg body weight/day @RISK software N = 10,000 iterations
Gauchi et al. 2002a,b	France	SCOOP task 3.2.7	ASPCC survey Children N = 232	MP-P method (gamma, lognormal) NP-NP method Bootstrap confidence intervals SA high quantiles SAS software
Kuiper-Goodman et al. 2010	Canada	Health Canada CGC 1994–2006 ND = lognormal imputation ND = 0.5LOD	CSFII (24 h recall) survey (United States) 1994–1998 Validated for Canada All age-sex strata	MP method P method Adjustments to "usual" exposure SAS software N = 1000 iterations

(Continued)

Table 17.1 (Continued) Overview of Characteristics, Databases and Probabilistic Models Used in a Selection of Studies Performing Probabilistic Exposure Assessments of OTA

Reference	Country	Occurrence Data	Consumption Data	Probabilistic Exposure Assessment Models Details
Boon et al. 2011	Europe	VWA 2002–2006 ND = 0.5LOD	Concise DB EFSA	OIM, BBN, ISUF MCRA software
Boon et al. 2011	Netherlands	VWA 2002–2006 ND = 0.5LOD	DNFCS (2-dWDR) 1997–1998 16–64 years (n = 4285)	OIM, BBN, ISUF MCRA software

2-dWDR, 2 days weighted dietary record; ASPCC, Association Sucre-Produirs Sucrés Communication Consommation; BBN, Betabinomial-normal model; DNFCS, Dutch National Food Consumption Survey; FFQ, food frequency questionnaire; ISUF, Iowa State University Foods model; LOD, limit of detection; ND, not detected; MP method, mixed parametric method; OIM, observed individual means method; P method, parametric method; SA, sensitivity analysis; TDS, total diet study; VWA, Dutch Food and Consumer Product Safety Authority.

was completed with a sensitivity analysis based on a quadratic poly-nomial regression model for the quantiles of OTA exposure, including 32 main factors, their 496 two-factor interactions and their 32 quadratic terms. Through 6561 exposure simulations, the authors found that the parameters "Cereals", both consumption and contamination, exhibited the largest influence on the models. A similar MP-P approach was imple-mented in Catalonia, in the north-eastern region of Spain, where the authors integrated contamination data from six major food commodi-ties (e.g., breakfast cereals and loaf bread) with consumption data gath-ered from a food frequency questionnaire specially designed to evaluate the consumption patterns of food susceptible to being contaminated by mycotoxins in different age-groups (Coronel et al., 2011). The authors found that the probabilistic mean estimates were similar to the esti-mates from the deterministic approach but slightly higher for children. The median daily intake estimates were mostly below the provisional maximum tolerable daily intake (PMTDI) of 17 and 14 ng/kg bw/day. Similarly in China, a probabilistic exposure assessment was restricted to the adult population and combined distributions of food consumption data with distributions of OTA in food (Han et al., 2013). The authors concluded that no health risk should be expected derived from OTA based on the EFSA and Joint Food and Agriculture Organization/World Health Organization (FAO/WHO) Expert Committee on Food Additives (JECFA) safety reference values, but the highest quantiles (i.e., the 97.5th percentile) could exceed the level of 5 ng/kg body weight/day proposed by the Scientific Committee.

Another famous and comprehensive risk assessment study of OTA was conducted in Canada by Kuiper-Goodman et al., (2010), includ-ing probabilistic (partial and full) exposure assessments and a novel adjustment method for the "usual" exposure. In this case, the partial probabilistic exposure model was built by combining the full individual distributions of food consumption with the average occurrence data for each vector of food. The full probabilistic model combines both full dis-tributions: occurrence of OTA and consumption of each food commodity. The adjustment method is motivated by the need to capture the long-term exposure over long periods, representing the usual or habitual exposure profile, accounting for the "within"- and "between"-individual variance. In order to adjust for this, the authors applied a two-stage model, whereby they first estimated the partial probabilistic exposure (Stage 1), weight-ing each food consumption vector to its average level of OTA. The adjust-ment factors were derived using variance reduction procedures. Next, the adjustment factors were implemented to adjust each single iteration for the partial probabilistic and the full probabilistic exposures, resulting in the "adjusted" probabilistic exposures (Stage 2).

A probabilistic refinement of deterministic exposure assessment of OTA was also conducted in the Netherlands, comparing three different long-term exposure probabilistic models: observed individual means (OIM), Iowa State University Foods (ISUF) and betabinomial-normal (BBN) models (Boon et al., 2011). The authors used concentration data of OTA from the Dutch Food and Consumer Product Safety Authority (VWA) covering the years 2002–2006. The food consumption was gathered from the EFSA's Concise database for different countries, and individual data came from the Dutch National Food Consumption Survey (DNFCS) of 1997–1998 for the total population. The results showed that the OIM outperformed the BBN and ISUF estimates. The authors concluded that the BBN model should be the preferred approach, as it can model exposure distributions that depend on covariates.

17.4.2.2 Quantitative Assessment of Risk Derived from Dietary Exposure to Deoxynivalenol

Probabilistic exposure assessment of deoxynivalenol (DON) was performed within the context of the Total Diet Study of mycotoxins in Catalonia (in the north-eastern region of Spain). The occurrence of DON in high-consumption commodities appeared to be widespread in the region, including cereal-based foods such as bread, pasta, breakfast cereals and snacks. The chemical analysis was performed with highly sensitive and validated gas and liquid chromatographic methods for a total of 556 individual or composite samples. Food consumption data was obtained by means of a food frequency questionnaire including 38 specific food items consumed in large quantities in the region and known to be potentially contaminated with mycotoxins. The population groups covered were seniors ($n = 76$), adults ($n = 720$), adolescents ($n = 235$), children ($n = 69$), infants ($n = 164$), celiac sufferers ($n = 70$) and adults with ethnic dietary patterns ($n = 56$) (Cano-Sancho et al. 2011).

Following the methodology proposed by Gauchi and Leblanc (2002) for OTA, the authors applied a parametric-parametric (P-P) method, choosing the gamma distribution to fit the *pdf*s.

The P-P method used to estimate the normalized exposure k of the S simulation set was built as follows:

$$\hat{E}_i = \sum \tilde{C}_{i(j)} \tilde{t}_j \tag{17.2}$$

where $\tilde{C}_{i,(j)}$ is a random normalized consumption i for the foodstuff j, drawn from Cj, the corresponding adjusted gamma pdf, and t_j is a random contamination for the foodstuff j, drawn from t_j, the corresponding adjusted gamma pdf.

The mean of normalized exposures over the simulation set S was then estimated using the following equation, where n was the number of random deviates drawn (10,000 in the study):

$$\hat{E}_S = \frac{1}{n} \sum \hat{E}_k \tag{17.3}$$

Inspired by the comprehensive evaluation of different bootstrap methods to build CIs performed by Gauchi and Leblanc (2002), "pseudo-parametric CIb", referred to as Type 1, were built by randomly drawing B samples of size n in the exposure simulation set S. The boundaries of the 95% confidence interval were calculated by taking the 0.025th and 0.975th empirical quantiles of the final bootstrap distribution. The results revealed higher summary estimates computed through probabilistic models when compared with the deterministic ones. In general, both gamma and lognormal distributions fitted the exposure pdfs well, with the exception of adult females and ethnic populations. The bootstrap CIs supported the estimates obtained from both lognormal and gamma fitted exposure simulations, exhibiting a substantial lack of precision for the extremes of the right tail (P95 and P99). Given the relevant contribution of pasta to the overall exposure estimates, a complementary study modelled the impact of cooking on the exposure estimates through a probabilistic approach, adding a novel term in the equation to correct for eventual mycotoxin losses during the boiling process (Cano-Sancho et al. 2013). The effect of cooking and especially food processing has been widely recognized; however, these modifying factors have not commonly been considered in the probabilistic models to estimate the impact of this source of uncertainties on the final estimates.

17.4.2.3 Quantitative Assessment of Risk Derived from Dietary Intake of T2-HT2

T2 and HT2 toxins are the most representative type A trichothecenes that may be found in cereal-based foods. Their high toxicity combined with the large uncertainties related to their exposure assessment has led to conflicting risk assessments involving some probabilistic approaches. The similar structure of both trichothecenes and common modes of action has favoured the combined risk assessment. A re-assessment of chronic exposure with updated data on the occurrence (2011–2016) of T2 and HT2 in food (19,505 analytical results) resulted in a much higher proportion of left-censored data than previous estimates. The consequence of this large proportion of censored data was a substantial increase of the upper bound estimates (worst-case) compared with the previous assessments (e.g., three-fold for infants with an estimated upper bound of 146 ng/kg bw/day) (EFSA, 2017).

The EFSA's CONTAM Panel has extended the chronic exposure assessment through a probabilistic assessment of acute exposure. The experts used a total of 41 most recent dietary surveys carried out in 23 different European countries. Probabilistic acute exposure was assessed for each reporting day by multiplying the total consumption amount for each food category by one occurrence level randomly drawn among the individual results available for that food category. Average acute exposure ranged from a minimum of 13.4 ng/kg bw per day, estimated in "Elderly", up to a maximum of 64.7 ng/kg bw per day, estimated in "Toddlers". The highest 95th percentile acute dietary exposure was estimated for a dietary survey within the age class "Infants" (170 ng/kg bw per day) (EFSA, 2017).

The occurrence of T2 and HT2 in Catalonia (Spain) was also reported to be highly censored, entailing large uncertainties on risk assessment. The authors noticed that most of the contamination in samples was in the range between the limit of detection and the limit of quantitation, with a few samples above the upper bound. The probabilistic methodology, based on a parametric model previously described for DON, was further coupled to three different approaches to manage the left-censored data: a) a substitution method, b) a parametric method based on the maximum likelihood estimation (MLE) and c) an NP method based on the Kaplan–Meier estimation (KM). Children, the most exposed group, exhibited median values computed from the gamma fitted simulation *pdf* of exposure to T2-HT2 of 0.05, 0.07 and 0.04 µg/kg bw/day for the substitution, MLA and KM method, respectively. Additionally, the authors were able to compute the bootstrap CIs to quantify the accuracy and reliability of the simulation estimates (Cano-Sancho et al. 2012).

17.4.2.4 (Integrative) Quantitative Assessment of Carcinogenic Risk Derived from Dietary Intake of Aflatoxin B1

Probabilistic approaches in the QRA of mycotoxins have been commonly focused on modelling exposure assessment and in a few cases, the hazard characterization, resulting in a probabilistic reference dose distribution (e.g., TDI). In order to quantify the variability and reduce the uncertainty in both exposure assessment and hazard characterization, the integrative probabilistic risk assessment (IPRA) models combine the probabilistic procedures in the same model. The overall framework was first proposed by van der Voet and Slob (2007) and applied to carcinogens using afla- toxin B1 (AFB1) as a case study (RIVM 2011). Probabilistic risk assessment of AFB1 in young children from the Netherlands was considered appro- priate after performing lower-tier assessments resulting in an MOE lower than 100 (Boon et al. 2009).

The authors developed the IPRA through the following steps:

Step 1. Estimate the distributions of the individual exposure, of the individual critical effect dose and subsequently, of the *individual MOE* (IMoE).

Step 2. From the latter distribution, derive the population character- istics of interest, such as the *probability of critical exposure* (PoCE) or a particular percentile of the IMoE distribution.

Step 3. Quantify the uncertainties of model inputs and evalu- ate the resulting uncertainty in the estimated PoCE (or other characteristic).

The authors used the same exposure assessment data and methodology for children aged 2–6 years, as reported by Boon et al. (2009).

The carcinogenic dose–response was based on three different stud- ies, combined using "study" as a covariate of the potency parameter, and adjusted with the PROAST software. The benchmark dose (BMD) distri- butions were derived by the parametric bootstrap method, i.e., by generat- ing datasets from the fitted model, refitting the model to the artificial data and recalculating the BMD in each run (number of runs = 1000).

Four different approaches were applied for the probabilistic risk assessment:

Approach A. MOE, expressed as a ratio between a point of depar- ture (PoD, e.g., benchmark dose) obtained from the critical dose– response data and the estimated human exposure level.

Approach B. This approach adapts the methodology for non-carcino- gens to cancer, using the median effective dose (ED50) from quan- tal dose–response on the basis that the ED50 is an estimate of the tolerance dose for the median animal, and the inter-individual variability is reflected in the intra-species distribution.

Approach C. This is a linear extrapolation whereby a straight line is "drawn" from the dose with an observable tumour incidence (the PoD) to the tumour incidence in the controls, with the aim of obtaining the dose at which a low, acceptable (but unobservable) incidence would occur. A series of models is fitted to the dose–response data, and for those models that fit the data sufficiently well, the BMD10 is assessed, together with an uncertainty distribution (using bootstrapping).

Approach D. A model extrapolation (and model-averaging) approach, whereby instead of linear extrapolation, extrapolation is performed from a fitted model or rather, from a series of fitted models that were found to result in a reasonable fit. The advantage of Approach D over linear extrapolation (Approach C) is that it provides a more realistic confidence interval for the dose at a given (low) risk, because it takes model uncertainty into account.

These results illustrate the problem that simple linear extrapolation as currently applied by some countries does not allow for potential inter- and intra-species differences. In IPRA, potential inter- and intra-species differences are fully taken into account.

Overall, the authors found a similar low MOE as Boon et al. (2009) did for children, gaining some insights into the cancer risk uncertainties. In the case of AFB1, the uncertainties involved (i.e., in exposure and in dose–response) are relatively small.

- For instance, instead of concluding that risks cannot be excluded, this study suggests that a risk as large as 1 in 200 cannot be excluded.
- Approach C (linear extrapolation) does not provide information on the most important uncertainty in cancer risk assessment, i.e., the uncertainty related to low-dose extrapolation.
- Approach D (model extrapolation) gives some more specific information than Approach B by providing risk in two dimensions, such as: between 1% and 80% of the population would be subject to a chance of developing cancer of 1 in 1000 (or smaller).

17.5 QUANTITATIVE EXPOSURE ASSESSMENT OF MYCOTOXINS BASED ON FOOD CHAIN DATA

Growing evidence has shown that different parameters may be good predictors of fungal growth and mycotoxin production in raw food materials, either in the field, postharvest or later in the production chain, and by extension, predictors of mycotoxin occurrence in final products and eventual exposure of consumers. Hence, risk management strategies can be applied at different stages along the food chain, from primary production, processing and manufacturing, transport and distribution, storage and retail to preparation and consumption of the food, which may result in a decreased exposure to mycotoxins. An integrative food chain risk assessment strategy (also known as a farm-to-fork strategy) is required to assess the impact of the risk management options (e.g., mycotoxin reduction strategies) on the final concentration in food and related exposure in the population. Methodologically, such risk assessment approaches may be either deterministic (based on mycotoxin point estimations for the different steps in the food chain) or probabilistic, if probability distributions for mycotoxin contamination are established (in this way, the probability distribution of exposure, in particular the percentiles, is more accurately estimated). In any case, most frameworks are developed on the basis of simulated scenarios where the parameters of interest (e.g., storage temperature) are modelled on the basis of realistic applications. To assess the efficacy and usefulness of management strategies focused on reducing mycotoxin burden in food, a complete exposure assessment might not be required, as the mycotoxin concentration in the finished product can be compared with the reference processing system to check whether a lower concentration results. There are many different types of strategies or control measures, instigated by regulation or chosen by the operators (e.g., good agricultural and animal production practices, good hygiene practices [GHPs] during manufacture and processing, and good consumer handling practices).

Control measures in the food industry regarding mycotoxins may comprise some of the following activities:

- Ensuring control of initial levels of hazards (e.g., avoiding nuts and spices from certain origins; avoiding raw materials from primary producers who do not adhere to good agricultural practices; establishing requirement specifications with suppliers and requiring verifiable documentation, e.g., letters of guarantee or certificates of analysis attesting the safe level of mycotoxins; using sampling and analyses as necessary; and using appropriate methods based on established criteria to reject unacceptable ingredients or products).
- Preventing an unacceptable increase of hazards.
  a) Preventing contamination, for example adopting GHPs that minimize mycotoxin contamination from transport, drying and storage facilities, establishments or equipment, and from

aqueous solutions in fruit and nut processing due to excessive reuse. For the particular case of mycotoxins, GHPs are also important to prevent contamination by mycotoxigenic fungi, which may further develop and produce mycotoxin in subsequent process stages.

b) Preventing fungal growth during transportation, storage and processing, for example cold storage of apples; adjusting a_w in stored cereals, nuts, coffee or spices; adding preservatives to stored fruits and cereals; controlling temperature and moisture/a_w in dehydrating fruits; adjusting storage times; using packaging techniques and materials to protect food from contamination; or implementing effective controls within the food processing environment (e.g., pest control).

- Reducing or eliminating mycotoxins

a) Selecting ingredients (e.g., applying electronic sorters to reject nuts that are likely to contain AFs; culling fruits for fruit juice production that are likely to contain patulin; rejecting rotten grape bunches that are likely to contain OTA; cleaning of cereals, resulting in the separation of mouldy grains, which account for most of the *Fusarium* toxins and AFs in a lot).

b) Additionally some measures that are not implemented to control mycotoxins, or that are intrinsic to the food process, may exert a certain control on mycotoxins, inactivating mycotoxins to some extent, because mycotoxins are quite heat stable (e.g., heat treatments, like roasting, frying and baking; commercial sterilization; fermentation processes), physical segregation of the most contaminated fractions of raw materials (e.g., milling of cereals, must extraction from grapes or malt, pressure washing of apples, centrifugation, filtration) (García-Cela et al. 2012).

The application of simulation-based models using data collected in the preliminary phases of the food-production chain appears to be a promising approach to evaluate risk management strategies focused on reducing the burden of mycotoxins in the food-web. These types of models, mainly borrowed from the microbiological risk assessment field, may efficiently integrate probabilistically a vast list of parameters gathered from the different steps that may affect the final consumer exposure. For QRA to become a useful decision support system, the impact of all these control measures on mycotoxin concentration should be modelled in such a way that simulations should be run easily, providing robust evaluations for risk

501

assessment. Moreover, it is important to address variability and uncertainty. Variability refers to quantities that are distributed within a defined population, such as food consumption rates, raw material contamination, mycotoxin production, etc.). These are inherently variable and cannot be represented by a single value, so that we can only determine their moments (e.g., mean, variance, skewness, etc.) with precision. In contrast, true uncertainty or model-specification error (e.g., statistical estimation error) refers to a parameter that has a single value, which cannot be known with precision due to measurement or estimation error (WHO, 1995). Two-dimensional Monte-Carlo simulations allow an efficient, separate evaluation of both variability (Dimension 1) and uncertainty (Dimension 2), which may be easily performed by multiple statistical packages.

Some examples of QRA have been taken from the literature, and the main points are summarized here:

17.5.1 Case Study 1: Simulation of Consumer Exposure to Deoxynivalenol According to Wheat Crop Management and Grain Segregation (Le Bail et al. 2005)

DON is mainly produced in wheat in the field and rarely during postharvest. According to Le Bail et al. (2005), assuming a consumption of 175 g of wheat flour per day (WHO, 2003) and a PMTDI of 60 µg/day for an adult weighing 60 kg, the maximum permissible concentration in wheat is 340 µg/kg. However, it is difficult to respect this limit every year at field scale. Thus, two risk management strategies were tested to reduce DON exposure through the consumption of wheat derivatives: the use of tillage methods as a cropping system was tested against no-tillage, and on the other hand, once the wheat was harvested, different crop segregation strategies were tested during postharvest. Point estimations were carried out for the different scenarios tested, while a distribution of consumption was applied for modelling the exposure.

17.5.1.1 Preharvest
It is known that no-tillage practices lead to an increase in DON in cropping wheat (Champeil, 2004). Thus, three scenarios were considered and applied to experimental plots:

Scenario 1: all the plots of the area were ploughed (0% no-tillage)
Scenario 2: half the area was ploughed (50% no-tillage)
Scenario 3: all the plots were direct-drilled (100% no-tillage)

Simulations were carried out assuming an initial DON concentration of 100–800 µg/kg for Scenario 1, and different ratios for Scenario 3/Scenario 1 from 1 to 8, and their impact on exposure was assessed. It was shown that PMTDI was not exceeded even for the highest ratio if the initial contamination level was 100 µg/kg for Scenario 1. In contrast, a ratio of 2 was sufficient to attain the PMTDI of mean exposure for contamination values of 400 µg/kg and over.

17.5.1.2 Postharvest

Segregation of cereal crops from different field plots in different batches according to predicted mycotoxin contamination may be a suitable risk management alternative. The authors considered the following scenarios:

Scenario (a): all wheat crops in the area are blended as a single batch.

Scenario (b): the plots are divided into two classes according to the preceding crop: Batch M (maize) for the plots with maize as preceding crop and Batch O (others) for the other plots.

Scenario (c): the plots of the area are divided in two classes according to the real grain contamination value to segregate two batches: Batch L (L) for the plots with contamination <1250 µg/kg and Batch H for the plots exceeding this value.

Two experimental datasets were used in this case, the first one ($n = 17$ plots) with median 15 µg/kg (15–2250 µg/kg) and the second one ($n = 21$ plots) with median 135 µg/kg (10–16,685 µg/kg). For the first dataset, levels of exposure below PMTDI for Scenario (a) were obtained; thus, the two alternative scenarios, (b) and (c), gave no advantage over the simple blending of grain from all the fields, and the segregation of some batches even resulted in exposure levels above the PMTDI. For Dataset 2, Scenario (a) always resulted in exposure above the PMTDI. In this case, Scenarios (b) and (c) made it possible to create batches – Batch O for Scenario (b) and Batch L for Scenario (c) – for which the estimated exposure was below 1 µg/day kg. Scenario (c) was the most useful, as it not only reduced exposure but also corresponded to 90% of the fields, so that only a small proportion had to be rejected; however, it is difficult to implement, as it requires analysis of grain for DON content in each field (which is too costly) or a prediction tool to be applied per field. Alternatively, Scenario (b) can be applied; however, in this case, wheat from 53% of the fields should be rejected for human consumption, which is also economically costly.

17.5.1.3 Processing

No scenarios were tested at this point. DON content in flour was determined using a corrective coefficient of 0.44, applied to grain content, to take into account the effects of processing. Regarding consumption data, the percentage of flour in the recipes of the different finished products was taken into account.

In recent years, a number of publications have provided useful data for a better modelling of the effect of processing wheat to wheat products on DON content, including operations such as sorting, cleaning, milling, fermentation, baking and extrusion cooking (Generotti et al. 2015; Schwake-Anduschus et al. 2015; Vidal et al. 2014, 2016).

Transport and distribution, storage, retail, and preparation and consumption of the finished wheat products are not expected to have an effect on the presence of DON.

17.5.2 Case Study 2: Evaluation of Strategies for Reducing Patulin Contamination of Apple Juice Using a Farm-to-Fork Risk Assessment Model (Baert et al., 2012)

Patulin is a mycotoxin usually found in apple derivatives, mainly apple juice, due to the processing of low-quality, *Penicillium expansum* infected fruits. Almost 100% of *P. expansum* isolates are able to produce patulin (Morales et al 2010). In 2003, the EC issued a recommendation including a Code of Practice for the prevention and reduction of patulin contamination in apple juice and apple juice ingredients in other beverages, which points out a number of strategies for patulin risk management.

Baert et al. (2012) developed a QRA model for patulin in apple juice as a function of apple storage and processing conditions. The model is developed on the basis of a comprehensive list of parameters identified all along the food processing chain, from the collection of apples until storage of the produced apple juice. The model was used to test the influence of different risk management measures: duration of the deck storage for fresh apples, either using or not using a controlled atmosphere (CA) during storage, duration of refrigerated storage before and after CA storage, duration of deck storage at the juice plant, segregation of damaged or mouldy apples prior to processing, etc. The model was validated against empirically generated data ($n = 177$) of apple storage and juice production using the different possible scenarios. The simulations ($n = 10,000$) were performed with the @RISK software using Latin Hypercube sampling, setting the random generator seed at 1 to guarantee the stability of the estimations. An exposure assessment was not carried out, and the suitability of the different scenarios was assessed in terms of final contamination in the apple juice.

17.5.2.1 Preharvest

Although present in the field, *P. expansum* usually develops postharvest, during deck or cool storage, and produces patulin. Consequently, preharvest strategies aiming to reduce *P. expansum* inoculum in the field could be applied, but they are not as effective as postharvest strategies.

17.5.2.2 Postharvest

Baert et al (2012) built a probabilistic QRA model for apple postharvest, from picking to deck or cold storage prior to processing to juice, including all possible scenarios of duration, temperature and %O_2 through their probability distributions. Probability of infection, lag phase duration and colony diameters were estimated for *P. expansum*, as well as probability of patulin production and concentration, and were included in the risk assessment model from previously published models for the different storage conditions. Apples stored under CA conditions contained more patulin than fresh or cold stored ones; for example, a reduction from 40% to 20% in CA stored apples in a batch reduced the mean patulin concentration by almost 50%. Reduction of the wound frequency had a limited impact on final patulin concentration, as did short storage at 1 °C before or after CA storage. By contrast, the time between delivery at the juice plant and processing led to a large increase in patulin concentration, mostly when apples had been CA stored.

17.5.2.3 Processing

The fate of patulin throughout the production of cloudy and clear apple juice was modelled in a deterministic way by taking point values for the different percentages of reduction during washing, milling, pasteurization, filtration, refrigeration, etc. from the literature. Sorting of apples with lesion surfaces larger than 10 cm² led to fewer than 0.5% of juice samples containing over 25 µg/kg.

From all strategies tested, a joint strategy was refined by including a maximum of 20% of CA stored apples and removing apples with lesions bigger than 10 cm²; the model predicted in this case a 88% reduction of the patulin content and only 0.3% of apple juices over 25 µg/kg.

Transport and distribution, storage, retail, and preparation and consumption of the finished apple juices are not expected to have an effect on the presence of patulin.

Overall, the model showed realistic concentration estimates in apple juice; however, an overestimation at the higher concentrations was recognized. The authors pointed out the need for further raw data and models predicting the production of patulin in order to refine the predictions.

17.5.3 Case Study 3: A Stochastic Simulation Model for the Quantitative Assessment of the Concentration of Mycotoxins in Milk and the Related Human Exposure (Signorini et al. 2012)

Aflatoxin M1 (AFM1) is an increasing concern for the dairy industry in European countries. It is thought that the increase may be linked to the increase in AFB1 in maize and maize by-products used for feed production in certain areas due to climate change as well as to the inclusion of cottonseed and rapeseed in the feed.

Signorini et al. (2012) built a QRA model by combining probability distributions for the AFB1 concentration in the different raw materials included in cattle feed in Argentina and the amount of each ingredient in the diet, which was affected by the season and by the time from lactation initiation for each animal. In this way, four different simulations were created (2 seasons × 2 lactation periods), for which the model estimated AFB1 probability distributions for each diet. Monte-Carlo simulations ($n = 5000$ iterations) were also performed with the @RISK package, and the simulated statistics showed adequate convergence. Carry-over equations from the literature were applied to these distributions, and the distributions for AFM1 in milk was obtained as a single distribution per year. The concentration of AFB1 in feed was estimated as 4.7 µg/kg (95%CI 0.832–23.14), and mean AFM1 in milk was 0.059 µg/kg (95%CI 0.032–0.323). The AFM1 level in milk was sensitive to AFB1 in concentrate feed, carry-over, AFB1 in corn silage, season and AFB1 in cotton seed, in this order, which indicates the priorities in risk management, e.g., the need for better segregation of maize used for concentrate feed production and improved management of silage production. The total daily intake estimated for AFM1 was 0.00122 ng/kg bw (95%CI 0.007–0.633).

The last summarized study exemplifies that modelling may be applied to a whole food chain as, in this case, preharvest studies could have been included, as well as processing of milk to final products. In particular, the processing of milk may affect exposure, although in the study, the authors considered that effect negligible and estimated exposure from data estimated for raw milk. A recent review by Campagnollo et al. (2016) summarized the effects of the main unit operations for dairy product processing in AFM1. An overview of the whole process and its impacts on exposure is presented in Figure 17.3; scattered data exist on most of the steps, but model building is still a challenge in some of them.

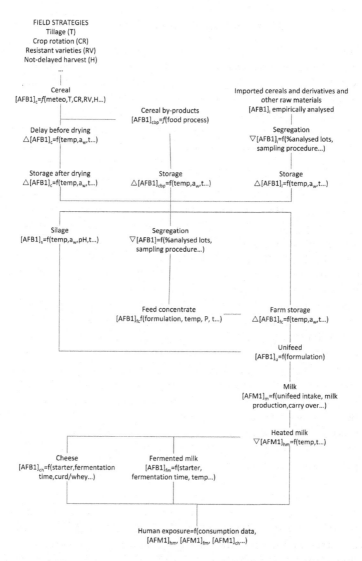

Figure 17.3 Overview of the farm-to-fork stages that determine the exposure of population to AFM1. [AFM1]$_c$: concentration of AFM1 in cereals; [AFM1]$_{cbp}$: concentration of AFM1 in cereal by-products; [AFM1]$_i$: concentration of AFM1 in imported raw materials; [AFM1]$_s$: concentration of AFM1 in silage; [AFM1]$_{fc}$: concentration of AFM1 in feed concentrate; [AFM1]$_u$: concentration of AFM1 in unifeed; [AFM1]$_m$: concentration of AFM1 in milk; [AFM1]$_{hm}$: concentration of AFM1 in heated milk; [AFM1]$_{ch}$: concentration of AFM1 in cheese; [AFM1]$_{fm}$: concentration of AFM1 in fermented milk.

17.6 CONCLUDING REMARKS

- There is an acknowledged lack of standardized probabilistic approaches and guidelines allowing the direct comparability of results. A normalized protocol addressing the major cornerstones on the probabilistic risk assessment pipeline could help with a more standardized and comparable body of studies. For instance, there are no thresholds of risk and related uncertainties estimated deterministically at which the probabilistic approach should be applied.
- The lack of data, in terms of quality and quantity, is a major limitation to conducting reliable and accurate QRA. In case of mycotoxins, the seasonal and year-to-year variability increases the uncertainty when combining consumption datasets with discordant timing. From a conservative point of view, the probabilistic input datasets could be complemented with average contamination data from other regions where the historical patterns of occurrence are comparable, with the aim of avoiding underestimations of risk estimates.
- There is room for enlargement and improvement of QRA models, including new scientific data; however, there is still a need to model many steps from field to consumer that have a potential impact on the final exposure. In this sense, Bayesian models based on Monte-Carlo simulations may accommodate the integration of multiple modules along the food chain, accounting for the variability and uncertainty of the parameters (Albert et al. 2008).
- Beyond the need for raw data to integrate into the farm-to-fork models, the development and validation of accurate prediction models for fungal growth and mycotoxin production become a priority for the integrative framework.
- It is urgent to perform probabilistic exposure assessments in developing countries. The high occurrence of mycotoxins in developing countries, combined with the high contribution of cereals to the overall dietary intake among their populations, places developing countries at the top of the list of populations whose health is at risk due to mycotoxin exposure. Unfortunately, the lack of official regulations and control systems for mycotoxins in many developing countries limits the availability of quality data to perform QRA. Coordinated action between governments, international agencies and scientists could achieve substantial refinements in exposure assessments in regions where the mycotoxin occurrence is suspected to trigger health risks (e.g., African countries).

508

REFERENCES

Albert, I., Grenier, E., Denis, J.-B., Rousseau, J. 2008. Quantitative risk assessment from farm to fork and beyond: A global Bayesian approach concerning food-borne diseases. *Risk Anal.* 28(2): 557–571.

Baert, K., Devlieghere, F., Amiri, A., De Meulenaer, B. 2012. Evaluation of strategies for reducing patulin contamination of apple juice using a farm to fork risk assessment model. *Int. J. Food Microbiol.* 154(3): 119–129.

Boon, P.E., Bakker, M.I., Van Klaveren, J.D., Van Rossum, C.T.M. 2009. Risk assessment of the dietary exposure to contaminants and pesticide residues in young children in the Netherlands, Bilthoven, the Netherlands: RIVM report 350070002/2009.

Boon, P.E., Bonthuis, M., van der Voet, H., van Klaveren, J.D. 2011. Comparison of different exposure assessment methods to estimate the long-term dietary exposure to dioxins and ochratoxin A. *Food Chem. Toxicol.* 49(9): 1979–1988.

Campagnollo, F.B., Ganev, K.C., Khaneghah, A.M., et al. 2016. The occurrence and effect of unit operations for dairy products processing on the fate of aflatoxin M1: A review. *Food Control* 68: 310–329.

Cano-Sancho, G., Gauchi, J.P., Sanchis, V., Marín, S., Ramos, A.J. 2011. Quantitative dietary exposure assessment of the Catalonian population (Spain) to the mycotoxin deoxynivalenol. *Food Addit. Contam. A* 28(8): 1098–1109.

Cano-Sancho, G., Marín, S., Ramos, A.J., Sanchis, V. 2012. Exposure assessment of T2 and HT2 toxins in Catalonia (Spain). *Food Chem. Toxicol.* 50(3–4): 511–517.

Cano-Sancho, G., Sanchis, V., Ramos, A.J., Marín, S. 2013. Effect of food processing on exposure assessment studies with mycotoxins. *Food Addit. Contam. A* 30(5): 867–875.

Champeil, A., Fourbet, J.F., Doré, T. 2004. Influence of cropping system on *Fusarium* head blight and mycotoxin levels in winter wheat. *Crop Prot.* 23/7: 635–645.

Commission Recommendation 2003/598/EC of 11 August 2003 on the prevention and reduction of patulin contamination in apple juice and apple juice ingredients in other beverages ((OJ L 203, 12.8.2003, p. 54–59).

Coronel, M.B., Marín, S., Cano-Sancho, G., Ramos, A.J., Sanchis, V. 2011. Exposure assessment to ochratoxin A in Catalonia (Spain) based on the consumption of cereals, nuts, coffee, wine, and beer. *Food Addit. Contam. A* 29: 979–993.

Council, E., Verger, P., Volatier, J.-L. 2005. Handling of contamination variability in exposure assessment: A case study with ochratoxin A. *Food Chem. Toxicol.* 43(10): 1541–1555.

EFSA (European Food Safety Authority). 2008. *Guidance Document for the Use of the Concise European Food Consumption Database in Exposure Assessment (EFSA/DATEX/2008/01)*. Parma: European Food Safety Authority.

EFSA (European Food Safety Authority). 2012. Guidance on the use of probabilistic methodology for modelling dietary exposure to pesticide residues. *EFSA J.* 10(10): 2839.

EFSA (European Food Safety Authority). 2017. Scientific report: Human and animal dietary exposure to T-2 and HT-2 toxin. *EFSA J.* 15(8): 4972.

509

García-Cela, E., Ramos, A.J., Sanchis, V., Marín, S. 2012. Emerging risk management metrics in food safety: FSO, PO. How do they apply to the mycotoxin hazard? *Food Control* 25(2): 797–808.

Gauchi, J.-P., Leblanc, J.-C. 2002. Quantitative assessment of exposure to the mycotoxin ochratoxin A in food. *Risk Analysis* 22(2): 219–234.

Generotti, S., Cirlini, M., Malachova, A., et al. 2015. Deoxynivalenol & deoxynivalenol-3-glucoside mitigation through bakery production strategies: Effective experimental design within industrial rusk-making technology. *Toxins* 7(8): 2773–2790.

Han, Z., Nie, D., Yang, X., Wang, J., Peng, S., Wu, A. 2013. Quantitative assessment of risk associated with dietary intake of mycotoxin ochratoxin A on the adult inhabitants in Shanghai city of P.R. China. *Food Control* 32(2): 490–495.

Kuiper-Goodman, T., Hilts, C., Billiard, S.M., Kiparissis, Y., Richard, I.D.K., Hayward, S. 2010. Health risk assessment of ochratoxin A for all age-sex strata in a market economy. *Food Addit. Contam.* 27(2): 212–240.

Le Bail, M., Verger, P., Doré, T., Fourbet, J.-F., Champeil, A., Ioos, R., Leblanc, J.-C. 2005. Simulation of consumer exposure to deoxynivalenol according to wheat crop management and grain segregation: Case studies and methodological considerations. *Regul. Toxicol. Pharmacol.* 42(3): 253–259.

Morales, H., Marín, S., Ramos, A.J., Sanchis, V. 2010. Influence of post-harvest technologies applied during cold storage of apples in *Penicillium expansum* growth and patulin accumulation: A review. *Food Control* 21(7): 953–962.

RIVM. 2011. *Integrated Probabilistic Risk Assessment (IPRA) for Carcinogens: A First Exploration*. Bilthoven, the Netherlands: RIVM Report 320121002/2011.

Schwake-Anduschus, C., Proske, M., Sciurba, E., Muenzing, K., Koch, M., Maul, R. 2015. Distribution of deoxynivalenol, zearalenone, and their respective modified analogues in milling fractions of naturally contaminated wheat grains. *World Mycotoxin J.* 8(4): 433–443.

Signorini, M.L., Gaggiotti, M., Molineri, A., et al. 2012. Exposure assessment of mycotoxins in cow's milk in Argentina. *Food Chem. Toxicol.* 50(2): 250–257.

van der Voet, H., Slob, W. 2007. Integration of probabilistic exposure assessment and probabilistic hazard characterization. *Risk Anal.* 27(2): 351–371.

Vidal, A., Morales, H., Sanchis, V., Ramos, A.J., Marín, S. 2014. Stability of DON and OTA during the breadmaking process and determination of process and performance criteria. *Food Control* 40: 234–242.

Vidal, A., Sanchis, V., Ramos, A.J., Marín, S. 2016. The fate of deoxynivalenol through wheat processing to food products. *Curr. Opin. Food Sci.* 11: 34–39.

WHO. 2003. *Gems/Food Regional Diets, Food Safety Issues, Food Safety Department Revision September 2003*. Geneva: WHO.

WHO (Food and Agriculture Organization/World Health Organization). 1995. *Uncertainty and Variability in the Risk Assessment Process*. 7. Geneva, Switzerland: World Health Organization. Application of risk analysis to food standards issues.

WHO (Food and Agriculture Organization/World Health Organization). 2008. *Microbiological Risk Assessment*. 7. Rome, Italy: World Health Organization. Exposure assessment of microbiological hazards in food: Guidelines. 92pp.

INDEX

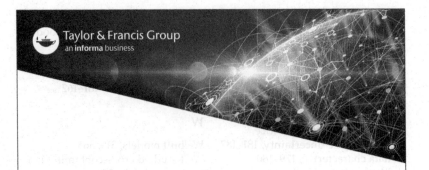

Printed in the United States
by Baker & Taylor Publisher Services